T0350963

Design for Safety

Wiley Series in Quality & Reliability Engineering

Dr Andre Kleyner

Series Editor

The Wiley series in Quality & Reliability Engineering aims to provide a solid educational foundation for both practitioners and researchers in Q&R field and to expand the reader's knowledge base to include the latest developments in this field. The series will provide a lasting and positive contribution to the teaching and practice of engineering.

The series coverage will contain, but is not exclusive to,

- statistical methods;
- physics of failure;
- reliability modeling;
- functional safety;
- six-sigma methods;
- lead-free electronics;
- warranty analysis/management; and
- risk and safety analysis.

Design for Safety

Edited by

Louis J. Gullo
Raytheon Missile Systems, Arizona, USA

Jack Dixon
JAMAR International, Inc., Florida, USA

Registered Office(s)
John Wiley & Sons, Inc., 111 River Street, Hoboken, NJ 07030, USA
John Wiley & Sons Ltd, The Atrium, Southern Gate, Chichester, West Sussex, PO19 8SQ, UK

Editorial Office
The Atrium, Southern Gate, Chichester, West Sussex, PO19 8SQ, UK

For details of our global editorial offices, customer services, and more information about Wiley products visit us at www.wiley.com.

Wiley also publishes its books in a variety of electronic formats and by print-on-demand. Some content that appears in standard print versions of this book may not be available in other formats.

Library of Congress Cataloging-in-Publication data applied for

ISBN: 9781118974292

Cover Design: Wiley
Cover Images: (Left to right) © 3DSculptor/Gettyimages; © ms. Octopus/Shutterstock; © prosot-photography/iStock; © gali estrange/Shutterstock

Set in 10.5/13pt Palatino by SPi Global, Pondicherry, India
Printed and bound in Malaysia by Vivar Printing Sdn Bhd.

10 9 8 7 6 5 4 3 2 1

To my wife, Diane, and my children, Louis, Jr., Stephanie, Catherine, Christina, and Nicholas.
Louis J. Gullo
To my wife, Margo.
Jack Dixon

And

to all the heroes of the world, especially all the safety heroes that make the world a safer place.
Louis J. Gullo
Jack Dixon

Contents

Series Editor's Foreword

The Wiley Series in Quality and Reliability Engineering aims to provide a solid educational foundation for researchers and practitioners in the field of dependability, which includes quality, reliability, and safety, and expand the knowledge base by including the latest developments in these disciplines.

It is hard to overstate the effect of quality and reliability on system safety. A safety-critical system is a system whose failure or malfunction may result in death or serious injury to people. According to Federal Aviation Administration (FAA), system safety is the application of engineering and management principles, criteria, and techniques to optimize safety by the identification of safety-related risks, eliminating or controlling them by design and/or procedures, based on acceptable system safety precedence.

Along with continuously increasing electronics content in vehicles, airplanes, trains, appliances, and other devices, electronic and mechanical systems are becoming more complex with added functions and capabilities. Needless to say, this trend is making the jobs of design engineers increasingly challenging, which is confirmed by the growing number of safety recalls. These recalls are prompting further strengthening of reliability and safety requirements and a rapid development of functional safety standards, such as IEC 61508 Electrical/Electronic/Programmable systems, ISO 26262 Road Vehicles, and others, which have increased the pressure on improving the design processes and achieving ever higher reliability as it applies to system safety.

There are no do-overs in safety. You cannot undo the damage to a human caused by an accident caused by an unsafe system; therefore it is extremely important to design a safe system the first time. This book *Design for Safety*, written by Louis J. Gullo and Jack Dixon explores the safety engineering and takes the concept of design and system safety to a new level. The book takes you step by step through the process of designing for safety. These steps include the development of system requirements, design for safety checklist, and application of the critical design tools, such as fault tree analysis, hazard analysis, FMEA, system integration, testing, and many others.

Both authors have lifelong experience in product design, safety, and reliability, and sharing their knowledge will be a big help to the new generation of design engineers as well as to the seasoned practitioners. This book offers an excellent mix of theory, practice, useful applications, and commonsense engineering, making it a perfect addition to the Wiley Series in Quality and Reliability Engineering.

Despite its obvious importance, quality, reliability, and safety education are paradoxically lacking in today's engineering curriculum. Very few engineering schools offer degree programs, or even a sufficient variety of courses, in quality or reliability methods, and the topic of safety only receives minimal coverage in today's engineering student curriculum. Therefore, the majority of the quality, reliability, and safety practitioners receive their professional training from colleagues, professional seminars, publications, and technical books. The lack of opportunities for formal education in these fields emphasizes too well the importance of technical publications like this one for professional development.

We are confident that this book, as well as this entire book series, will continue Wiley's tradition of excellence in technical publishing and provide a lasting and positive contribution to the teaching and practice of quality, reliability, and safety engineering.

Dr. Andre Kleyner,
Editor of the Wiley Series in Quality
and Reliability Engineering

Preface

Anyone who designs a product or system involving hardware and/or software needs to ask the following questions and seek answers to:

- Will my designs be safe for the users of the product or system that I design for them?
- Will my designs be safe for people affected by the users of the product or system that I design for them?
- Are there applications that my designs may be used for that are not safe even though it is not the original intentions of my design?
- Can anyone die or be harmed by my designs?

The designers and engineers that fully answer these questions and take action to improve the safety features of a design are heroes. These engineering heroes are usually unsung heroes who don't receive nor seek any reward or recognition.

When you think of heroes, you might conjure up the image of a US Army Medal of Honor recipient, or a brave firefighter willing to sacrifice his or her life to rescue people from a towering inferno, or a policeman cited for courage in the line of duty, but you probably won't imagine an engineer willing to sacrifice his or her job or career to prevent a potential catastrophic hazard from occurring within a product or system. Every day and throughout the world, multitudes of engineers working in numerous development and production engineering career fields within a global marketplace discover and analyze safety-critical failure modes and assess risks of hazards to the user or customer, which may cause loss of life or severe personal injury. These engineers display a passion for their work with consideration for the safety aspects of their products or systems realizing the ultimate impacts to the health and well-being of their user community. The passion of these engineers usually goes unnoticed, except by other engineers or managers who work closely with them. The passion of these engineers may be recognized in extreme or unusual circumstances with an individual or team achievement award, but they most certainly would not become hailed as heroes. Why not? Does our engineering society place value of those willing to display courage in managing challenging technical problems? Of course, there is value in this characteristic of an engineer, but only when it results in making the organization or company more money, not reducing or eliminating the potential of dangerous hazards that could harm the user community. The engineers demonstrating courage in tackling the challenging technical problems to keep people safe are just doing their jobs as system safety engineers or some other related job function, but they would not be considered as heroes.

When you think of heroes in engineering, you might say Nikola Tesla or Thomas Edison made significant contributions to the advancement of a safe world in terms of developing commercial power to light homes at night and prevent fires due to lit candles igniting window dressings or draperies. We are sure you will agree that commercial power saves lives, indirectly. As a result of commercial power, most home fires caused by candles lighting a home at night have been prevented, but fires at home will still occur regardless of the use of commercial power replacing candles. There are other mitigating factors that have a direct correlation to causes of fires at home, such as smoking cigarettes in bed or poor insulation of electrical wiring or overloaded electrical circuits.

A direct application of saving lives is a preventive action to design out an explosive hazard in an automobile due to the fuel tank during an automobile collision. As a result of an engineer's diligence, persistence, and commitment to mitigate the risk of a fuel tank explosion in a car during normal operation or during a catastrophic accident, it is clear that the engineer's actions would have saved lives, directly. This direct application of design improvements that result in no deaths or personal injuries caused by automobile fuel tank explosions should warrant the title of "Engineering Hero" to the ones worthy of such distinction.

Engineering needs more heroes [1]. Engineers with the biggest paychecks get the widest acclaim. To be a hero today, you must be considered financially successful and ahead of your peers. There must be other ways to recognize engineering heroes on a broad scale, but how? There is no Nobel Prize for engineering. There is no engineering award with similar global status and prestige. Engineers cannot routinely recognize their heroes in a similar fashion as do physicists, economists, and novelists. To be fair, in lesser known circles than Nobel, engineers are recognized by their peers through the Kyoto Prize, Charles Stark Draper Prize of the US National Academy of Engineering, and the IEEE's own Medal of Honor, to name a few engineering honors. We agree with G. Pascal Zachary when he states that a valid criterion for an engineer to be considered an engineering hero is when one overcomes adversity. Engineering heroism appears when an engineer overcomes personal, institutional, or technological adversity to do their best job possibly while realizing what is ethically or morally right, contributing to the social and cultural well-being of all humanity.

Anyone who convinces a product manufacturer to install a safety feature on an existing product should be praised as a hero. One example of a design for safety feature that was installed on an existing product is the "safety mechanism" designed for firearms. A safety catch mechanism or safety switch used for pistol and rifle designs was intended to prevent the accidental discharge of a firearm, helping to ensure safe handling during normal use. The safety switch on firearms has two positions: one is "safe" mode and the other is "fire" mode. The two-position safety toggle switch was designed on the military grade firearm, M16 automatic rifle. In "safe" mode, the trigger cannot be engaged to discharge the projectile in the firing assembly. Other types of safety mechanisms include manual safety, grip safety, decocker mechanism, firing pin block, hammer block, transfer bar, safety notch, bolt interlock, trigger interlock, trigger disconnect, magazine disconnect, integrated trigger safety mechanism, loaded chamber indicator, and stiff double-action trigger pull. "Drop safety mechanisms" or "trigger guards" are passive safety features designed to reduce the chance of an accidental firearm discharge when the firearm is dropped or handled in a rough manner. Drop safeties generally provide an obstacle in the firing mechanism, which can only be removed when the trigger is pulled, so that the firearm cannot otherwise discharge. Trigger guards provide a material barrier to prevent inadvertent trigger pulls. Many firearms that were manufactured in the late 1990s were designed with mandatory integral locking mechanisms that had to be deactivated by a unique key before the firearm could be fired. These are intended as child-safety devices during unattended storage of firearms. These types of locking mechanisms were not intended as safety mechanisms while carrying. Other devices in this category are muzzle plugs, trigger locks, bore locks, and firearm safes.

Accidents decreased tremendously over the years as a result of safety features. Accidental discharges were commonplace in the days of the "Ole West," circa 1850–1880. Those were the days before safety switches were designed into rifle and pistol designs. Now accidental discharges only occur when a loaded firearm is handled when the safety position is off. Since the implementation of this safety switch design, gunshots caused by accidental firing have been significantly reduced. There was a designer behind this safety switch design who thought about saving lives. In our minds, this designer was an unsung hero, one of many heroes in the development of safe firearms.

We propose these unsung heroes deserve immense credit for preventing unnecessary injury or death from accidental discharge of firearms. There are many more examples of this.

The idea for this book was conceived as a result of publishing our first book, *Design for Reliability*. We saw the need for additional books discussing various topics associated with the design process. As a result, we are planning to create a series of *Design for X* books with this one, *Design for Safety*, being the second in the series. Our book fills the gap between the published body of knowledge and current industry practices by communicating the advantages of designing for safety during the earliest phase of product or system development. This volume fulfils the needs of entry-level design engineers, experienced design engineers, engineering managers, and system safety engineers/managers who are looking for hands-on knowledge of how to work collaboratively on design engineering teams.

Reference

[1] Zachary, G. P. (2014), Engineering Needs More Heroes, *IEEE Spectrum*, 51, 42–46.

Louis J. Gullo
Jack Dixon

Acknowledgments

We would like to thank Dev Raheja for his contributions to this book and for his co-editing of *Design for Reliability*, the first book in our planned *Design for X* series. Without the inspiration from Dev Raheja, only a few of these words would have been written. We have been humbled by his knowledge and grateful for his contributions to this book in offering us a cohesive framework using the ten paradigms in which to tie the pages together. We also are indebted to Nancy Leveson and her publishers. Her contributions to the field of system software safety are immense and greatly appreciated. There are many others who have made this work possible, adding to the body of knowledge from which we have drawn on. Among them, we especially want to thank Mike Allocco, Brian Moriarty, Robert Stoddard, Joseph Childs, and Denis W. Stearns.

Louis J. Gullo
Jack Dixon

Introduction: What You Will Learn

Chapter 1 Design for Safety Paradigms (Raheja, Gullo, and Dixon)
This chapter introduces the concept of design for safety. It describes the technical gaps between the current state of the art and what it takes to design safety into new products. This chapter introduces ten paradigms for safe design that help you do the right things at the right times. These paradigms will be used throughout the book as guiding themes.

Chapter 2 The History of System Safety (Dixon)
This chapter provides a brief history of system safety from the original "fly-fix-fly" approach to safety, to the 1940s' hints at a better way of doing aircraft safety, to the 1950s' introduction of the term "system safety," and to the Minuteman program that brought the systematic approach to safety to the mainstream. Next, the development of and history of MIL-STD-882 is discussed. The growth of system safety and various hazard analyses techniques over the years are covered in detail. The expansion of system safety into the nonmilitary, commercial arena is discussed along with numerous industry standards. Tools of the trade, management of system safety, and integration of system safety into the business process are summarized.

Chapter 3 System Safety Program Planning and Management (Gullo and Dixon)
This chapter discusses the management of system safety in detail. It describes how system safety fits into the development cycle, how it is integrated into the systems engineering process, and what the key interfaces are between system safety and other disciplines. The System Safety Program Plan is described in detail as well as how it is related to other management plans. Another important document, the Safety Assessment Report, is also outlined in detail.

Chapter 4 Managing Risks and Product Liabilities (Gullo and Dixon)
In this chapter, the importance of product liability is emphasized beginning with some financial statistics and numerous examples of major losses due to bad design. The importance of risk and risk management is described. This chapter includes a brief summary of product liability law and what it means to the safety engineer and the organization developing the product or system.

Chapter 5 Developing System Safety Requirements (Gullo)
This chapter's main emphasis is on developing safety requirements including why we need them and why they are so important. We discuss what requirements are and how they enter into various types of specifications. This chapter covers in detail how to develop good safety requirements and provides examples of both good and bad requirements.

Chapter 6 System Safety Design Checklists (Dixon)
This chapter introduces various types of checklists and why they are an important tool for the safety engineer. It covers procedural, observational, and design checklists and provides examples of each type. The uses of checklists are also discussed, and several detailed checklists are provided in the appendices of the book.

Chapter 7 System Safety Hazard Analysis (Dixon)
This chapter introduces some terminologies and discusses risk in detail as an introduction to hazard analyses. After that, it covers several of the most widely used hazard analysis techniques including preliminary hazard list, preliminary hazard analysis, subsystem hazard analysis, system hazard analysis, operating and support hazard analysis, and health hazard analysis. The chapter ends with a discussion of hazard tracking and its importance.

Chapter 8 Failure Modes, Effects, and Criticality Analysis for System Safety (Gullo)
This chapter describes how the Failure Modes and Effects Analysis (FMEA) and Failure Modes, Effects, and Criticality Analysis (FMECA) are useful for system safety analysis. It discusses various types of FMEAs including Design FMECA, Software Design FMECA, and Process Failure Modes, Effects, and Criticality Analysis (PFMECA) and how they may be applied in a number of flexible ways at different points in the system, hardware, and software development life cycle.

Chapter 9 Fault Tree Analysis for System Safety (Dixon)
Fault Tree Analysis (FTA) is covered in this chapter. It is a very popular type of analysis used in system safety. It is a representation in tree form of the combination of causes (failures, faults, errors, etc.) contributing to a particular undesirable event. It uses symbolic logic to create a graphical representation of the combination of failures, faults, and errors that can lead to the undesirable event being analyzed. The purpose of FTA is to identify the combinations of failures and errors

that can result in the undesirable event. This chapter provides a brief history of the development of FTA and provides a detailed description of how the analyst creates and applies FTA.

Chapter 10 Complementary Design Analysis Techniques (Dixon)
This chapter covers several additional popular hazard analysis techniques including event trees, sneak circuit analysis, functional hazard analysis, barrier analysis, and bent pin analysis. It also provides brief introductions to a few additional techniques that are less often used including Petri nets, Markov analysis, management oversight risk tree, and system-theoretic process analysis.

Chapter 11 Process Safety Management and Analysis (Dixon)
This chapter introduces Process Safety Management (PSM). It is an effort to prevent catastrophic accidents involving hazardous processes that involve dangerous chemicals and energies. It applies management principles and analytic techniques to reduce risks to processes during the manufacture, use, handling, storage, and transportation of chemicals. A primary focus of PSM is on hazards related to the materials and energetic processes present in chemical production facilities, but it can also be applied to facilities that handle flammable materials, high voltage devices, high current load devices, and energetic materials, such as rocket motor propellants. In this chapter we discuss the regulatory requirement for PSM, elements of PSM, hazard analysis techniques, and related regulations and end with a discussion of inherently safer design.

Chapter 12 System Safety Testing (Gullo)
In this chapter we discuss the purpose and importance of safety testing. The different types of safety tests are described along with the test strategy and test architecture. The development of safety test plans is covered. This chapter contains a section on testing for regulatory compliance and discusses numerous national and international standards. The topic of Prognostics and Health Monitoring (PHM) is introduced along with a discussion of the return on investment associated with PHM. We also discuss how to leverage reliability test approaches for safety testing. Safety test data collection is covered along with what to do with test results. The chapter is ended with a discussion on designing for testability and test modeling.

Chapter 13 Integrating Safety with Other Functional Disciplines (Gullo)
In this chapter, we cover several ways of integrating safety with other engineering and functional disciplines. We discuss the many key interfaces to system safety engineering, and we define the cross-functional teams. We have touched on modern decision-making in a digital world and on knowing who are your friends and your foes. The importance of constant communication is emphasized. We talk about a code of conduct and values. This chapter introduces paradigms from

several different sources and how they relate to system safety and how their application can make you a better engineer and help make you and your organization more successful.

Chapter 14 Design for Reliability Integrated with System Safety (Gullo)

The integration with all functional disciplines is very important for effectively and efficiently practicing system safety engineering, but the most important of these functional discipline interfaces is the interface to reliability engineering. This chapter builds on and applies the lessons from Chapter 13 to establish a key interface with reliability engineering. In this chapter we discuss what reliability is and how it is intertwined with system safety. Specifically we discuss how system safety uses reliability data and how this data is used to help determine risk. We conclude the chapter with examples of using reliability data to design for safety.

Chapter 15 Design for Human Factors Integrated with System Safety (Dixon and Gullo)

In starting this chapter, we refer back to the previous two chapters where we discussed the ways system safety engineers should integrate and interface with other types of engineers and functional disciplines and, in particular, with reliability engineering. Another important engineering interface for a system safety engineer is Human Factors Engineering (HFE). System safety benefits greatly from a well-established and reinforced interface to HFE. In this chapter we define HFE and its role in design of both hardware and software. We discuss the Human–Machine Interface (HMI), the determination of manpower and workload requirements, and how they influence personnel selection and training. We detail how human factors analysis is performed and how the various tools are used. Also discussed is how the human in the system influences risk, human error and its mitigation, and testing to validate human factors in design.

Chapter 16 Software Safety and Security (Gullo)

This chapter introduces the subjects of software safety and security. Many of today's systems are software-intensive systems, and it is necessary to analyze, test, and understand the software thoroughly to ensure a safe and secure system and to build system trust that the system always works as intended without fear of disruptions or undesirable outcomes. This chapter provides a detailed discussion of cybersecurity and software assurance. Next we discuss the basic software system safety tasks and how software safety and cybersecurity are related. Software hazard analysis tools are discussed along with a detailed discussion of Software FMECA. This chapter ends with a discussion of software safety requirements.

Chapter 17 Lessons Learned (Dixon, Gullo, and Raheja)

A lesson learned is the study of similar types of data and knowledge discovered from past events to prevent future traumatic recurrences or enable great successes. This chapter focuses on the importance of using lessons learned to prevent future accidents. It discusses the importance of capturing lessons learned and how to analyze failures to learn from them. We also discuss the importance of learning from successes and near-misses. Throughout this chapter pertinent examples are provided along with analysis of why lessons learned are so important. We also cover the process of continuous improvement.

Chapter 18 Special Topics on System Safety (Gullo and Dixon)

This final chapter delves into several special topics and applications to consider in relation to the future of system safety. There is no better industry marketplace to address the future of system safety features than the commercial aviation and automobile industries. We examine the historical and current safety data of both industries to see what it tells us about the historical trends and the future probabilities of fatal accidents. We explore the safety design benefits from commercial air travel that could be leveraged by automobile manufacturers and developers of new ground transportation systems. This chapter also discusses future improvements in commercial space travel.

1

Design for Safety Paradigms

Dev Raheja, Louis J. Gullo, and Jack Dixon

1.1 Why Design for System Safety?

Only through knowledge of a specific system's performance can a person understand how to design for safety for that particular system. Anyone designing for safety should realize that there is no substitute for first-hand knowledge of a system's operating characteristics, architecture, and design topology. The most important parts of this knowledge is understanding the system—learning how it performs when functioning as designed, verifying how the system performs when applied under worst-case conditions (including required environmental stress conditions), and experiencing faulty conditions (including mission-critical failures and safety-critical failures).

1.1.1 What Is a System?

A system is defined as a network or group of interdependent components and operational processes that work together to accomplish the objectives and requirements of the system. Safety is a very important aim of a system while executing and accomplishing its objectives and requirements. The design process of any system should ensure that everybody involved in using the system or developing the system gains something they need, avoiding the allure to sacrifice one critical

Design for Safety, First Edition. Edited by Louis J. Gullo and Jack Dixon.
© 2018 John Wiley & Sons Ltd. Published 2018 by John Wiley & Sons Ltd.

part of the system design in favor of another critical part of the system. This context includes customers, system operators, maintenance personnel, suppliers, system developers, system safety engineers, the community, and the environment.

1.1.2 What Is System Safety?

System safety is the engineering discipline that drives toward preventing hazards and accidents in complex systems. It is a system-based risk management approach that focuses on the identification of system hazards, analysis of these system hazards, and the application of system design improvements, corrective actions, risk mitigation steps, compensating provisions, and system controls. This system-based risk management approach to safety requires the coordinated and combined applications of system management, systems engineering, and diverse technical skills to hazard identification, hazard analysis, and the elimination or reduction of hazards throughout the system life cycle.

1.1.3 Organizational Perspective

Taking a systems approach enables management to view its organization in terms of many internal and external interrelated organization and company business connections and interactions, as opposed to discrete and independent functional departments or processes managed by various chains of command within an organization. (Note: The term "organization" will be used throughout the book to refer to all system developer and customer entities to include businesses, companies, suppliers, operators, maintainers, and users of systems.) When all the connections and interactions are properly working together to accomplish a shared aim, an organization can achieve tremendous results, from improving the safety of its systems, products, and services to raising the creativity of an organization to increasing its ability to develop innovative solutions to help mankind progress.

1.2 Reflections on the Current State of the Art

System safety is defined as the application of engineering and management principles, criteria, and techniques to achieve acceptable risk within the constraints of operational effectiveness and suitability, time, and cost throughout all phases of the system life cycle [1]. We have come a long way since the early days of system safety in the 1960s. System safety in many organizations has been successfully integrated into the mainstream of systems engineering and is vigorously supported by management as a discipline that adds value to the product development process. Many analysis techniques have been created and revised numerous times to make them more effective and/or efficient. The application of system safety in product design and development has proven valuable in reducing accidents and product liability.

However, there are still many challenges facing system safety engineers. First and foremost, even after over 50 years, system safety is still a small and somewhat obscure discipline. It needs more visibility. While many organizations success- fully implement system safety, many continue to ignore its benefits and suffer the consequences of delivering inferior, unsafe products.

Other challenges include the continually increasing complexity of systems being developed. Now, instead of only worrying about one system at a time, we must worry about building safe systems of systems. This additional complexity has introduced new challenges of how to address the interactions of all the systems that might make up a system-of-systems.

Inadequate specifications and requirements continue to plague the discipline. Too often weak, generic specifications are provided to the designers leading to faulty designs because the requirements were vague or ill defined.

The management of change is often another weakness in the product life cycle. As changes are made to the product or system, system safety must be involved to ensure that the changes themselves are safe and that they do not cause unintended consequences that could lead to accidents.

The human often causes safety problems by the way he uses, or abuses, the product. All too often the user can be confused by the complexity of a product or system or by the user interface provided by the software that operates it. Taking the human into consideration during the design process is paramount to its suc- cessful deployment.

The goal of this book is to help remedy some of these problems and build upon the many years of success experienced by system safety. In this chapter we pre- sent 10 paradigms we believe will lead to better and safer product designs. Throughout this chapter, and the book, we provide both good and bad examples so the reader can identify with real-world cases from which to learn.

1.3 Paradigms for Design for Safety

Forming an ideal system's approach to designing new systems involves developing paradigms, standards, and design process models for a developer to follow and use as a pattern for themselves in their future design efforts. These paradigms are often called "words of wisdom" or "rules of thumb." The word "paradigm," which originated from the Greek language, is used throughout the content of this book to describe a way of thinking, a framework, and a model to use to conduct yourself in your daily lives as a system safety engineer, or any type of engineer. A paradigm becomes the way you view the world, perceiving, understanding, and interpreting your environment and helping you formulate a response to what you see and understand.

This book starts by focusing on 10 paradigms for managing and designing systems for safety. These 10 paradigms are the most important criteria for designing for safety. Each of these paradigms is listed next and is explained in detail in separate clauses of this chapter following the list:

- Paradigm 1: Always aim for zero accidents.
- Paradigm 2: Be courageous and "Just say no."
- Paradigm 3: Spend significant effort on systems requirements analysis.
- Paradigm 4: Prevent accidents from single as well as multiple causes.
- Paradigm 5: If the solution costs too much money, develop a cheaper solution.
- Paradigm 6: Design for Prognostics and Health Monitoring (PHM) to minimize the number of surprise disastrous events or preventable mishaps.
- Paradigm 7: Always analyze structure and architecture for safety of complex systems.
- Paradigm 8: Develop a comprehensive safety training program to include handling of systems by operators and maintainers.
- Paradigm 9: Taking no action is usually not an acceptable option.
- Paradigm 10: If you stop using wrong practices, you are likely to discover the right practices.

These paradigms, which are referenced here, are cited throughout the course of this book. Table 1.1, at the back of this chapter, provides a guide to where in this book the various paradigms are addressed.

1.3.1 Always Aim for Zero Accidents

Philip Crosby, the former Senior Vice President (VP) at ITT and author of the famous book *Quality is Free*, pioneered the zero defects standard. Philip Crosby considered "zero defects" as the only standard you needed. This applies even more to safety. It is a practice that aims to prevent defects and errors and to do things right the first time. The ultimate aim is to reduce the level of defects to zero. The overall effect of achieving zero defects is the maximization of profitability.

To experience high profitability, an organization has to compare the life cycle costs of designing the product using current methods versus improving the design for zero accidents using creative solutions. Such creative solutions are usually simple. Because of this, they are often called elegant solutions. It may be a cheaper option to develop an elegant solution over a complex solution. As the great Jack Welch, former CEO of General Electric (GE), said in his book *Get Better or Get Beaten* [2], doing things simply is the most elegant thing one can do. In a company in Michigan, a shaft/key assembly for a heavy-duty truck transmission was designed for zero failures for at least 20 years by changing the heat-treating method. Heat treatment (e.g., annealing) is a process where heating and cooling of a metallic item alter the material properties for the purpose of improving the design strength and reducing the risk of hazards or failures. In this case, the temperature range and the heating/cooling rates were varied to achieve the optimum design strength. The cost of the new heat treatment method was the same as the previous method. The only additional cost was the cost to run a few experiments to determine the temperature range and the heating/cooling rates that were needed to get the desired strength. The new heat treatment method became a cheaper method when the company eliminated warranty costs, risk of safety-related lawsuit costs, and projected maintenance costs. They received more business as a result of customer satisfaction through achieving high quality and safety. The Return on Investment (ROI) was at least 1000%. This was an elegant solution. This solution was much cheaper than paying legal penalties and maintenance costs for accidents.

Similarly, consider another example involving de Havilland DH 106 Comet aircraft. A redesign of the de Havilland DH 106 Comet aircraft resulted from the first three de Havilland DH 106 Comet aircraft fuselage failures in 1952. The failures were located around the perimeter of the large square windows on the fuselage, which were manufactured without increasing the thickness of the fuselage in this area. The metal fatigue from high stress concentration on the sharp corners was causing the failures. They eliminated the sharp corners by designing oval-shaped windows. This redesign was much cheaper than paying for the accidents that would have resulted if the design was not changed [3].

1.3.2 Be Courageous and "Just Say No"

Paradigm 2 is to be courageous and "Just say no" to those who want to rush designs through the design review process without exercising due diligence and without taking steps to prevent catastrophic events. Say "No" at certain times during the system development design process to prevent future possible catastrophic events as they are discovered. Many organizations have a Final Design Review. Some call it the Critical Design Review. This is the last chance to speak up if anyone is concerned about anything in the design. A very important heuristic to remember is "Be courageous and just say no." The context here is that if the final design is presented with known safety design issues, and everyone votes "yes" to the design approval without seriously challenging it, then your answer should be "no." Why? Because there are almost always new problems lingering in the minds of the team members, but they don't speak up at the appropriate time. They are probably thinking that it is too late to interfere or they want to be a part of the groupthink process where everyone thinks alike. No matter how good the design is, an independent facilitator can find many issues with it. Ford Motor Company hired a new VP during the design of the 1995 model of the Lincoln Continental car. The company had been making this car for years, and everyone on the team had at least 10 years of experience. The design was already approved, but the new VP insisted on questioning every detail of the design with a cross-functional team composed of engineers from each subsystem.

Though its redesign began four months later than had been intended, the 1995 Lincoln Continental was available on the market one month ahead of schedule. The team made over 700 design changes. Since they made these improvements while the design was still on paper, the team completed their project using only a third of their budgeted 90 million dollars, resulting in savings of 60 million dollars [4].

It takes courage to be a real change agent and a true believer that a change has critical importance. Jack Welch stated in his book *Winning* [5], that real change agents comprise less than 10% of all business people. They have courage—a certain fearlessness about the unknown. As Jack says, "Change agents usually make themselves known. They're typically brash, high energy, and more than a little bit paranoid about the future. Very often, they invent change initiatives on their own or ask to lead them. Invariably, they are curious and forward looking. They ask a lot of questions that start with the phrase: "Why don't we....?"'

To recommend design for safety changes at critical points in the system design cycle, and be successful in implementing the design changes, you will need to win people to your way of thinking. As Dale Carnegie stated in his book *How to Win Friends and Influence People* [6], you need to exercise 12 principles to win people to your way of thinking. Paraphrasing these principles here, we have:

Principle 1: Get the best of an argument by avoiding it.
Principle 2: Respect the other person's opinions and avoid saying, "You are wrong."
Principle 3: If you realize that you are wrong, admit you are wrong immediately.
Principle 4: Begin a discussion in a friendly way and nonconfrontational.
Principle 5: Provoke "yes, yes" responses from the other person immediately.
Principle 6: Allow the other person to do the most talking.
Principle 7: Let the other person think your idea is also their idea.
Principle 8: See things from the other person's point of view and perspective.
Principle 9: Consider and sympathize with the other person's ideas and desires.
Principle 10: Appeal to nobler motives.
Principle 11: Dramatize your ideas to display your passion.
Principle 12: Throw down a challenge when nothing else works.

1.3.3 Spend Significant Effort on Systems Requirements Analysis

Most accidents or system failures originate from bad requirements in specifications. The sources of most requirement failures are incomplete, ambiguous, and poorly defined device specifications. They result in making expensive engineering changes later, one at a time called "scope creep." Often robust changes cannot be made because the project is already delayed and there may not be resources for implementing new features.

Look particularly hard for missing functions in the specifications. Usually there are insufficient requirements for modularity (to minimize interactions), reliability, safety, serviceability, logistics, human factors, testability, diagnostics capability, and the prevention of old failures in current designs. Specifications need to address inter-operability functions in more detail such as for the requirements for internal interfaces, external interfaces, user–hardware interfaces, and user–software interfaces. Specifications also need to address how the product should behave if and when there is an unexpected behavior such as the device shutdown from a power failure or from an unexpected human error. Those who are trying to design around a faulty specification should only expect a faulty design. Unfortunately, most companies still discover design problems when the design is already in production. At this stage there are usually no resources and no time available for major design changes.

To identify missing functions in a specification, performing requirements analysis is necessary. This analysis is always conducted by a cross-functional team. At least one member from each discipline should be present, such as R&D, design, quality, reliability, safety, manufacturing, field service, marketing, and if possible, a customer representative. New products have missing and vague requirements as much as 60% [7]. Therefore, writing accurate and comprehensive performance specifications is the prerequisite for a safe design. Personal experiences during interviews with various people attending safety training reveal that the troubleshooting technicians on complex electronic products are unable to diagnose about 65% of the problems (i.e., cannot duplicate the fault). Obviously, improved fault isolation requirements in the specifications are necessary for such products, but the cost may be too prohibitive. In large complex systems, it is incredibly expensive to obtain 100% test coverage and harder still to obtain 100% fault coverage isolating to a single item that caused a failure.

To find the design flaws early, a team has to view the system from different angles. You would not buy a house by just looking at the front view. You want to see it from all sides. Similarly, a system concept has to be viewed from at least the following perspectives:

- Functions of the product
- What undesired functions the product should never do (such as sudden acceleration)
- Range of applications

- Range of environments
- Active safety (inherent safety controls during the use of the device)
- Duty cycles during life
- Reliability for lifetime
- Robustness for user/servicing mistakes
- Logistics requirements to avoid adverse events
- Manufacturability requirements for defect-free production
- Internal interface requirements
- External interface requirements
- Installation requirements to assure safe functioning (an MRI was found inaccurate after the installation)
- Shipping/handling capabilities to keep the device safe (electronic components fail in humid environments)
- Serviceability/diagnostics capabilities
- PHM to warn users in case of an anomaly
- Interoperability with other products
- Sustainability
- Potential accidents and abuses
- Human factors

Most designers are likely to miss some of the aforementioned requirements. Almost all of them affect safety. This knowledge is not new; it can be incorporated by inviting experts in these areas to brainstorm. Requirements should include how the system should behave when a sneak failure occurs. A tool called Sneak Circuit Analysis (SCA) can be used for predicting sneak failures. A good specification will also address what the system shall not do, such as "there shall be no sudden fires in automobiles caused by any reason." The point here is, if we do everything right the first time, we avoid pain and suffering to users.

1.3.4 Prevent Accidents from Single as well as Multiple Causes

Many Failure Modes and Effects Analysis (FMEA) assess designs to determine the possibility of accidents that can happen from a single point failure cause, such as a component failure or a human error, or from multiple failure causes. According to the system safety theory [8], at least two events happen in any accident. There has to be a latent hazard in the system and a trigger event, such as human error, to activate an accident. A latent hazard can be an oversight in design, poor supervision, poor specification, inadequate risk analysis, or inadequate procedure. The Fault Tree Analysis (FTA) tool can reveal the combinations of causes for the hazard and the trigger event. To design out the accident, you could either prevent the latent hazard or prevent the trigger event.

An example concerning the Space Shuttle Challenger accident makes this clear. The NASA Challenger carrying eight astronauts exploded shortly after lift-off in 1996, causing the deaths of all personnel on board. The rubber o-ring seal that was not capable of functioning properly below 40°F caused the hazard. The trigger event was that management decided to fly at a lower temperature (human error) when the specification was clear about not launching below 40°. The whole accident could have been prevented if either the hazard (seal weakness) was designed out or the decision-making process by management was able to prevent errors in decisions to launch. Since human errors are very hard to control, the best strategy would be to design the seal that is good at any temperature feasible for a launch. This is the change NASA made after the accident. They added a heating wire around the seal so that the seal would be flexible enough to provide a good seal at any temperature.

The most difficult thing is to know the latent hazards prior to design release and to prevent hazards by design. We can predict many hazards if we use the tools such as Preliminary Hazard Analysis (PHA), FTA, SCA, Operating & Support Hazard Analysis (OS&HA), and FMEA.

1.3.5 If the Solution Costs Too Much Money, Develop a Cheaper Solution

The previous example of the Challenger shows that at very little cost, the seal would have been designed to fly at any temperature. The cost of heating the wire was probably less than $1000, while the cost of the accident was in the millions of dollars and cost eight lives. The lesson to be learned is that a cross-functional team should not accept any design without challenging the design. Apparently this design (not to fly below 40°) was not challenged enough. If someone would have challenged the seal as a hazard, they had a chance of preventing the accident at very low cost. Another example of an inexpensive change was covered previously in Paradigm 1 where just changing the heat treatment method resulted in at least 20 years life for a heavy-duty truck transmission.

The solutions are simple and not costly if a cross-functional team engages in creative brainstorming and comes up with at least 10 possible solutions for every problem. One of them is likely to be very cost effective. In the previous example where just changing the heat-treating method resulted in failure-free long life, this approach was used. This organization required 500% ROI even on safety to encourage very robust and cost-effective designs. It required system hazard analysis before approving the specifications and required design mitigation for all catastrophic hazards. The trick is to make all the big safety changes early during the concept stage where the cost of change is insignificant. The precedence for mitigating risk during the concept stage is as follows:

- Change the requirements to avoid the hazard.
- Introduce fault tolerance.

- Design to complete the mission safely.
- Provide early prognostic warnings.

Note that inspecting and testing are not included in the previous list. One cannot depend on inspection and testing for safety. However, it is always wise to inspect or test just to be confident in the solutions and to watch out for new defects inadvertently introduced during engineering changes.

These strategies almost always result in drastic reductions in life cycle costs, such as reduction in warranty costs, reduction in fatalities, reduction in maintenance/repair costs, reduction in accidents costs, and reductions in environmental damage costs and several other costs. Safety must always make a good business case if a best solution is chosen from creative brainstorming.

1.3.6 Design for Prognostics and Health Monitoring (PHM) to Minimize the Number of Surprise Disastrous Events or Preventable Mishaps

PHM technology and application enhances system safety, efficiency, availability, and effectiveness. In complex systems such as telecommunications and aerospace systems, most of the system failures are caused by fundamental limitations in the design strength and mechanical degradation mechanisms that propagate with time and lead to physical wear-out effects. Design strength must be able to withstand worst-case application and environmental stress conditions. There is a need for innovative solutions for discovering hidden problems, which usually turn up in rare events as probabilistic nondeterministic faults. Through the use of embedded sensors for health monitoring and predictive analytics within embedded processors, prognostic solutions for predictive maintenance are a possibility. PHM is an enabler for system reliability and safety. We need innovative tools for discovering hidden problems, which usually turn up in rare events, such as an airbag that does not deploy when needed in a crash. In the case of an airbag design, some brainstorming needs to be done on questions such as "Will the air bag open when it is supposed to?" "Will it open at the wrong time?" "Will the system give a false warning?" or "Will the system behave fail-safe in the event of an unknown component fault?"

The bottom line is that no matter how much analysis we do, it is impossible to analyze millions of combinations of events or faults that might occur. These faults might be due to incomplete test coverage or fault coverage because of limited time and funds to provide 100% fault detection capability through Built-In-Test (BIT) or external support test equipment. There will always be a certain percentage of faults that are unknown unknowns. There are an unknown number of failures that will be remain hidden, latent, and unknown until they occur and are detected, most likely during a mission when a function that is critical to a mission is disabled. These are called unpredicted failures. The following data on a major airline, announced at a FAA/NASA workshop [9], shows the extent of unpredicted failures:

- Approximately 130 problems known to FAA
- Approximately 260 actual problems in airline files
- Approximately 13,000 problems reported confidentially by the employees of the airline

The sneak failures are more likely to be in the embedded software where it is impractical to do a thorough analysis. Frequently the specifications are faulty because they are not derived from the system performance specification or they are not based on the system-of-systems. Peter Neumann, a computer scientist at SRI International, highlights the nature of damage from software defects [10]:

- Wrecked a European satellite launch
- Delayed the opening of the new Denver airport by one year
- Destroyed NASA Mars mission
- Induced a US Navy ship to destroy an airliner
- Shut down ambulance systems in London leading to several deaths

To counter such risks, we need an early prognostic warning, with health monitoring through embedded sensors to prevent a major mishap. The design process to incorporate these early warnings consists of postulating all the possible mishaps and designing intelligence to detect unusual behavior of the system. The intelligence may consist of measuring important features and making a decision on their impact. For example, a sensor input occasionally occurs after 30 ms instead of 20 ms as the timing requirement states. The question is: "Is this an indication of a disaster?" If so, the sensor should be replaced before the failure manifests itself to a critical state.

1.3.7 Always Analyze Structure and Architecture for Safety of Complex Systems

The systems today are connected directly and indirectly to many other systems. In large systems such as weapon systems, safety is linked to other systems, vehicles, soldiers, and satellites. There are almost an unlimited number of interactions possible. Tweaking in one place is bound to create some change in another place. As a result, all latent hazards are impossible to predict with high confidence. Therefore the organizations following a good process and doing the right things need to rethink how to deal with enormous complexity. It is hard enough to do the right things, but it is even harder to know what the right things are!

So, what can we do to control complexity? An organization must analyze not just the safety of its own system but also the safety of the system-of-systems made up of interconnecting systems. Safety needs to pay a lot more attention to hazard analysis on the structure and architecture of the complex system. The architecture must have modularity and traceability to safety-related interactions. Unfortunately, by the time safety engineers get involved, the structure and architecture are usually already chosen. Therefore, early involvement by safety engineering is critically important when dealing with large, complex systems.

1.3.8 Develop a Comprehensive Safety Training Program to Include Handling of Systems by Operators and Maintainers

Development of a complete safety training program for certifying the operators and maintainers requires not only recognizing the components and subsystems but also understanding the total system. Many safety training programs are focused only on the subsystem training. When this occurs, it means that the certification of the person operating or maintaining the equipment is limited. The operating and maintenance personnel may therefore not realize that the total system can be affected by hidden hazards. For example, having only one power source may negatively affect a maintainer's work. For safety-critical work, a redundant source would be a good mitigation to implement. However, if all sources of power are lost (prime, secondary, and emergency), the total system will not work until a correction is made. Safety training programs must include the process of making safe corrections. It should include total inter-connected system training addressing the worst potential secondary effect of hazards. Scenarios must be developed that provide instructors and students with a realistic understanding of the whole system and how to protect people from harm. Safety training represents a major mitigation method with complex systems; therefore it is imperative that complete training be provided for correctly operating and maintaining the system at all times and for all likely scenarios.

1.3.9 Taking No Action Is Usually Not an Acceptable Option

Sometimes, the teams cannot come up with a viable solution. They take no action or postpone the action hoping that a problem will be solved over time. This may be done out of denial, or because of fear to take action, or the fear of upsetting the superiors in management. Meanwhile, a product with a potential for fault finds its way to the market. Whatever be the case, all stakeholders including the customer become victims. The goal is to prevent the customers from becoming victims and casualties from poor design practices. Paradigm 2 earlier in this chapter provides helpful guidance on how to influence change. Doing nothing about a known problem is unacceptable.

1.3.10 If You Stop Using Wrong Practices, You Are Likely to Discover the Right Practices

The cause of many product recalls is insufficient knowledge of what needs to be done before the product development begins. Some wrong things discussed in this chapter are worth repeating:

- The design team starts designing the product without a thorough requirements analysis. They accept requirements without much challenging. They pay very little attention to missing and vague requirements.
- The project is approved without proper selection of the right team. A good design team must, as a minimum, include a customer representative from marketing or field service, a person with a thorough knowledge of the science of designing for safety, a person from manufacturing to assure the ability to produce the defect-free product, and a reliability engineer to guide the team on failure-free performance over the expected product life.
- The design team trusts 100% testing and 100% inspection in production. They do not understand that 100% inspection is less than 80% effective over time according to the quality gurus such as Deming and Crosby. Testing rarely represents the actual use environment.
- The design team frequently make costly design changes instead of following a structured approach for robust designs that lower life cycle costs.

Encourage everyone on a design team to identify wrong things. Ask every team member to identify at least five wrong things. They are always able to do so. With a good leader, this can be done with each team member. Then, create a plan to stop working on the wrong things, and replace them with the right things. In one case, employees came up with 22 solutions for a single design problem. About five of them had more than a 600% ROI. Almost always, the right things seem to just appear by themselves. You just have to look for them.

1.4 Create Your Own Paradigms

The advantage of creative brainstorming for potential mishaps and potential solutions is that teams will have occasional "aha" moments. Someone will have insight into what should have been done instead of what was done. This situation relates to a personal experience related to a system requirements analysis. In one particular example, at least 100 missing requirements were identified in the system specification during a brainstorming session on a complex product. At that time, a particular quote came to mind. This quote was "If you don't know where you are going, you'll probably end up somewhere else." This can be a nice paradigm for R&D engineers and specification writers. Such ideas can come while reading a good book also. Dr. Deming, the father of the Japanese quality system, used to say, "Working hard won't help if you are working on wrong things." This is a very useful paradigm. This was realized in another example in which a company had 400-page document on FMEA, but they did not make a single design change. That is nonproductive! They relied on inspection and testing. They worked hard on this FMEA in one meeting per week for six months. It surely did not help the company. One can create a new paradigm from this incidence: *Design out the problems, it is cheaper than inspecting and testing in production.*

1.5 Summary

In conclusion, these paradigms help you in doing the right things at the right times. If we prevent hazards and failures by doing the right things right the first time, there is no such thing as the cost of safety since the cost is part of the initial design and cannot be separated. These hazard and failure prevention actions include doing the right safety analyses, using the 10 paradigms, and incorporating robust design risk mitigations early. The ROI is usually at least 1000-fold comparing the cost of doing nothing against the cost of the design for safety analysis and resultant design change preventive actions. With ROI this high, there is no reason a safety-critical design change to preventive hazards should not be accomplished. Table 1.1 provides a guide to where in this book the various paradigms are addressed.

References

[1] MIL-STD-882E, Department of Defense Standard Practice System Safety, Washington, DC: U.S. Department of Defense, May 11, 2012.
[2] Slater, R., *Get Better or Get Beaten*, New York: McGraw-Hill, Second Ed, 2001, page 97.
[3] Wikipedia Encyclopedia, de Havilland Comet, https://en.wikipedia.org/wiki/De_Havilland_Comet (Accessed on October 4, 2015).
[4] Goleman, D., *Working with Emotional Intelligence*, London: Bloomsbury, 1998.
[5] Slater, R. *Winning*, New York: Harper Business, 2005, page 139.
[6] Carnegie, D., *How to Win Friends and Influence People*, New York: Pocket Books, a division of Simon & Schuster, Inc., 1936.

[7] Raheja, D. and Allocco, M., *Assurance Technologies Principles and Practices*, Hoboken: Wiley-Interscience, 2006.

[8] Raheja, D., *Preventing Medical Device Recalls*, Boca Raton: CRC Press, Taylor & Francis, 2014.

[9] Farrow, D. R., Speech at the Fifth International Workshop on Risk Analysis and Performance Measurement in Aviation sponsored by FAA and NASA, Baltimore, August 19–21, 2003.

[10] Mann, C. C., Why Software Is So Bad, MIT Technology Review, July, 2002, http://www.technologyreview.com/featuredstory/401594/why-software-is-so-bad/ (Accessed October 4, 2015).

Table 1.1 Paradigm locations

Paradigm number	Paradigm	Chapter																	
		1	2	3	4	5	6	7	8	9	10	11	12	13	14	15	16	17	18
Paradigm 1	Always aim for zero accidents	X	X		X							X	X	X	X				
Paradigm 2	Be courageous and "just say no"	X			X									X				X	
Paradigm 3	Spend significant effort on systems requirements analysis	X			X	X									X	X		X	
Paradigm 4	Prevent accidents from single as well as multiple Causes	X			X				X	X	X				X				
Paradigm 5	If the solution costs too much money, develop a cheaper solution	X			X			X				X							
Paradigm 6	Design for Prognostics and Health Monitoring (PHM) to minimize the number of surprise disastrous events or preventable mishaps	X			X								X		X				
Paradigm 7	Always analyze structure and architecture for safety of complex systems	X			X						X	X							
Paradigm 8	Develop a comprehensive safety training program to include handling of systems by operators and maintainers	X			X								X			X		X	
Paradigm 9	Taking no action is usually not an acceptable option	X			X			X						X				X	
Paradigm 10	If you stop using wrong practices, you are likely to discover the right practices	X			X		X							X				X	

2

The History of System Safety

Jack Dixon

2.1 Introduction

When designing for safety, a person's main focus is accident prevention—preventing injuries, loss of life, damage to equipment, and loss of valuable resources. The most effective means to accomplish accident prevention is by applying the engineering discipline known as system safety.

MIL-STD-882E defines safety as "Freedom from conditions that can cause death, injury, occupational illness, damage to or loss of equipment or property, or damage to the environment" [1].

The standard goes on to define system safety as "the application of engineering and management principles, criteria, and techniques to achieve acceptable risk within the constraints of operational effectiveness and suitability, time, and cost throughout all phases of the system life-cycle" [1].

The practice of system safety requires several elements in order to be successful. First, a system safety program must be established. This establishes the framework for a successful program by laying out the groundwork. Next, hazards must be identified. As hazards are identified, various types of hazard analyses must be accomplished to further definitize the hazards, develop recommendations for their control, and eliminate risk or reduce the risk to acceptable levels. Later in development, safety testing is conducted to ensure that the fixes, which were implemented to control the hazards, are effective. After safety testing, an updated

Design for Safety, First Edition. Edited by Louis J. Gullo and Jack Dixon.
© 2018 John Wiley & Sons Ltd. Published 2018 by John Wiley & Sons Ltd.

hazard analysis must be performed to assess the risks that remain. Management must accept any remaining risks.

In this chapter we provide a top-level introduction to system safety starting with its origins and a brief history of the discipline. Next we will highlight some of the tools used in the practice of system safety and the benefits reaped from their use. We will also briefly discuss how system safety is managed during a product development effort and how it is integrated into the business process. All of these topics will be further developed in later chapters.

2.2 Origins of System Safety

From the earliest of times, humans have been concerned with their safety. From seeking shelter to fighting off wild animals and other humans, safety has been very important. Safety was survival. To be safe meant you had a chance to survive. Safety also meant security. If your conditions were secure, you were protected and safe in your surroundings and environment. You satisfied the basic need to survive.

In the 1940s Abraham Maslow created a hierarchy of needs [2]. In his famous paper, he developed a theory that humans have five levels of needs. These are shown in Figure 2.1. As the lower level needs are satisfied, the person moves up the pyramid, focusing on satisfying the next higher level needs. Notice that once the person's basic physiological needs for food, water, air, sleep, and shelter are satisfied, the next most important need is safety. Safety includes security, health, and social stability. While safety is our focus here, other higher levels can be pursued after our safety needs are met. The person's social needs include love and belonging, embracing family, friends, and other connections. Higher up the

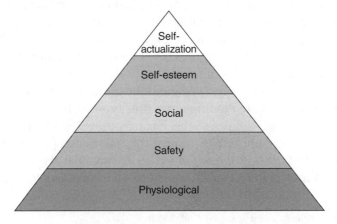

Figure 2.1 Maslow's hierarchy

pyramid is self-esteem. Here the person cultivates self-respect, pursues achievement and respect of others, and gains confidence. At the top of the hierarchy is self-actualization. At this level the person is pursuing meaning and purpose in their lives and may follow more creative activities.

It is obviously not known exactly when humans first began to worry about product safety, but the first record related to the safety of a product is found in the *Code of Hammurabi* [3], which is dated about 1754 BC. There are 282 laws in this code of laws, and several of them deal with product safety for houses and boats. The following are a paraphrased version of several of the safety-related laws in the Code:

229. If a builder builds a house for someone and does not construct it properly, and the house which he built falls in and kills its owner, then that builder shall be put to death.
230. If it kills the son of the homeowner, then the son of that builder shall be put to death.
231. If it kills a slave of the owner, then he shall pay slave for slave to the owner of the house.
232. If it ruins goods, he shall make compensation for all that has been ruined, and inasmuch as he did not construct properly this house which he built and it fell, he shall re-erect the house from his own means.
235. If a shipbuilder builds a boat for someone and does not make it tight, and if during that same year that boat is sent away and suffers injury, the shipbuilder shall take the boat apart and put it together tight at his own expense. The tight boat he shall give to the boat owner.

While some of the punishments may seem a little harsh by today's standards, it is the first recorded recognition of a product developer's responsibility to the customer. Product safety has evolved considerably since the times of the Babylonians.

2.2.1 History of System Safety

Once upon a time in a land not so far away, system safety as we know it today did not exist. Instead, when an accident occurred, an investigation was undertaken to see what might have gone wrong, and a fix was implemented in the system or product. This was known as the "fly-fix-fly" approach. In more general terms, this is a "trial-and-error" approach to system or product development.

System safety was born out of the dissatisfaction with the "fly-fix-fly" approach, from which the evolution of system safety started after World War II with the military's concern for aircraft safety. Amos L. Wood made the first presentation to the aviation industry of a system safety-type processes in January 1946 although the term "system safety" was not used. In his paper, "The Organization of an Aircraft Manufacturer's Air Safety Program" [4], he emphasized the need for a continuous focus of safety in design, post-accident analysis, safety education, and designs that minimize personnel errors. In September 1947, a paper published by William Stieglitz entitled "Engineering for Safety" [5] outlined a vision for system safety. He stated, "Safety must be designed and built into airplanes, just as are performance, stability, and structural integrity..." and "...A safety group must be just as important a part of a manufacture's organization as a stress, aerodynamics, or weights group...." and "Safety is a specialized subject just as are aerodynamics and structures." Here he is recognizing that safety is a specialty in and of itself. Prior to this safety was just another thing for which the engineers were responsible. This was the beginning of an interdisciplinary approach to viewing the system as a whole.

These farsighted visions formed the basis for the development of system safety. Waiting for an accident to happen to find design deficiencies was irresponsible. The "fly-fix-fly" philosophy was becoming an unacceptable way to identify and control hazards.

Over the next decade, the Department of Defense (DoD), led by the Air Force, began to stress a systematic, disciplined approach to system development focusing on proactive efforts to identify and analyze hazards, assess them for risk, and control them. The focus moved away from the "fly-fix-fly" philosophy toward the establishment of an acceptable level of safety designed into the system before being placed into operation. Hazards were being evaluated and controlled before an accident caused losses.

During this time frame, in the early 1950s, organizations began to form that were concerned with safety and with safety as a discipline. In 1950, the Air Force created the USAF Directorate of Flight Safety Research (DFSR). They began sponsoring industry conferences addressing safety issues of aircraft systems. The Air Force lead was soon followed by the creation of safety centers for the Navy in 1954 and the Army in 1957. The Flight Safety Foundation was also formed in this period, and seminars were held, bringing together engineering and management personnel. It was at one of these seminars in 1954 that Chuck Miller first introduced the term "system safety" in a paper entitled "Applying Lessons Learned from Accident Investigations to Design Through a Systems Safety Concept" [6].

At this time, the Air Force began developing Intercontinental Ballistic Missiles (ICBMs). System safety was not yet established as a separate discipline or as a system requirement. Everybody was responsible for safety; therefore, nobody was. These programs were becoming more hazardous. They were becoming more complex, using new technologies and many more interfaces, so the old way of doing safety was no longer effective. The ICBM program was plagued by numerous failures. According to the Air Force *System Safety Handbook for the Acquisition Manager* [7]:

…the Space Division (then the Ballistic Missile Division (BMD)) was engaged in the operational testing and site activation of our first ballistic missile systems. In the process, we had lost two Titan missile silos, three Atlas silos, and at least five people and we had an extremely low launch success rate. The significant factor prevalent in a large percentage of these incidents, was that causes could be traced to deficiencies in design, operational planning, and ill-conceived management decisions.

A Titan silo was destroyed because the counterweights, used to balance the silo elevator on the way up and down in the silo, were designed with consideration only to raising a fueled missile to the surface for firing. There was no consideration that, when you weren't firing in anger, you had to bring the fueled missile back down to defuel. The first operation with a fueled missile was nearly successful. The drive mechanism held it for all but the last 5′ when gravity took over and the missile dropped. Very suddenly, the 40′ diameter silo was altered to about 100′ diameter.

During operational tests on another silo, the decision was made to continue a test against the advice of the safety engineer when all indications were that, because of high oxygen concentrations in the silo, a catastrophe was imminent. The resultant fire destroyed a missile and caused extensive silo damage. In another accident, five people were killed when a single point failure in a hydraulic system caused a 120-ton door to fall.

Launch failures were caused by reversed gyros, reversed electrical plugs, bypass of procedural steps, and even launch with a known critical component failure. The BMD Commander issued a directive which resulted in a concentrated effort of accident prevention and elimination of the "Fly-Fix-Fly" approach. These directives eventually evolved into the system safety concept.

As a result of all these problems, the programs were canceled, and the Minuteman program was put on an accelerated deployment schedule. Major design flaws were the cause of many of these accidents. The "fly-fix-fly" approach was obviously not working. With systems that were this hazardous and that ultimately would carry nuclear warheads, it became apparent that a more systematic approach to preventing accidents was needed. This awareness ultimately led to the widespread use of the system safety approach.

We have come a long way since then.

2.2.2 Evolution of System Safety and Its Definitions

The history of system safety parallels the development of MIL-STD-882.

The definition of system safety has evolved over many years. While there were earlier attempts to include safety in the development of systems, "system safety" really evolved in the early 1960s during the Minuteman ICBM program. It was not until the adoption by DoD of MIL-S-38130A [8] in June of 1966 from an earlier Air Force version that the first official definition of system safety appeared. This specification defined system safety as "The optimum degree of safety within the constraints of operational effectiveness, time and cost, attained through specific application of system safety engineering throughout all phases of a system."

In 1969, the DoD published the first version of MIL-STD-882 based on the earlier MIL-S-38310A specification. By evolving from a specification to a standard, the awareness and importance of system safety was elevated. Building on the MIL-S-38130A definition, the system safety concept was expanded to include management, not just engineering, and extended coverage to the entire life cycle—"The optimum degree of safety within the constraints of operational effectiveness, time and cost, attained through specific application of system safety management and engineering principles throughout all phases of a system life cycle" [9].

In 1977, the next version of the standard was released—MIL-STD-882A. This time hazard identification and risk management were added: "The optimum degree of safety within the constraints of operational effectiveness, time, and cost attained through specific application of system safety management and engineering principles whereby hazards are identified and risk minimized throughout all phases of the system life cycle" [10].

MIL-STD-882B was introduced in 1984. The new definition was now "The application of engineering and management principles, criteria, and techniques to optimize safety within the constraints of operational effectiveness, time, and cost throughout all phases of the system life cycle" [11]. This iteration made system safety an "activity" rather than a "state."

In 1993 MIL-STD-882C made only a minor change by including "all aspects of safety" in the definition—"The application of engineering and management principles, criteria, and techniques to optimize all aspects of safety within the constraints of operational effectiveness, time, and cost throughout all phases of the system life cycle" [12].

After numerous draft revisions, MIL-STD-882D was held up by DoD's acquisition streamlining efforts and the elimination of many military standards. During this time, due largely to the threat of MIL-STD-882 being eliminated altogether by DoD, The International System Safety Society attempted to create an industry standard that could take the place of the military standard if it were eliminated. This draft standard was SSS-STD-882 [13]. The definition they chose was the same as MIL-STD-882C. The Society's standard was never released.

Fortunately, DoD decided to keep the MIL-STD-882 system safety standard. It was finally released in 2000 as a much slimmed down guidance document. All the detailed tasks that had been included over the years were removed. This time in the definition of system safety the words "to optimize safety" were replaced with "to achieve acceptable mishap risk." The definition in MIL-STD-882D became "The application of engineering and management principles, criteria, and techniques to achieve acceptable mishap risk, within the constraints of operational effectiveness and suitability, time, and cost, throughout all phases of the system life cycle" [14].

In the mid-2000s another attempt to revise the military standard for system safety was mounted. Again, numerous drafts were released in 2005 that put the detailed task descriptions back in, but it took until the spring of 2012 to have MIL-STD-882E [1] released. This revision was greatly expanded and resembles the earlier MIL-STD-882C more than MIL-STD-882D. It was expanded, not only to reintroduce the tasks that had been eliminated but also to add more. It also emphasized that system safety was a key part of Systems Engineering (SE), and it strengthened the integration of other functional disciplines into SE to ultimately improve consistency of hazard management practices across programs. As shown in the introduction to this chapter, the system safety definition was changed to "The application of engineering and management principles, criteria, and techniques to achieve acceptable risk within the constraints of operational effectiveness and suitability, time, and cost throughout all phases of the system life-cycle." This time risk was emphasized by the replacement of "to achieve acceptable mishap risk" with "to achieve acceptable risk."

Table 2.1 shows the evolution of MIL-STD-882 over the years and summarizes the changes in the definition of system safety.

2.2.3 The Growth of System Safety

While the birth of system safety was in the defense industry, over the years many other fields have adopted it. While we are not trying to cover all the industries that now use the system safety approach, or a modification of it, a sampling of industries and their standards is presented in an attempt to show the breadth of coverage that system safety now spans.

Since system safety had its origin in defense aviation, it is only natural that over the years the approach would be adopted by the space industry as well as by the commercial aviation industry. For the space industry, NASA standards developed along a path similar to MIL-STD-882 mainly because DoD and NASA use essentially the same contractors. According to *NASA Facility System Safety Guidebook*, like MIL-STD-882, system safety is defined as "the application of engineering and management principles, criteria, and techniques to optimize all aspects of safety within the constraints of operational effectiveness, time, and cost throughout all phases of the system life cycle" [15].

For the commercial aircraft industry, the Federal Aviation Administration (FAA) developed its own *System Safety Handbook* [16] for the use of FAA employees,

Table 2.1 Evolution of MIL-STD-882

Version	Published	Contents	System safety definition (changes indicated in bold type)
882	15 July 1969	• Emphasized SSPP • Provided definitions • Included "hazard level" (categories) • Required hazard analysis (Preliminary Hazard Analysis (PHA), Subsystem Hazard Analysis (SSHA), System Hazard Analysis (SHA), and Operating & Support Hazard Analysis (O&SHA)) • Included order of precedence • 24 pages	"The optimum degree of safety within the constraints of operational effectiveness, time and cost, attained through specific application of system safety management and engineering principles throughout all phases of a system life cycle"
882A	28 June 1977	• Updated development phases • Expanded requirements for each phase • Expanded government versus contractor responsibilities • Refined order of precedence • Introduced risk acceptance • Changed "hazard level" to "hazard severity" and added a table for "hazard probability" • Added Fault Hazard Analysis (FHA), Fault Tree Analysis (FTA), and Sneak Circuit Analysis (SCA) • 23 pages	"The optimum degree of safety within the constraints of operational effectiveness, time, and cost attained through specific application of system safety management and engineering principles **whereby hazards are identified and risk minimized** throughout all phases of the system life cycle"
882B	30 Mar 1984	• Substantially expanded to 98 pages • Continued expanding detailed guidance • Added tasks that could be called out contractually and referenced applicable Data Item Descriptions (DIDs) • Added system safety consideration of facilities and off-the-shelf equipment • First US standard to mention software and actually had a task on software safety hazard analysis	"The **application** of engineering and management principles, criteria, and techniques to optimize safety within the constraints of operational effectiveness, time, and cost throughout all phases of the system life cycle"

Table 2.1   (Continued)

Version	Published	Contents	System safety definition (changes indicated in bold type)
882C	19 Jan 1993	• Expanded to 116 pages • Reordered and expanded the tasks • Deleted the software hazard analysis task, but defined system safety engineering tasks and activities to be performed, but not specifically assigning those tasks to software or hardware • More emphasis on software • Added software hazard risk assessment process including software control categories and software hazard criticality matrix	"The application of engineering and management principles, criteria, and techniques to optimize **all aspects of** safety within the constraints of operational effectiveness, time, and cost throughout all phases of the system life cycle"
882D	10 Feb 2000	• This version was a result of the DoD's acquisition streamlining philosophy • Written more as a mishap risk management contractual stand-ard. Did away with prescribing particular safety tasks to be performed on contract and left it to the contractor and managing authority to "do what was required." This responsibility for tailoring and negotiating tasks on contract left in the hands of the program office often led to less than satisfactory results • Lacked software safety guidance • Shrunk to 31 pages	"The application of engineering and management principles, criteria, and techniques **to achieve acceptable mishap risk**, within the constraints of operational effectiveness and suitability, time, and cost, throughout all phases of the system life cycle"
882E	11 May 2012	• Expanded to 104 pages • Similar to "C," but better • Reinstated task descriptions… finally!	"The application of engineering and management principles, criteria, and techniques **to achieve acceptable risk** within the constraints of operational effectiveness and suitability, time, and cost throughout all phases of the system life-cycle"

supporting contractors and any other entities that are involved in applying system safety policies and procedures throughout FAA. Its definition of system safety is similar to the MIL-STD where it defined system safety as

> …a specialty within systems engineering that supports program risk management. It is the application of engineering and management principles, criteria and techniques to optimize safety. The goal of System Safety is to optimize safety by the identification of safety related risks, eliminating or controlling them by design and/or procedures, based on acceptable system safety precedence.

More recently, the FAA has recently implemented the Safety Management System (SMS). The International Civil Aviation Organization (ICAO) has defined SMS as "…a systematic approach to managing safety, including the necessary organizational structures, accountabilities, policies, and procedures" [17]. In implementing SMS, FAA Order 8000.369, *Safety Management System Guidance* [18], also defines system safety, much like the MIL-STD, as "The application of engineering and management principles, criteria, and techniques to optimize all aspects of safety within the constraints of operational effectiveness, time, and cost throughout all phases of the system lifecycle." In both cases, the FAA implementation of system safety follows the path set by the defense industry.

Related to the commercial aviation industry is the safety standard SAE ARP 4761, *Guidelines and Methods for Conducting the Safety Assessment Process on Civil Airborne Systems and Equipment* [19], which describes the system safety techniques for use during the development of aviation systems. In this document are guidelines and methods of performing the safety assessment for certification of civil aircraft. The concept of aircraft level safety assessment is introduced, and the tools to accomplish this task are outlined. Many of the methods described come directly from the defense world of system safety:

- Functional Hazard Analysis (FuHA)
- Preliminary System Safety Assessment (PSSA)
- System Safety Assessment (SSA)
- Fault Tree Analysis (FTA)
- Failure Modes and Effects Analysis (FMEA)

The railroad industry also has embraced the practice of system safety. For example, Amtrak has published their own safety document, "Amtrak System Safety Program" [20]. While Amtrak's definition is slightly different, it captures the essence of system safety. Amtrak defines system safety as "a detailed method of applying scientific, technical, operating, and management techniques and principles for the timely identification of hazard risk, and initiation of actions to prevent or control these hazards throughout the system life cycle and within the constraints of operational effectiveness, time, and cost."

In the area of mine safety, the system safety process has been adopted. The National Institute for Occupational Safety and Health (NIOSH) has published Information Circular 9456, "Programmable Electronic Mining Systems: Best Practice Recommendations" [21]. This provides recommendations and guidance on the functional safety of processor-controlled mining equipment. It is part of a risk-based system safety process encompassing hardware, software, humans, and the operating environment for the equipment's life cycle. The reports in this series address the various life cycle stages of inception, design, approval and certification, commissioning, operation, maintenance, and decommissioning. These recommendations were developed as a joint project between the NIOSH and the Mine Safety and Health Administration. The guidelines are intended for use by mining companies, original equipment manufacturers, and aftermarket suppliers to mining companies.

Even the food industry has gotten into the act. The food industry does not use the term "system safety," but they follow a somewhat similar approach. Certain process plants must meet requirements set forth in 9 CFR Part 417. To do this, they require Hazard Analysis and Critical Control Point (HACCP) plans. The United States Department of Agriculture (USDA) has published a guidebook for the preparation of HACCP plans [22], which are similar to SSPP's, and hazard analysis/ risk assessment is a major requirement for the analysis of meat and poultry plants operations. The HACCP approach is mandatory for the meat and juice industries, but is now being applied to other food processing operations as well as other related industries such as water quality, cosmetics, and pharmaceuticals.

Numerous generic standards (i.e., non-industry-specific standards) have been developed that have the essential characteristics of system safety. Again, without being totally exhaustive all of these standards, the following paragraphs highlight some of the most common ones.

One such standard is the *Standard for Software Safety Plans* published by the Institute of Electrical and Electronics Engineers (IEEE) [23]. This standard establishes the minimum acceptable requirements for the content of a software safety plan to address the processes and activities intended to improve the safety of safety-critical software.

Another popular generic standard, whose development began in the mid-1980s and has been adapted to various industries, is an international standard published by the International Electrotechnical Commission (IEC). IEC 61508 is entitled Functional Safety of Electrical/Electronic/Programmable Electronic (for E/E/PE) Safety-related Systems [24], and it comes in eight parts:

- IECTR 61508-0, Functional safety and IEC 61508
- IEC 61508-1, General requirements

- IEC 61508-2, Requirements for E/E/PE safety-related systems
- IEC 61508-3, Software requirements
- IEC 61508-4, Definitions and abbreviations
- IEC 61508-5, Examples and methods for the determination of safety integrity levels
- IEC 61508-6, Guidelines on the application of IEC 61508-2 and IEC 61508-3
- IEC 61508-7, Overview of techniques and measures

Prior to the 1980s, many regulatory bodies prohibited the use of software-based equipment in safety-critical applications. This international standard provides guidance for the functional safety of electric, electronic, and programmable electronic equipment that are used to perform safety functions. The standard focuses on risk-based safety-related system design much like system safety does. Although the standard does not specifically use the term "system safety" in the usual sense, it uses the same approach—applying system safety engineering analyses throughout the life cycle and assessing risks. The standard requires that hazard and risk assessment be carried out and provides guidance on a number of approaches. It uses hazard severity and hazard probabilities similar to MIL-STD-882 and also uses Safety Integrity Levels (SILs), which are a performance measurement required for a safety-instrumented function.

The IEC 61508 standard is a basic safety publication of the commission. It is an "umbrella" document with no industry-specific language, so it can cover multiple industries and applications. The standard provides guidelines, which, in turn, allow industries to develop their own standards specifically tailored to their particular industry. Several industry-specific standards based on the original IEC 61508 standard have been developed. Some examples are shown in Table 2.2.

Closely allied with these system safety standards is another family of standards, ISO 31000 [25], related to risk management and published by the International Organization for Standardization. This family of standards provides principles, framework, and a process for managing risk. We will discuss risk as it relates to product safety and liability in Chapter 4. ISO 31010 is a standard on risk management assessment techniques [26]. This international standard provides guidance on selection and application of systematic techniques for risk assessment. While the standard avoids using the term "system safety," many of the recommended analysis techniques are from the system safety world, including

- Checklists
- Preliminary Hazard Analysis (PHA)
- Hazard and Operability Study (HAZOP)
- Scenario analysis
- Root cause analysis

Table 2.2 A sampling of IEC 61508 standards

Standard	Title	Application
IEC 61511	Functional Safety: Safety Instrumented Systems for the Process Industry Sector	Establishes practices in the engineering of systems that ensure the safety of an industrial process through the use of instrumentation. The process industry sector includes many types of manufacturing processes, such as refineries, petrochemical, chemical, pharmaceutical, pulp and paper, and power
ANSI/ISA S84	Application of Safety Instrumented Systems for the Process Industries	Provides for the safe application of safety-instrumented systems including emergency shutdown systems, safety shutdown systems, and safety interlock systems for the process industry
IEC 62061	Safety of Machinery: Functional Safety of Electrical, Electronic and Programmable Electronic Control Systems	It is the machinery-specific implementation of IEC 61508. It provides requirements that are applicable to the system level design of all types of machinery safety-related electrical control systems and also for the design of noncomplex subsystems or devices
IEC 61513	Functional Safety: Safety Instrumented Systems for the Nuclear Industries	Provides requirements and recommendations for the instrumentation and control for systems important to the safety of nuclear power plants
ISO 26262	Road Vehicles: Functional Safety	An adaptation of IEC 61508 for Automotive Electric/Electronic Systems
IEC 62279	Railway Applications: Communication, Signalling and Processing Systems— Software for Railway Control and Protection Systems	Provides a specific interpretation of IEC 61508 for railway applications
IEC 61800-5-2	Adjustable Speed Electrical Power Drive Systems	A product-specific standard. Provides safety requirements for variable speed motor controllers
IEC 62304	Medical Device Software: Software Life Cycle Processes	Provides a framework of life cycle processes with activities and tasks necessary for the safe design and maintenance of medical device software

(Continued)

Table 2.2 (Continued)

Standard	Title	Application
EN 50128	Railway Applications: Communication, Signalling and Processing Systems— Software for Railway Control and Protection Systems	Specifies procedures and requirements for the development of programmable electronic systems for use in railway control and protection applications
EN 50129	Railway Applications: Communication, Signalling and Processing Systems— Safety-Related Electronic Systems for Signalling	Describes safety-related electronic systems, including subsystems and equipment, for railway signaling applications
EN 50402	Electrical Apparatus for the Detection and Measurement of Combustible or Toxic Gases or Vapour or of Oxygen—Requirements on the Functional Safety of Fixed Gas Detection Systems	Requirements on the functional safety of fixed gas detection systems applicable to fixed gas detection systems for the detection and measurement of flammable or toxic gases or vapors or oxygen

- FMEA
- FTA
- Event tree analysis

We will discuss risk and its relationship to system safety and to various types of system safety analyses techniques in greater detail in Chapter 7.

2.3 Tools of the Trade

MIL-STD-882E defines safety as "Freedom from conditions that can cause death, injury, occupational illness, damage to or loss of equipment or property, or damage to the environment" [1].

 Paradigm 1: Always Aim for Zero Accidents

As has been pointed out earlier, the standard defines system safety as "The application of engineering and management principles, criteria, and techniques to achieve acceptable risk within the constraints of operational effectiveness and suitability, time, and cost throughout all phases of the system life-cycle" [1]. One can think of system safety as the product safety function within the broader discipline of SE.

The goal of system safety is to identify risks inherent in a design and suggest risk mitigation measures as the design progresses.

So, how do we achieve these noteworthy goals? Over the years, many documents and techniques have been developed in support of building safe systems and products. We will highlight a few of them here and discuss them further in subsequent chapters.

System safety is a process that consists of eight elements. Figure 2.2 depicts the typical logical sequence of the process. Iteration between steps may be required.

First, and most important, is the System Safety Program Plan (SSPP), which documents the system safety methodology for the identification, classification, and mitigation of hazards as part of the overall SE process. The SSPP provides the detailed tasks and activities that are required to implement a systematic approach of hazard analysis, risk assessment, and risk management. It defines how the system safety management effort will be integrated into the SE process.

An important part of system safety, and the SSPP in particular, is how risk is handled. Risk is defined as part of the system safety process and usually includes a method of combining the probability that an event will occur with the severity of its consequences if it does occur. Categories of both probability and severity are identified in the SSPP along with the level of acceptability of the risk and what level of management accepts particular risks if they are not fully mitigated.

As the design progresses, hazards are identified and tracked. These hazards are tracked throughout the design process, often with an automated Hazard Tracking System (HTS). Various types of hazard analyses are conducted during the product development cycle to identify hazards, to assess the risk involved, and to mitigate the hazards and hopefully eliminate them entirely. Some of the most commonly used hazard analysis techniques are shown in Table 2.3.

2.4 Benefits of System Safety

The biggest benefit of system safety, of course, is the prevention of accidents, thus saving lives and preventing injuries. In addition, when accidents are prevented, damage to equipment, systems, facilities, and the environment is also averted.

In this prevention of accidents, the costs associated with the accidents are also avoided. Not only are the direct costs of accidents caused by the loss of or damage to equipment avoided, but also so are the indirect costs that may result from an undesirable event. These indirect costs may include product liability losses, lawsuits, increases in insurance rates, and loss of customers, business, and reputation.

Somewhat less obvious, but nonetheless important, are the costs that are avoided during the development phase of a system or product. By identifying

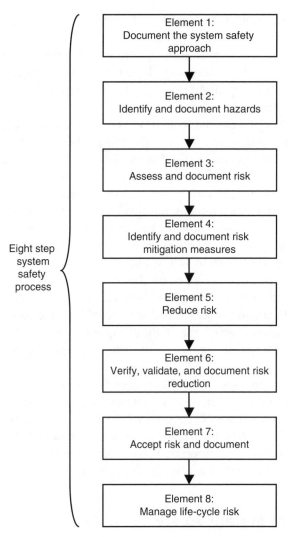

Figure 2.2 The system safety process [1]

hazards early in the design process, the costs associated with redesign are avoided. These costs may include fixing defects after a design is complete, redesigning a product due to the late discovery of a safety problem, retesting the product after a redesign, and the costs of delays caused by redesign.

Intangible benefits may also accrue as a result of a successful system safety program. For example, employee satisfaction may be improved due to participation in the process. The interfaces with customers and regulators may become more efficient given the transparent and participatory nature of the

Table 2.3 Common hazard analysis techniques

Analysis technique	Brief description
Preliminary Hazard List (PHL)	A compilation of potential hazards created early in development
Preliminary Hazard Analysis (PHA)	Identifies hazards, assesses the initial risks, and identifies potential mitigation measures
Subsystem Hazard Analysis (SSHA)	Used to verify subsystem compliance with requirements to eliminate hazards or reduce the associated risks. Also used to identify previously unidentified hazards associated with the design of subsystems and to make recommend actions to eliminate hazards or mitigate their risks
System Hazard Analysis (SHA)	Used to verify system compliance with requirements to eliminate hazards or reduce the associated risks. Also used to identify previously unidentified hazards associated with the subsystem interfaces and faults; to identify hazards associated with the integrated system design, including software and subsystem interfaces; and to make recommend actions to eliminate identified hazards or mitigate their risks
Operating & Support Hazard Analysis (O&SHA)	Used to identify and assess hazards introduced by operational and support activities and procedures and to evaluate the adequacy of operational and support procedures, facilities, processes, and equipment used to mitigate risks associated with these identified hazards
Health Hazard Analysis (HHA)	Used to identify human health hazards, to evaluate proposed hazardous materials and processes using such materials, and to propose measures to eliminate the hazards or reduce the associated risks when the hazards cannot be eliminated
Software Hazard Analysis (SWHA)	Software hazards should be identified as all the aforementioned hazard analyses are performed
Failure Modes and Effects Analysis (FMEA)	Used to determine the effects of failure modes of components, assemblies, subsystems, processes, or functions
Fault Tree Analysis (FTA)	A logical and graphical model of the system that shows various combinations of possible failures and non-failures that could lead to a previously defined undesirable event

system safety process. Customer satisfaction and confidence in the organization may be improved.

Therefore, system (or product) safety is an important aspect of design, and its implementation in any program will enhance the design process and result in better products being developed.

2.5 System Safety Management

System safety management includes all plans and actions taken to identify hazards, to assess and mitigate associated risks, and to track, control, accept, and document risks encountered in the design, development, test, acquisition, use, and disposal of systems, subsystems, equipment, and infrastructure.

The SSPP is the document that defines how system safety will be managed in a development program. The SSPP documents the system safety methodology for the identification, classification, and mitigation of hazards as part of the overall SE process. The SSPP provides the detailed tasks and activities that are required to implement a systematic approach of hazard analysis, risk assessment, and risk management. It defines how the system safety management effort will be integrated into the SE process.

At a minimum the following topics should be covered in an SSPP:

- Describe the risk management effort and how the program is integrating risk management into the SE process, the integrated product and process development process, and the overall program management structure.
- Identify and document the requirements applicable to the system to include standards, system specifications, design criteria, risk management requirements, and test requirements.
- Define how the appropriate risk acceptance authority will formally accept hazards and associated risks.
- Define how hazards will be documented and tracked with a closed-loop HTS.

A more detailed description of the SSPP and its contents will be discussed in Chapter 3.

2.6 Integrating System Safety into the Business Process

2.6.1 Contracting for System Safety

There are two aspects of contracting for system safety depending on your perspective. First, if you are the procuring agency, you are concerned with including system safety requirements in contracts. Second, if you are the contractor responding to a procuring agency, then you are concerned with meeting system safety requirements.

There are two main categories of contracts—fixed-price contracts and cost-reimbursement contracts. The major difference between these two types of contracts is who bears the financial risk and how much is borne by either party. With a fixed-price contract, the contractor assumes the financial risk in that the contractor has agreed to develop a particular product for an established price based on a specification provided by the customer. Should the contractor exceed the price originally estimated for the development, the contractor must still provide the product regardless of additional costs. With a cost-reimbursement contract, the customer assumes the financial risk in that the contractor develops the product based on the customer's specification and incurs costs, which are passed on to the customer regardless of whether the initial estimates of the cost are exceeded.

There are variations on these two categories of contract types, resulting in different balances of financial risk between the parties. The following are types of contracts that are typically used:

Increasing financial risk to seller

Firm Fixed-Price (FFP): These contracts require the contractor to deliver a product, as specified by the customer, for an agreed to price.

Fixed-Price with Award Fee (FP/AF): These contracts are fixed-price, but have award fees added as a way to motivate the contractor to perform above and beyond certain standards usually in schedule, quality, etc.

Fixed-Price with Economic Price Adjustment (FP/EPA): These contracts are also fixed price, but are used when markets are unstable to allow for unexpected variations in labor or commodities.

Fixed-Price Incentive (FPI): These contracts are fixed-price, but have incentives built into agreed to formulae to guarantee a certain profit.

Increasing financial risk to buyer

Cost-Plus Incentive Fee (CPIF): These contracts are used when a contract has little uncertainty.

Cost-Plus Award Fee (CPAF): These contracts are structured to provide a fixed award fee which is not tied to performance.

Cost-Plus Fixed Fee (CPFF): These contracts are used when there is uncertainty or development risk in the development of a product and provide a fixed fee in addition to the cost.

For all of these contracts, it is common practice for the procuring organization to issue a Request for Proposal (RFP), which will contain a Statement of Work (SOW) and a specification. The SOW provides in clear, understandable terms the work to be done in developing or producing the goods to be delivered or services to be performed by a contractor. It defines what is required in specific, performance-based, quantitative terms and is prepared in explicit terms that will enable the contractor to clearly understand the customer's needs. The specification provides the qualitative and quantitative design and performance requirements. It is typically referenced in the SOW, but the specific technical requirements are not in the SOW, but are in the specification.

From a system safety standpoint, the SOW will usually invoke top-level standards and requirements such as MIL-STD-882. It will define what is required to be in the system safety program for the particular development project. It will spell out documentation requirements for SSPP and various hazard analyses that may be required. The specification will generally provide more specific and detailed safety requirements specifically related to the hardware, equipment, or software being procured. For example, a detailed requirement in a specification might be "The on/off control mechanism shall not protrude beyond the equipment's surface."

It is important that both parties, the customer and the contractor, be aware of their roles in this arrangement. The customer must carefully determine what system safety tasks they want the contractor to perform, what system safety documentation they want produced by the contractor, and what safety requirements need to be met. The contractor must respond in their proposal to all of the customer's safety-related requirements, and later after the contract is awarded, they must perform the system safety tasks properly, conduct all the required analyses, provide the required documentation, and meet all of the customer's safety requirements.

References

[1] U.S. Department of Defense, *Standard Practice System Safety*, MIL-STD-882E, Department of Defense, Washington, DC, 2012.

[2] Maslow, A. H., A Theory of Human Motivation, *Originally Published in Psychological Review*, 50, 370–396, 1943.

[3] Code of Hammurabi, Circa 1754 BC. Based on the 1915 translation by L. W. King, http://avalon.law.yale.edu/ancient/hamframe.asp (Accessed on August 23, 2017)

[4] Wood, A. L., The Organization of an Aircraft Manufacturer's Air Safety Program, 14th Annual Meeting of the Institute of Aeronautical Sciences (IAS), New York City, 1948.

[5] Stieglitz, W., Engineering for Safety, Aeronautical Engineering Review, February 1948.

[6] Miller, C. O., Applying Lessons Learned from Accident Investigations to Design Through a Systems Safety Concept, Paper presented at the Flight Safety Foundation Seminar, Sante Fe, November 11, 1954.

[7] Olson, R. E., *System Safety Handbook for the Acquisition Manager*, Air Force Systems Command SD-GB-10, Space Division, Air Force Systems Command, Los Angeles, CA, 1982.

[8] U.S. Department of Defense, *Safety Engineering of Systems and Associated Subsystems and Equipment: General Requirements for*, MIL-S-38130A, Department of Defense, Washington, DC, June 1966.

[9] U.S. Department of Defense, *System Safety Program for Systems and Associated Subsystems and Equipment: Requirements for*, MIL-STD-882, Department of Defense, Washington, DC, 1969.

[10] U.S. Department of Defense, *System Safety Program Requirements*, MIL-STD-882A, Department of Defense, Washington, DC, 1977.

[11] U.S. Department of Defense, *System Safety Program Requirements*, MIL-STD-882B, Department of Defense, Washington, DC, 1984.

[12] U.S. Department of Defense, *System Safety Program Requirements*, MIL-STD-882C, Department of Defense, Washington, DC, 1993.

[13] SSS-STD-882, *System Safety Society Standard for System Safety Program Requirements*, The International System Safety Society, Unionville, VA, 1994.

[14] U.S. Department of Defense, *Standard Practice for System Safety*, MIL-STD-882D, Department of Defense, Washington, DC, 2000.

[15] NASA-STD-8719.7, *NASA Facility System Safety Guidebook*, National Aeronautics and Space Administration, Washington, DC, 1998.

[16] U.S. Federal Aviation Administration, *System Safety Handbook*, U.S. Department of Transportation, Federal Aviation Administration, Washington, DC, 2000.

[17] International Civil Aviation Organization, *Safety Management Manual*, ICAO SMM Doc. 9859, International Civil Aviation Organization, Montreal, Quebec, 2013.

[18] FAA Order 8000.369, *Safety Management System Guidance*, U.S. Department of Transportation, Federal Aviation Administration, Washington, DC, September 30, 2008.

[19] Society of Automotive Engineers, *Guidelines and Methods for Conducting the Safety Assessment Process on Civil Airborne Systems and Equipment*, ARP 4761, SAE International, The Engineering Society For Advancing Mobility Land Sea Air and Space, Warrendale, PA, 1996.

[20] Amtrak System Safety Program, Amtrak, Washington, DC, December 2007.

[21] Sammarco, J. J., *Programmable Electronic Mining Systems: Best Practice Recommendations*, Information Circular, 9456, U.S. Department of Health and Human Services, Public Health Service, Centers for Disease Control and Prevention, National Institute for Occupational Safety and Health, Pittsburgh Research Laboratory, Pittsburgh, PA, May 2001.

[22] U.S. Food Safety and Inspection Service, *Guidebook for the Preparation of HACCP Plans*, U.S. Department of Agriculture, Food Safety and Inspection Service, Washington, DC, September 1999.

[23] IEEE Computer Society; Software Engineering Standards Committee; Institute of Electrical and Electronics Engineers; IEEE Standards Board, *Standard for Software Safety Plans*, IEEE Standard 1228, IEEE, New York, 1994.

[24] International Electrotechnical Commission, *Functional Safety of Electrical/Electronic/ Programmable Electronic Safety-Related Systems (E/E/PE, or E/E/PES), Eight Parts*, International Electrotechnical Commission (IEC), Geneva, 2010.

[25] Standards Association of Australia; Standards New Zealand, *Risk Management: Principles and Guidelines*, ISO 31000, Standards Australia International, Sydney, 2009.

[26] South African Bureau of Standards, *Risk Management: Risk Assessment Techniques*, ISO 31010, SABS Standard Division, Pretoria, 2009.

Suggestions for Additional Reading

Raheja, D. and Allocco, M., *Assurance Technologies Principles and Practices*, John Wiley & Sons, Inc., Hoboken, NJ, 2006.

Raheja, D. and Gullo, L. J., *Design for Reliability*, John Wiley & Sons, Inc., Hoboken, NJ, 2012.

Roland, H. E. and Moriarty, B., *Systems Safety Engineering and Management*, John Wiley & Sons, Inc., New York, 1990.

3

System Safety Program Planning and Management

Louis J. Gullo and Jack Dixon

3.1 Management of the System Safety Program

System Safety Engineers (SSEs) apply various types of processes to the design and development of many types of products and systems. These products and systems provide a wide assortment of applications and solutions across different markets and industries. These products or systems offer benefits for customers, all within the military, government entities, and private industries. Some of these markets and industries, which are addressed in this book, are defense, aerospace, commercial avionics, automotive, telecommunication, medical, and retail consumer marketplaces. Formal planning and management for system safety is needed in order to effectively implement a System Safety Program (SSP) to ensure the development of safe products. There are different management styles and program requirements associated with diverse products in systems development, and there are different approaches to implementing an SSP. The key to implementing a successful SSP is to develop a system safety management plan that will reflect the details of how system safety is to be implemented. SSEs refer to this plan by several names depending on the industry and program. For the purposes

Design for Safety, First Edition. Edited by Louis J. Gullo and Jack Dixon.
© 2018 John Wiley & Sons Ltd. Published 2018 by John Wiley & Sons Ltd.

of this book, this plan is a System Safety Program Plan (SSPP), a terminology borrowed from the defense and aerospace industries.

Many other government organizations and industries have adopted the system safety approach to product/system development and have integrated system safety requirements into their safety management programs. The plans that are created and the requirements within the plans may differ depending on why, how, and where system safety engineering is being applied.

This chapter includes discussion as to what is necessary to establish appropriate safety management specifically related to implementing system safety engineering successfully throughout the system life cycle. The requirements for a successful SSP are emphasized. Flexibility is important in developing an SSP since its details may differ widely for simple systems, complex systems, System-of-Systems (SoS), and families of systems.

3.1.1 System Safety Management Considerations

In establishing a safety management program, there are many facets to contemplate including the various safety standards, the complexity of the product/system, the human element, and the risk involved.

Safety-related standards, codes, requirements, practices, rules, and procedures for particular industries or products have become extensive and must be addressed within safety management. Numerous guidelines and recommended practices have been established that detail program content, processes to be used, and procedures to follow.

Numerous other subjects that might be considered or involved in a particular SSP depending on the product type and its complexity include:

- Process safety
- Product safety
- Munitions safety
- Hazardous material safety
- Life safety
- Weapon safety
- Fleet safety
- Industrial hygiene
- Health hazards
- Aircraft/flight safety
- Environmental effects
- Fire safety
- Radiation safety
- Nuclear safety

- Range safety
- Construction/facility safety

The human side of the equation remains another important and most complex consideration. Humans are involved on many levels of product or system development, production, use, and misuse. Humans are often the initiators or contributors to accidents and mishaps, and their potential for making errors must be considered. The SSP must consider ergonomics in design, usability, and human limitations.

Understanding the risks involved with the product or system being developed requires the detailed analysis of hazards, the severity of their consequences, and the probability of their occurrence. Assessing this risk and finding and evaluating alternative designs or approaches that eliminate the hazard or reduce the risk imposed by it is the major objective of any SSP. The idealistic goal of system safety is to design systems that are totally free of risks.

All of these considerations lead to the conclusion that system safety management programs are needed in order to integrate, implement, and manage the many concerns related to developing a system or product that will be safe throughout its life cycle.

3.1.2 Management Methods and Concepts

The key elements within management may vary on the particular industry or entity. However, all management is concerned with cost, schedule, and performance. General management involves organizing, decision-making, planning, and control [1].

3.1.2.1 Organizing

One general aspect of management relates to the creation of the effective organization. Organizational design includes forming the organization and the organizational structure; defining authority, span of control, accountability, and line and staff relationships; achieving coordination among groups; and defining organizational boundaries.

From the system safety standpoint, organizing is an important facet of managing an SSP. An effective system safety organization and its functions must be defined along with how the system safety efforts fit within the systems engineering process. Staffing and other resources are planned to provide adequate support for the program. Roles and responsibilities of each person and each interfacing organizational unit involved in the SSP must be clearly delineated. If there is contractor and subcontractor involvement in the system safety effort, their roles, responsibilities, interfaces, and lines of communication must be defined.

Commitment is important. The safety engineers must be dedicated to producing a safe product. The systems engineering team and all the subject matter experts from all the various disciplines supporting the systems engineering effort must have the personal investment in the program to make it successful. Most importantly, management commitment is essential to having a successful SSP. Top management must believe that system safety is important and must stand behind their belief.

3.1.2.2 Decision-Making

One of management's central functions is making decisions. Management must establish a decision-making framework and define the decision-making process. They must also understand the differences between decision-makers, the rationality of decision-makers, the goals, the uncertainty, and the risk involved with making decisions.

The decision-making process always involves six steps:

- Setting goals
- Searching for alternatives
- Comparing and evaluating alternatives
- Choosing among alternatives
- Implementing the decision
- Follow-up, control, and feedback

From the system safety standpoint, goals and objectives for the SSP must be established early. Risk assessment and risk assessment criteria must be established in order to guide hazard analyses and risk mitigation. General design criteria, which include standards, guidelines, and top-level safety requirements, are part of the early decision-making process. This process also uses trade studies to suggest and evaluate preliminary design concepts. A Hazard Tracking System is defined early in the program's life cycle in order to provide tracking, control, and eventual mitigation of hazards.

3.1.2.3 Planning

Planning is the key management function for dealing with change. The major purposes of planning are to

- Anticipate and avoid problems
- Identify opportunities
- Develop courses of action that will help the organization reach its desired goals

Planning involves the identification and allocation of resources. These resources include money, personnel, materials, and equipment.

Most detailed planning uses a Program Evaluation and Review Technique (PERT). This method typically uses one of any number of commercially available computer programs that lay out the entire development and production schedule in a network. It provides for resource allocation, time and cost considerations, and dependencies among activities. The analysis of the network results in defining network paths and the critical paths within the network where delays will occur if the schedule is not adhered to. As the system matures, PERT becomes a control mechanism.

From the system safety standpoint, a schedule of system safety activities must be developed with adequate resources applied to the activities to ensure success. This schedule should include all the required system safety activities, their dependencies (i.e., what inputs are needed to accomplish the activity), and their output products (i.e., analysis, documents, etc.). All activities should have start and finish dates that mesh with the overall systems engineering effort and development plan. The system safety activities should also be properly integrated with other system-level activities such as technical reviews, testing, manual and training material development, and delivery milestones.

3.1.2.4 Control

Formal controls are a means to help the organization achieve its desired results. Who has the authority to do what to whom must be defined.

Control is used to

- Standardize performance
- Protect assets
- Standardize quality
- Limit the amount of authority that can be exercised by people or organizations
- Measuring and directing performance of personnel
- Achieve desired results through planning
- Coordinate workers and equipment toward the goals of the organization

From the system safety standpoint, the two main aspects of control are (1) to ensure that the SSP stays on track and (2) to ensure that hazards are identified, analyzed, and eliminated or controlled to an acceptable level of risk.

To ensure that the SSP stays on track, management must oversee the system safety activities to keep them on schedule, adequately resourced, and producing quality products. This is accomplished by following the plan that was established early on and reviewing the safety products as they are generated.

To ensure that hazards are identified and appropriately dispositioned, a Hazard Tracking System, with clearly defined authorities for closeout and acceptance of risk, must be in place and its use enforced.

3.2 Engineering Viewpoint

Before we delve into the processes used by system safety engineering, we must first understand systems engineering and the engineering viewpoint. Engineering concentrates on how to develop new systems and how to improve existing systems. There are several definitions of systems engineering, each having a slightly different meaning. Wikipedia [2] defines systems engineering as "an interdisciplinary field of engineering that focuses on how to design and manage complex engineering systems over their life cycles." It also defines a systems engineering V-Model [3] to explain systems engineering processes. The V-Model is discussed later in this chapter. INCOSE 2004 *Systems Engineering Handbook* [4] defines systems engineering as "an interdisciplinary approach and means to enable the realization of successful systems." The *NASA Systems Engineering Handbook* [5] defines it as "a methodical, disciplined approach for the design, realization, technical management, operations, and retirement of a system." The Systems Engineering Body of Knowledge (SEBoK) [6] defines four types of systems engineering:

1. Product Systems Engineering (PSE)
2. Enterprise Systems Engineering (ESE)
3. Service Systems Engineering (SSE)
4. SoS

PSE is the traditional systems engineering focused on the design of physical systems consisting of hardware and software. ESE pertains to the view of enterprises, organizations, or combinations of organizations, as systems. SSE is the engineering of service systems, such as Service Oriented Architecture (SOA). A service system may be a system designed for the intended purpose of servicing another system, such as many civil infrastructure systems.

Systems engineering combines and leads multiple engineering disciplines and areas of expertise driving toward a single objective. This objective is to design and manage a system over its entire system and/or product life cycle. Systems engineering on any particular systems should be a cradle-to-grave responsibility, which means that systems engineering begins at the concept stage—writing and analyzing requirements, continuing through systems design and development, proceeding to systems production and delivery to customers, following systems performance and improvements during operations and support in customer applications, and finishing when design life or service life is reached—and the systems is taken out of service. Disciplines such as requirements generation, requirements analysis, architectural design, detailed design, integration, verification and validation, project management, evaluation metrics and measurements, project and development team coordination, systems safety and reliability, life cycle logistics and supportability, and other management, technology, and engineering disciplines become more difficult as systems grow larger and more complex. Systems engineers work with development practices and processes, optimization models, methods and trade studies, and risk management tools and techniques within the scope of various programs and projects. Systems engineering maintains a holistic view of a system while focusing on system analyses and eliciting customer feedback in terms of user applications, system performance parameters, overall customer needs, problems to be resolved, and required functionality. This customer feedback is essential for system success and should include high-level system requirements documentation provided through program contractual channels. It should be collected as early as possible in the system development cycle. Once customer feedback is collected, and preliminary system analyses are conducted using the customer data, systems engineering begins to document system requirements, flowing down or decomposing requirements to lower levels of the system hierarchy. Then systems engineers proceed with design synthesis, simulation, tests, and system verification and validation while considering the complete system solution or end result, given the stress and environmental conditions the system may be exposed to over the system life cycle.

3.2.1 Software Tools

There are a multitude of tools that are used for systems engineering. There are books written on this subject. Our intent here is to make the reader aware that systems engineering uses software tools extensively and to focus on two types of software tools as a priority: requirements generation and tracking tools, such as IBM DOORS, and system architecture and design tools, such as IBM Rhapsody. Using these tools ensure efficient and effective systems design and engineering, prevent engineering mistakes and problems later in the development process, and ensure systems can be delivered on time and on budget with the level of quality and reliability expected by the customer.

3.2.2 Design Concepts and Strategy

System design concepts and strategies provide the engineering plans and vision for moving forward into system development with an eye on the goals and objectives. System development should begin with the end in mind and with an understanding of the approach and processes to be employed to reach the goals and objectives. These design concepts and strategies usually include high-level views of the system architecture. Use of IBM Rhapsody tool is very helpful in defining these system views and architecture through the use of System Modeling Language (sysML). Development of activity diagrams and sequence diagrams support articulation of the design concepts and strategies. Also, the Zachman Framework or Department of Defense Architecture Framework (DODAF) are useful means to develop the ultimate goals and visions for describing the way a system will function and perform all intended operations over the system life cycle. Object-oriented design is another means for planning and defining a system of interacting objects for the purpose of solving a software problem within a system design. System design concepts and strategies will be different from one type of system to another. Systems that include mission-critical or safety-critical functions will endure much more rigor in their development processes, compared with a system that does not include either.

3.2.3 System Development Process (SDP)

Systems engineering includes system life cycle process management. This book points to ISO/IEC/IEEE 15288 for the key processes used in the System Development Process (SDP). The system life cycle processes occur during the following stages: concept, development, production, utilization, support, and retirement, as defined in IEEE/IEC 15288 [7]. The concept stage is the initial steps in the SDP to define the system from a user-level perspective reflecting the system performance needs for the customer. One of the system definition documents is the Concept of the Operation (CONOPS). Requirements definition documents and architecture diagrams are created during the concept stage and completed during the development stage. During the development stage, detailed design, implementation, integration, test, verification, and validation processes are performed. The production stage occurs after validation and successful completion of system test. The utilization, support, and retirement stages are aligned with the operation and maintenance processes in the SDP. Risk management is a key process used throughout the stages of the system life cycle, and system safety is a discipline used within risk management to discover, predict, and prevent risks related to system design safety hazards. These stages in the system life cycle, which are shown in Figure 3.1, will be discussed in a sequential method in the subsequent chapters of the book as they relate to design for safety practices and functions.

3.2.4 Systems Engineering V-Model

The technical processes for systems engineering include stakeholder requirements definition, requirements analysis, architectural design, implementation, integration, verification, transition, validation, operation, maintenance, and disposal. These technical processes involve various types of engineering disciplines and functions, such as system safety. Most of these technical processes

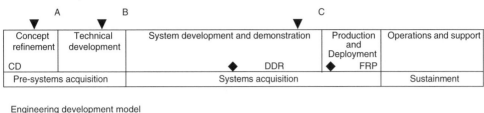

DoD 5000.02 acquisition model

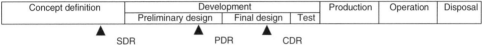

Engineering development model

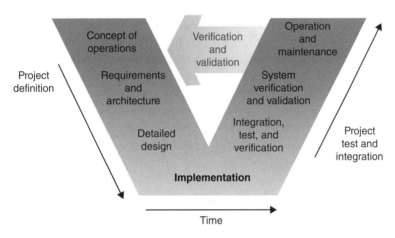

Figure 3.1 Phases of development

Figure 3.2 Systems engineering V-model [8]

are graphically depicted on the Systems Engineering V-Model. An illustration of the V-Model [8] is shown in Figure 3.2.

According to Wikipedia [3], the V-Model first appeared at Hughes Aircraft in 1982 during the proposal effort for the FAA Advanced Automation System (AAS) program. The V-Model formed the test strategy for the Hughes AAS program's Design Competition Phase (DCP) and was created to show the test and integration approach to address the new challenges in detection of latent software defects. The need for this new level of latent defect detection drove the goal to automate the planning processes of the air traffic controller as required by the Automated En route Air Traffic Control (AERA) program.

3.2.5 Requirements Generation and Analysis

Requirements definition and creation are critical in systems engineering. If the requirements are wrong, this has a ripple effect throughout the rest of the development stages. Requirements must be correct from the beginning of the system life cycle, or else costly rework will occur, and the later in the process a requirement defect is found and corrected, the more costly it is. Use of requirements generation and analysis tools, such as IBM DOORS, can certainly help the systems engineer in finding requirement gaps and defects early in the SDP. Many of the system level requirements involve specification of system functions. A typical system requirement may be stated as "the system shall perform the functions necessary to store and back up key system parametric data every 24 hrs." Systems engineering requirements are usually written in a positive form, such as "the system shall perform…." System safety requirements are usually written by SSEs in the negative form, such as "the system shall not…." A typical system safety requirement may be stated as "system operation shall not cause fatalities and bodily harm." It is very difficult; however, to verify these types of requirements, such as safety requirements written in a negative form. Since systems engineers want requirements that are verifiable, system requirements are written in a positive form. System safety requirements may be written by a systems engineer or SSE in a positive form, such as "the system operation shall prevent personal injury, avoid catastrophic damage, and allow only safe conditions" or "the system shall predict performance in an environment where no one gets hurt." Examples of sources for system safety requirements may be found in

- OSHA Law and Regulations (Standards—29 CFR)
- IEC 61508. IEC 61511
- ISA 84.00.01-2004
- Safety requirements spec. Fse-global.com

Following the creation and documentation of requirements in specifications, the systems engineer will analyze the requirements. This analysis is performed to fill any gaps in the requirements that may have been missed when requirements are flowed down from top-system specifications, such as in System-of-Systems Integration (SoSI), or in the decomposition of requirements that are apportioned or allocated to subsystem specifications. Requirements analysis will include the tracing of requirements through the specification tree where every specification in the system is identified in a tree structure configuration or hierarchy. In performance of traceability analysis, an engineer will look for not only requirements gaps but also requirement linkages with forward and reverse traceability. The purpose is to ensure there are no lower-level orphan requirements, which are requirements with no parent requirement, or no childless parent requirements. In other words, each top-level requirement must be either flowed down or decomposed to lower-level requirements. The analysis also looks for requirement redundancies, ambiguities, vague requirements, unverifiable requirements, or requirement defects that contradict one another.

3.2.6 System Analysis

Requirements analysis is the first form of system analysis. System analyses are performed throughout the stages of the V-Model and involve preparation of system block diagrams to assess performance of the system. Functional block diagrams are created for each system block. Analysis of the functions is critical to deciding if a function is to be allocated as hardware or software or a combination of both. System analysis includes assessment using models and simulations and execution of algorithms used in system performance processing to determine if the algorithm is performing as required, prior to building the first system prototypes and testing the system. Use of system architecture and design tools, such as IBM Rhapsody, certainly helps the systems engineer in performing system analyses, in finding performance gaps and defects early in the SDP.

3.2.7 System Testing

During the course of system analysis, system testing will begin. The primary reason for system test is to Verify and Validate (V&V) system requirements in a controlled lab or testing area where environmental conditions are controllable. This is not intended to mean that all system requirements are verified and validated by testing in a lab. Requirements V&V may be accomplished by inspection, analysis, test, or demonstration. Usually system analysis and system testing activities overlap in the SDP since the two system functions learn something in one application and attempt to duplicate it in another application. Correlation between analysis results and test results are important to build confidence in the system performance and ensure that when a system passes a critical test milestone, it is certainly ready to progress to the next step in development or ready to be released to the customer. Demonstrations with the customer are usually a formal level of system testing, which may be performed to ensure a system performs as intended within the customer use applications and environments. Use of IBM DOORS is critical to efficient V&V of system requirements, as well as V&V for all lower level specification requirements, such as hardware and software requirements. Hardware and software testings are performed independently and V&V is accomplished at the hardware and software levels of design, prior to starting system integration and testing. Once the hardware and the software are integrated, then the system testing is performed.

3.2.8 Risk Management

Risk management is a key practice within the program, systems, and project management processes, which is employed throughout all the stages of the system V-Model. Risk management processes are integrated throughout the system life cycle processes with all key interfaces involved. Risk management includes the identification and categorization of risks, the assessment of the probabilities of occurrences and consequences of the risks, risk mitigation strategies, risk status tracking and reporting, and risk action plans to ensure unacceptable risks will not occur or minimize their probabilities of occurrence. System safety engineering, as well as other key engineering disciplines, performs these types of risk management functions. Interfaces between program/project management, engineering, quality, operations, Information Systems (IS), and other professionals within an organization contribute to the risk management planning, processes, and execution throughout the system life cycle processes and stages. More on risk management processes will follow later in the book.

3.3 Safety Integrated in Systems Engineering

System safety is one of many elements of a good systems engineering program. As such, it must be an integral part of each of the system life cycle phases.

During the concept stage safety objectives and top-level safety requirements are developed and documented in the system-level specifications. The SSOO is developed during this phase. Typically, a Preliminary Hazard List (PHL) is developed based on the generic hazards that can be identified from conceptual designs. A Preliminary Hazard Analysis (PHA) is performed as concepts are definitized.

During the development stage more detailed hazard analyses are conducted as the product/system design becomes better defined. These may include Subsystem Hazard Analysis (SSHA), System Hazard Analysis (SHA), Operating & Support Hazard Analysis (O&SHA), and so on.

During the production stage, the various analyses may be updated as other hazards are identified, or the design is refined to accommodate changes necessary for production.

During the utilization, support, and retirement stages, the various analyses may again be updated as other hazards are identified during early deployment or use. In particular, O&SHA may need to be updated as experience is gained with the product or system.

For the purposes of supporting the Systems Engineering V-Model, an SSE may construct a similar V-Model, the System Safety V-Chart, to show how the SSE integrates with the SDP. This system safety process follows the systems engineering process as shown in the System Safety V-Chart in Figure 3.3.

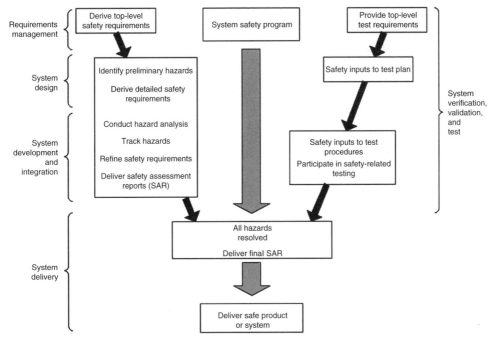

Figure 3.3 System safety V-Chart

3.4 Key Interfaces

Systems engineering has interfaces with multiple engineering specialties and technical and human-centered disciplines. Systems engineering ensures that all likely aspects of a project or systems are considered and integrated into a whole.

Key interfaces with systems engineering include, but are not limited to, the following disciplines:

- Business development (including customer engineering, application engineering, and marketing)
- Configuration management
- Control engineering
- Electrical engineering (including circuit designers for analog, digital, RF, power, and component devices)
- Human factors engineering (including Human System Integration (HSI))
- Industrial engineering
- Interface design
- Logistics and supportability engineering
- Manufacturing and production engineering

- Mechanical engineering
- Operations research
- Performance engineering
- Program and project management (including planning, scheduling, and proposal management)
- Quality engineering (including design assurance, design for six sigma engineering, mission assurance, systems assurance, and product assurance)
- Reliability, Maintainability, and Availability (RMA) engineering
- Risk management
- Environmental Health and Safety Engineering (EHSE)
- Supply chain engineering
- System Security Engineering (SSE) (including Information Assurance (IA), Operations Security (OPSEC), Communications Security (COMSEC), and anti-tamper engineering)
- Software engineering (including software assurance)
- Test engineering

System safety engineering interfaces with many of these disciplines within systems engineering process. SSEs are typically most involved with the disciplines of reliability, maintainability, quality, logistics, human factors, software engineering, and test engineering. These disciplines should be involved in the hazard analysis, hazard control, hazard tracking, and risk resolution activities.

3.5 Planning, Execution, and Documentation

Just as the key to any successful program is good planning, proper execution, and high-quality documentation, so it is with a successful SSP. Creating a comprehensive SSPP up front gets the safety program off to the right start and monitoring, and controlling the system safety activities keeps the safety program on track throughout the life cycle. Providing outstanding documentation for the various types of analyses and assessments helps ensure a safe product and customer satisfaction.

3.5.1 System Safety Program Plan

The SSPP is the guiding document for the management of any SSP. The tasks and activities of system safety management and engineering are defined in the SSPP. The SSPP as outlined in this section is patterned after MIL-STD-882E [9].

The goal is to develop a plan that is specifically designed to suit the particular product, project, process, operation, or system. A good plan is tailored to appropriately treat the complexities associated with the organizations involved and the product or system being developed. Tasks included in the SSPP should fit the product or system at hand, and the schedule of activities should be realistic. The major elements of the plan are highlighted in the following paragraphs.

3.5.1.1 Definition of the System Safety Program Plan

The SSPP details the task and activities of system safety management and system safety engineering required to identify, evaluate, and eliminate or control hazards throughout the changes from the baseline configuration. The SSPP describes fully the planned safety tasks and activities required to meet the SSP requirements.

3.5.1.2 Contents of the System Safety Program Plan

The SSPP typically contains the following information (adapted from DI-SAFT-81626 [10]):

System safety organization

- How the system safety organization fits within the organization of the total program.
- Lines of communication.
- Responsibility, authority, and accountability of system safety personnel, contractors, subcontractors, and system safety working groups.
- Identify the organizational unit responsible for executing each task.
- Identify the authority in regard to resolution of all identified hazards.
- Define the SSE staffing of the system safety organization for the entire effort.
- The process through which management decisions will be made regarding critical and catastrophic hazards, corrective action taken, mishaps or malfunctions, waivers to safety requirements, and program deviations.

SSP milestones

- Identify safety milestones.
- Provide a program schedule of safety tasks in relationship to other program milestones showing predecessor and successor tasks and whether tasks are specified on the critical chain.
- Identify integrated system activities (i.e., design analyses, tests, and demonstrations) applicable to the SSP, but specified in other engineering studies.

System safety requirements

- Describe or reference the methods that will be used to identify and apply safety/hazard control.
- List the safety standards and system specifications that are the sources of safety requirements that apply.
- Describe the risk assessment procedures.
- Define hazard severity categories, hazard probability levels, and system safety precedence to be followed in satisfying safety requirements.
- Describe the integration of subcontractor equipment safety information.

Hazard analysis

- The analysis technique and format that will be used in qualitative and quantitative analysis to identify hazards, their causes and effects, and recommended corrective action.
- The depth within the system to which each analysis technique will be used.
- The technique for establishing a single closed-loop Hazard Tracking System.

Safety verification

- The verification requirements for ensuring that safety is adequately demonstrated by analysis, inspection, and/or testing

Training

- Describe safety training for engineers, technicians, and operating and maintenance personnel.

Mishap reporting and investigation

- Describe the mishap and hazardous malfunction analysis process for mishaps prior to delivery of the product or system.

System safety interfaces

- Identify the interface between system and safety and all other applicable disciplines, such as maintainability, quality assurance, reliability, human factors engineering, transportability engineering, and medical support (health hazard assessments).

3.5.1.3 System Safety Analyses

To develop a safe system, numerous types of system safety analyses may need to be utilized. The goal of all safety analyses is to identify hazards associated with the product or system being developed and to assess the risk those hazards they present. The objective is to eliminate or control the risk to an acceptable level.

There are many types of system safety analyses. The approach to risk management and the various analyses planned for a particular development program should be delineated in the SSPP and the analyses scheduled at the appropriate time in the life cycle. A number of the analysis techniques will be described in later chapters of this book as shown in Table 3.1.

3.5.1.4 Tasks and Tailoring

In addition to the various safety analyses that will be applied, there are many other system safety-related tasks that should be included in the SSPP. MIL-STD-882E provides a good guideline for which safety analyses and which other system safety tasks should be applied during various phases of system development. Table 3.2 is an adaptation of a table from MIL-STD-882E showing typical

Table 3.1 Types of system safety analysis

Analysis	Chapter location
Risk management	Chapter 4
Preliminary Hazard List (PHL)	Chapter 7
Preliminary Hazard Analysis (PHA)	
Subsystem Hazard Analysis (SSHA)	
System Hazard Analysis (SHA)	
Operating & Support Hazard Analysis (O&SHA)	
Health Hazard Analysis (HHA)	
Failure Modes Effects, and Criticality Analysis (FMECA)	Chapter 8
Fault Tree Analysis (FTA)	Chapter 9
Event trees	Chapter 10
Sneak Circuit Analysis (SCA)	
Functional Hazard Analysis (FuHA)	
Barrier analysis	
Bent pin analysis	
Petri nets	
Markov analysis	
Management Oversight Risk Tree (MORT) analysis	
System-Theoretic Process Analysis (STPA)	
Human Factors Engineering (HFE) analysis	Chapter 15

Table 3.2 Task application matrix

Task	Title	Task type	MSA	TD	EMD	P&D	O&S
101	Hazard Identification and Mitigation Effort Using the System Safety Methodology	MGT	G	G	G	G	G
102	System Safety Program Plan	MGT	G	G	G	G	G
103	Hazard Management Plan	MGT	G	G	G	G	G
104	Support of Government Reviews/Audits	MGT	G	G	G	G	G
105	Integrated Product Team/Working Group Support	MGT	G	G	G	G	G
106	Hazard Tracking System	MGT	S	G	G	G	G
107	Hazard Management Progress Report	MGT	G	G	G	G	G
108	Hazardous Materials Management Plan	MGT	S	G	G	G	G
201	Preliminary Hazard List	ENG	G	S	S	GC	GC
202	Preliminary Hazard Analysis	ENG	S	S	S	GC	GC
203	System Requirements Hazard Analysis	ENG	G	G	G	GC	GC
204	Subsystem Hazard Analysis	ENG	N/A	G	G	GC	GC
205	System Hazard Analysis	ENG	N/A	G	G	GC	GC
206	Operating and Support Hazard Analysis	ENG	S	G	G	G	S
207	Health Hazard Analysis	ENG	S	G	G	GC	GC
208	Functional Hazard Analysis	ENG	S	G	G	GC	GC
N/A	Failure Modes, Effects, and Criticality Analysis	ENG	N/A	G	G	GC	GC
209	System-of-Systems Hazard Analysis	ENG	N/A	G	G	GC	GC
210	Environmental Hazard Analysis	ENG	S	G	G	GC	GC
301	Safety Assessment Report	ENG	S	G	G	G	GC
302	Hazard Management Assessment Report	ENG	S	G	G	G	S
303	Test and Evaluation Participation	ENG	G	G	G	G	S

		Task type	MSA	TD	EMD	P&D	O&S
304	Review of Engineering Change Proposals, Change Notices, Deficiency Reports, Mishaps, and Requests for Deviation/Waiver	ENG	N/A	S	G	G	G
401	Safety Verification	ENG	N/A	S	G	G	S
402	Explosives Hazard Classification Data	ENG	N/A	S	G	G	GC
403	Explosive Ordnance Disposal Data	ENG	N/A	S	G	G	S

Task type	Program phase	Applicability codes
ENG: Engineering	MSA: Materiel Solution Analysis	G: Generally Applicable
MGT: Management	TD: Technology Development	S: Selectively Applicable
	EMD: Engineering and Manufacturing Development	GC: Generally Applicable to Design Change
	P&D: Production and Deployment	N/A: Not Applicable
	O&S: Operations and Support	

selections of tasks (keyed to task descriptions in MIL-STD-882E) and the tailoring that might be used for each of them.

3.5.2 Safety Assessment Report

The Safety Assessment Report (SAR) is one of the most important system safety documents done for product/system development. The SAR is a comprehensive evaluation of the safety risks being assumed prior to test or operation of the system and/or at contract completion. It identifies all safety features of the system, design, and procedural hazards that may be present in the system being developed, and specific procedural controls and precautions that should be followed when using the product/system. The SAR delivery schedule should be planned early in the program and should be included in the SSPP.

3.5.2.1 Contents of the Safety Assessment Report

The SAR typically contains the following information (adapted from DI-SAFT-80102B [11]):

Introduction

- The purpose of the SAR

System description

- Reference other program documentation such as technical manuals, SSPP, system specification, etc.
- The purpose and intended use of the system
- A brief historical summary of system development
- A brief description of the system and its components
- Software and its roles
- A description of any other system(s), which will be tested or operated in combination with this system
- Include photos, charts, flow/functional diagrams, sketches, or schematics to support the system description, test, or operation

System operations

- A description of the procedures for operating, testing, and maintaining the system
- The safety design features and controls incorporated into the system as they relate to the operating procedures

- A description of any special safety procedures needed to assure safe operations, test, and maintenance, including emergency procedures
- A description of anticipated operating environments, and any specific skills required for safe operation, test, maintenance, transportation, or disposal
- A description of any special facility requirements or personal equipment to support the system

Systems safety engineering

- Summarize the safety criteria and methodology used to classify and rank hazardous conditions.
- A description of the analyses and tests performed to identify hazardous conditions inherent in the system.
- Include a list of all hazards by subsystem or major component level that have been identified and considered from the inception of the program.
  - A discussion of the hazards and the actions that have been taken to eliminate or control these items.
  - A discussion of the effects of these controls on the probability of occurrence and severity level of the potential mishaps.
  - A discussion of the residual risks that remain after the controls are applied or for which no controls could be applied.
- A discussion of the results of tests conducted to validate safety criteria requirements and analyses.

Conclusions and recommendations

- Include a short assessment of the results of the safety program efforts.
- List all significant hazards along with specific safety recommendations or precautions required to ensure the safety of personnel and property. The list of hazards should be categorized as to whether or not they may be expected under normal or abnormal operating conditions.
- For all hazardous materials generated by or used in the system, the following information shall be included.
  - Material identification as to type, quantity, and potential hazards.
  - Safety precautions and procedures necessary during use, storage, transportation, and disposal.
  - A copy of the Material Safety Data Sheet (OSHA Form 20 or DD Form 1813).
- Include a statement that the system does not contain or generate hazardous materials (i.e., explosive, toxic, radioactive, carcinogenic, etc.).

- A statement signed by the system safety manager and the program manager stating that all identified hazards have been eliminated or controlled and that the system is ready to test, operate, or proceed to the next acquisition phase.
- Include recommendations applicable to the safe interface of this system with the other system(s).

References

- A list of all pertinent references such as test reports, preliminary operating manuals, and maintenance manuals

3.5.3 Plans Related to System Safety

There are other plans that may appear in a particular system development effort that are related to system safety. It is important for the system safety engineer to be aware of these plans and to participate in their creation and implementation.

Systems Engineering Management Plan (SEMP)
SEMP is the overarching systems engineering plan that describes the entire systems engineering program for the product/system under development. It will typically address the SSP as one of its sections or may actually include a SSPP as an appendix.

Hazardous Materials Management Plan (HMMP)
HMMP is a document that lists all hazardous materials used on the program. It contains information about the material, where it is used, how much is used, and how it is to be disposed of.

Reliability Program Plan (RPP)
RPP is the reliability engineering's equivalent of the SSPP. It will describe the management of the reliability program, what analyses are planned, and how failures will be dealt with.

Test plan
Test planning documents are very important to system safety engineering. There may be numerous test plans for tests that will be conducted at various points during the program. Safety engineering should participate in the development of test plans to ensure that safety testing is conducted at the appropriate time to verify compliance with safety requirements.

3.6 System Safety Tasks

There are numerous safety-related tasks to be done during a development program many of which were described earlier in this chapter.

We have discussed those most closely associated with the various types of hazard analyses and have provided some guidelines on when those analyses should be accomplished. We will discuss many of these hazard analyses in much greater detail in later chapters.

Other safety-related tasks that will be discussed include the following.

Using design checklists will be covered in Chapter 6. While checklists should not be relied upon as the only safety tool to be used, checklists provide valuable clues and guidelines that can help the system safety analyst to conduct hazard analyses. They can also provide a tool that can be used during safety testing.

Risk data collection, risk assessment, risk mitigation, and risk management are key functions for the proper application of system safety. They help ensure that the risk associated with using the product or system being developed will be minimized. Chapter 7 will discuss these topics in great detail.

Safety testing has been alluded to several times in this chapter. Chapter 12 will expand upon the subject of testing as it relates to safety and to proving that the product or system under development is going to be safe to operate and maintain. Topics discussed will include the types of testing that should be performed, when it is conducted, and what to do with the results of the various types of testing.

References

[1] Helllriegel, D. and Slocum, J. W., *Management: Contingency Approaches*, Addison-Wesley, Reading, MA, 1978.

[2] Wikipedia, Systems Engineering, http://en.wikipedia.org/wiki/Systems_engineering (Accessed on July 31, 2017).

[3] Wikipedia, V-Model, http://en.wikipedia.org/wiki/V-Model (Accessed on July 31, 2017).

[4] INCOSE, *Systems Engineering Handbook*, 2004, INCOSE, Seattle, WA.

[5] NASA, *NASA Systems Engineering Handbook*, NASA/SP-2007-6105 Rev 1, NASA, Washington, DC, 2007.

[6] BKCASE Editorial Board, *The Guide to the Systems Engineering Body of Knowledge (SEBoK), v. 1.3.1, R.D. Adcock (EIC)*, The Trustees of the Stevens Institute of Technology, Hoboken, NJ, 2014, http://www.sebokwiki.org (Accessed on February 4, 2015). BKCASE is managed and maintained by the Stevens Institute of Technology Systems Engineering Research Center, the International Council on Systems Engineering, and the Institute of Electrical and Electronics Engineers Computer Society.

[7] ISO/IEC 15288, *Systems and Software Engineering: System Life Cycle Processes*, IOS/IEC, Geneva, 2008.

[8] Osborne, O., Brummond, J., Hart, R., Zarean, M., and Conger, S., *Clarus: Concept of Operations*, Publication No. FHWA-JPO-05-072, Federal Highway Administration (FHWA), 2005, http://ntl.bts.gov/lib/jpodocs/repts_te/14158.htm (Accessed on July 31, 2017).

[9] MIL-STD-882E, *System Safety*, U.S. Department of Defense, Washington, DC, 2012.

[10] DI-SAFT-81626, *Data Item Description, System Safety Program Plan*, U.S. Department of Defense, Washington, DC, 2001.

[11] DI-SAFT-80102B, Data Item Description: Safety Assessment Report, U.S. Department of Defense, Washington, DC, 1995.

Suggestions for Additional Reading

Checkland, P., *Systems Thinking, Systems Practice: Includes a 30-Year Retrospective*, John Wiley & Sons, Inc., Hoboken, NJ, 1999.

Raheja, D. and Allocco, M., *Assurance Technologies Principles and Practices*, John Wiley & Sons, Inc., Hoboken, NJ, 2006.

Raheja, D. and Gullo, L. J., *Design for Reliability*, John Wiley & Sons, Inc., Hoboken, NJ, 2012.

4

Managing Risks and Product Liabilities

Louis J. Gullo and Jack Dixon

4.1 Introduction

Product liability can present a substantial problem for organizations involved in the development, production, and sale of products and systems. Product defects can result in liability of the manufacturer for injuries or damage caused by its products.

To put product liability in perspective, the following table (Table 4.1) extracted from the Bureau of Justice Statistics shows the number of federal product liability cases and their awards between 1990 and 2003 [1]. In reviewing these statistics, it is obvious how product liability can be extremely costly. For instance, notice that the estimated median award for product liability trials in 2000 was over $1 million. Between the years 1990 and 2003, there is quite a bit of fluctuation in the estimated median awards. The lowest estimate for median awards was in 2002 with $306K. It should be mentioned that these estimates are only for those cases brought to federal court and do not include the many other cases settled in state, district, and local courts.

If we look at a sample of state court statistics from just one state, we find even more reason to worry about product liability. See Table 4.2 for 2014 product liability data from Missouri [2]. Here we see over $17 million was paid out in settlements.

Design for Safety, First Edition. Edited by Louis J. Gullo and Jack Dixon.
© 2018 John Wiley & Sons Ltd. Published 2018 by John Wiley & Sons Ltd.

Table 4.1 Product liability in federal court

Plaintiff winners in non-asbestos product liability trials terminated in US district courts, 1990–2003

Fiscal year	Number of non-asbestos product liability trials[1]	Percent of plaintiff winners[2]	Number of plaintiff		
			Winners	Awarded damages[3]	Estimated median award[4] ($)
1990	279	35.5	99	89	783,000
1991	284	33.1	94	85	923,000
1992	267	34.8	93	76	847,000
1993	237	29.5	70	66	673,000
1994	255	27.1	69	64	341,000
1995	229	29.7	68	61	355,000
1996	201	28.4	57	45	433,000
1997	233	29.2	68	59	805,000
1998	177	32.2	57	51	339,000
1999	165	26.7	44	36	783,000
2000	100	28.0	28	24	1,024,000
2001	79	38.0	30	22	702,000
2002	107	33.6	36	30	306,000
2003	87	33.3	29	25	450,000

Data Source: Federal Judicial Center, Integrated Data Base (Civil), fiscal years 1990–2003.
Note: Damage awards are adjusted for inflation in 2003 dollars. Award data are rounded to the nearest thousand.
[1] The number of non-asbestos product liability trails is limited to those with a known judgment.
[2] The plaintiff winner statistic does not include tort trials in which both the plaintiff and the defendant won.
[3] Award data were not available for all plaintiff winners because the award field is not mandatory for data entry. In addition, some plaintiff winners were awarded attorneys' fees and court costs, while others were awarded a judgment in the form of an injunction. These plaintiffs were also not counted as award winners.
[4] Monetary damage awards are estimates rather than exact award amounts. These awards include both compensatory (economic and noneconomic) and punitive damage awards. Punitive damages could not be calculated separately from the actual monetary damage award because punitive award data are not available in the integrated federal data file.

The same report summarized data for the 10 years between 2004 and 2014 and shows a total of $361 million paid out in settlements as shown in Table 4.3.

Another important area to consider in product liability is product safety recalls. Recalls may be optional or mandatory, depending on the nature of the problem, the risks, the costs, and the safety implications and effects. If products have substantial risks of catastrophic effects and damages, the manufacturer often recalls

Table 4.2 Product liability in Missouri 2014

Product liability by business classification for 2014

Business classification	Percentage of all paid claims	All paid claims	Average indemnity ($)	Total indemnity paid ($)	Average loss expense on paid claims ($)	Average initial reserve on paid claims ($)	Average no. of months report to close
Subcontractor to manufacturer	7.58	46	42,766	1,967,218	10,326	4,561	14
Manufacturer	64.74	393	23,791	9,350,042	6,860	6,667	13
Wholesaler	0.82	5	20,233	101,165	2,609	20,154	21
Retailer	15.49	94	31,223	2,934,967	2,869	28,554	8
Servicer–repairer	7.91	48	34,022	1,633,050	10,734	6,749	12
Distributor	3.46	21	87,857	1,844,997	15,526	43,985	13
Total	**100.00**	**607**	**29,376**	**17,831,439**	**7,076**	**11,306**	**13**

Table 4.3 Product liability in Missouri: 10-year summary 2005–2014

Product liability 10-year summary by business classification for 2005–2014

Business classification	Percentage of all paid claims	All paid claims	Average indemnity ($)	Total indemnity Paid ($)	Average loss expense on paid claims ($)	Average initial reserve on paid claims ($)	Average no. of months report to close
Subcontractor to manufacturer	7.89	601	38,666	23,238,417	8,731	7,287	16
Manufacturer	57.01	4,341	56,881	246,921,779	10,464	9,968	17
Wholesaler	0.93	71	78,542	5,576,499	17,198	15,554	31
Retailer	10.64	810	28,061	22,729,157	4,503	9,665	10
Servicer–repairer	10.33	787	35,737	28,124,764	6,644	8,600	11
Distributor	13.20	1,005	35,052	35,227,156	6,982	12,261	24
Total	**100.00**	**7,615**	**47,514**	**361,817,772**	**8,901**	**9,937**	**17**

them in an urgent manner, and sometimes, the government requires the recalls according to a time schedule. Probably, the high visibility product safety recalls that are most often in the news are automobile-related recalls.

The National Highway Traffic Safety Administration (NHTSA) marked a record year in 2014, with the highest number of vehicle recalls in one year due to the highest number of NHTSA investigations and enforcement efforts in more than three decades. The annual recall report shows that, in 2014, there were 803 vehicle recalls affecting 63.9 million vehicles, which included two of the largest vehicle recalls in history. NHTSA investigation and enforcement efforts influenced 123 of these vehicle recalls that affected 19.1 million vehicles. Table 4.4 shows the 10 largest recalls in 2014 [3].

After that banner year for recalls, news broke in 2015 of what would become the largest recall ever involving the Takata-made airbag inflators. The airbag inflators were failing and were suspected of causing explosive ruptures that have caused 7 deaths and nearly 100 injuries in the United States. Not only was the recall instituted by NHTSA, but the agency imposed the largest civil penalty ever of $200 million on Takata. Because the safety defect repairs were likely to put Americans at risk, the NHTSA for the first time invoked its legal authority granted under the 2000 TREAD Act to impose an accelerated recall. At that time the recalls involved more than 23 million inflators, 19 million vehicles, and 12 automakers [4].

Later, in 2016, after NHTSA's confirmation of the root cause behind the inflators' propensity to rupture, and the ruptures of the Takata inflators have been tied to 10 deaths and more than 100 injuries in the United States, NHTSA expanded the recall by an additional estimated 35–40 million inflators, adding to the already 28.8 million inflators previously recalled [5].

Table 4.4 Ten largest vehicle recalls

Manufacturer	Recall campaign	Component	No. of vehicles recalled
Ford Motor Company	81V008	Parking gear	21,000,000
Ford Motor Company	96V071	Ignition	7,900,000
General Motors LLC	71V235	Engine mounts	6,682,084
General Motors LLC	14V400	Ignition switch	5,877,718
General Motors LLC	81V025	Control arm	5,821,160
Honda (American Honda Motor Co.)	14V351	Air bags	5,394,000
Ford Motor Company	05V388	Deactivation switch	4,500,000
Ford Motor Company	09V399	Deactivation switch	4,500,000
Toyota Motor Corporation	09V388	Pedal entrapment	4,445,056
Ford Motor Company	72V160	Shoulder belts	4,072,000

This Takata recall just keeps growing. As of October 2016 Toyota is recalling an additional 5.8 million vehicles in Japan, Europe, China, and elsewhere to replace the defective airbag inflators. According to Toyota, this brings the recall to 231 million vehicles it has recalled globally to fix the inflators. There have been 16 deaths reported globally linked to Takata with 17 automakers recalling approximately 69 million inflators in the United States and 100 million worldwide [6].

FINAL UPDATE: On June 25, 2017, Takata filed for bankruptcy [7]. After the auto industry's largest product recall ever and 17 deaths due to defective airbags, it is estimated that Takata's total liabilities are currently $17 billion and may increase further as negotiations with carmakers continue.

So, as can be seen from this small sampling of data, product liability is widespread, covering many products, and costing the automotive industry a lot of money. Product liability is risky business! Safe design is good business.

4.2 Risk

Risk is defined by the Merriam-Webster Dictionary as "the possibility that something bad or unpleasant (such as an injury or a loss) will happen" [8].

There are many types of risk. Risk from a safety standpoint is usually associated with an injury or an accident of some kind. As the definition implies, however, risk can be associated with any kind of loss. Most loss can be considered economic; even accidents and injuries have economic loss in addition to the obvious pain and suffering. Some typical categories of risk include:

- Safety risk
- Health risk
- Environmental risk
- Program risk
- Security risk
- Financial risk
- Insurance risk
- Political risk
- Technological risk
- Portfolio risk
- Ecological risk [9]

There are many risks associated with product liability. Obviously, there is the risk of an injury or accident. In order to manage the risk associated with injuries and accidents, we use the tools of system safety to eliminate or control the risk to an acceptable level. Hazards associated with the product must be identified early in the design process. These hazards must be evaluated for their likelihood of occurrence and the consequences of their occurrence. Ways to eliminate the hazard must be determined and if the hazard cannot be eliminated completely, and then ways to mitigate the risk involved must be found.

Other types of risk that are relevant to the discussion of product liability include the economic impact of a product liability event:

- How much is defending a product liability case going to cost?
- How much is a product recall going to cost?
- How much product will have to be scrapped or reworked?
- How much will redesign cost?
- How many customers will the organization lose?
- What damage will be done to the company's brand?
- How will customer confidence and loyalty be affected by the bad press coverage?

4.3 Risk Management

Risk management is recognized as an essential contributor to business and project success. Risk management focuses on addressing uncertainties in a proactive manner to minimize threats, maximize opportunities, and optimize achievement of objectives. There is a wide convergence and international consensus on the necessary elements for a risk management process. This joint vision of a risk management process is supported by a growing range of capable tools and techniques, an accepted body of knowledge, an academic and research base, and a wide experience of practical implementation across many industries. Despite this vision, risk management often fails to meet expectations, as demonstrated by the continued history of business failures. Foreseeable threats materialize into problems and crises, and achievable opportunities are missed, leading to lost revenue and bankruptcies. The mere existence of accepted principles, well-defined processes, and widespread documented practice is not sufficient to guarantee success and avoid costly liabilities. Other essential ingredients are missing [10].

Risks escalate and costly product liabilities surface when a designer's knowledge of a specific system's performance is lacking a deep understanding of how and why it works, and the person does not understand how to design for safety for that particular system. Anyone designing for safety should realize that there is no substitute for first-hand knowledge of a system's operating characteristics, architecture, design topology, and design margin or safety factors proven through analysis, simulation, or empirical testing. When this knowledge is lacking in the mind of the responsible design engineer, bad things happen. When an organization does not take an appropriate systems approach to system design and development and omits many internal and external interrelated company business connections and interactions, then a system is prone to cause undesirable events and uncontrolled circumstances beyond the control of the organization. When all the connections and interactions are not properly working together to accomplish a shared aim, a business can achieve terrible results, from degrading the safety posture of its systems, products, and services to decreasing the creativity of an organization and contributing to its inability to develop innovative solutions to help mankind progress.

Risk management strives to reduce risk of all types. Managing risk is a normal task performed by managers and designers. The increasing complexity of projects and systems increases the risk of failures and their consequences. Therefore, it is imperative that all risks are identified, assessed, and mitigated.

Risk management offers many benefits:

- Improve the ability to meet or exceed schedule, budget, and technical performance requirements
- Improve the detailed understanding of the product or system
- Reduce business loses
- Reduce product liability concerns
  - Identify and correct hazards and failures
  - Develop design trade-offs among alternatives or options based on their relative risk
  - Establish priorities for improving safety

Management must, of course, strike a reasonable balance between risk and benefit. The cost of eliminating all risks could be prohibitively expensive and time-consuming. Successful risk management requires thoughtful planning and provisioning of adequate resources. Risk management should be implemented as early as possible in the life cycle of every product development effort. The goal is to identify risks involved, to help decision-making, and to develop strategies to handle the risks before they become critical issues.

The process of risk management can be handled in various ways, but the *Department of Defense Risk, Issue, and Opportunity Management Guide for Defense Acquisition Programs* provides a good, generic approach to the risk management process. Figure 4.1 provides a snapshot of this process [11].

While risk management has many applications, for our purposes, we will focus on design risk in this book. This involves the risk inherent in product design. For design risk as it is related to product liability, risk management attempts to control risk in product design by reducing the probability of having an accident and by reducing the consequences of an accident if an accident does occur. The purpose of risk management in this situation is to reduce risk to an acceptable level.

Figure 4.1 Risk management process

To determine what is an acceptable level of risk depends on many factors, including the type of hazard, the risk tolerance of the ultimate user, and the particular industry involved, to name a few. Typically though, accepting a risk boils down to a cost-benefit analysis where the level of risk is considered in conjunction with the effort and cost necessary to reduce the risk further. In other cases, such as a heavily regulated industry, an independent authority may reserve the ultimate decision to accept a risk. Design risk and risk assessment will be covered in greater detail in Chapter 7.

4.4 What Happens When the Paradigms for Design for Safety Are Not Followed?

Forming an ideal system's approach to designing new systems involves developing paradigms, standards, and design process models for developers, and sometimes users, to follow and patterns for themselves in their future design efforts. Chapter 1 focused on 10 paradigms for managing and designing systems for safety. These paradigms are referenced throughout the course of this book and are often called "words of wisdom" or "rules of thumb." The antithesis of these 10 paradigms are the most critical risks that need to be eliminated or mitigated for designing for safety to avoid expensive product liabilities, court cases, and legal costs.

- Paradigm antithesis 1: Never aim for zero accidents
- Paradigm antithesis 2: Don't speak up, don't be courageous and "just say yes"
- Paradigm antithesis 3: Spend minimal effort on systems requirements analysis
- Paradigm antithesis 4: Allow escapes from single as well as multiple causes
- Paradigm antithesis 5: If the solution costs too much money, do nothing
- Paradigm antithesis 6: Design for systems that provide no advanced warning and react to disastrous events or preventable mishaps as they are found by customers
- Paradigm antithesis 7: Never analyze structure and architecture for safety of complex systems
- Paradigm antithesis 8: Training? Who needs training? Ever heard of OJT?
- Paradigm antithesis 9: No action is the best option; don't pay attention to it and it will go away
- Paradigm antithesis 10: Don't waste your time documenting practices

A company or person who does not follow the safety paradigms and succumbs to the easy way out, by following any of the paradigm antitheses, may be headed for serious difficulties and challenges in court cases involving tort liabilities. A company or person who decides to avoid the safety paradigms makes the conscious decision to take their chances that events will not occur and result in costly litigation. The next section will explain the basics of tort liability and product liability from a legal viewpoint. We will introduce negligence as a cause of product liability that results in unfavorable judgments against companies and individuals that are held liable for customer injury or property damage.

4.5 Tort Liability

Definition of tort liability: Legal duty or obligation of one party to another party who is a victim as a result of a civil wrongdoing or injury. Tort is a wrongful act or an infringement of a right (other than under contract) that may lead to civil legal liability. The liability may be caused by a combination of a violation of a person's rights and an act that goes against a law, rule, or code of conduct causing damage, personal injury, or a private wrongdoing. Evidence should be evaluated in a court hearing to identify who the liable party is in the case.

Definition of strict liability: Strict liability applies when a defendant places another person in danger, even in the absence of negligence, simply because the person (plaintiff) possesses a dangerous product. The plaintiff does not need to prove negligence.

Strict liability torts: Strict product liability is liability without fault for an injury proximately caused by a product that is defective and not reasonably safe.

Therefore, in establishing strict liability, the injured plaintiff needs only to prove that (1) the product was defective and (2) the product defect was the cause of the injury. In other words, the focus at trial is on the product, not the conduct of the manufacturer, because it does not matter whether the manufacturer took every possible precaution. If the product was defective and caused an injury, the manufacturer is liable.

There are instances when a person becomes responsible for things that may go wrong even if the person did not intend for the wrong to occur. Some actions hold a person strictly liable regardless of the circumstances. For example, if a person owns an exotic dangerous animal that is trained as a household pet, such as a lion, and that pet escapes from the house of its master and bites a neighbor, the pet owner may be held liable. If the lion escapes out of the house by pushing open a partially closed front door, the pet owner will probably be held liable for personal injuries even though the lion owner did not let the lion out the door. If the lion escapes out of the house by crashing through the living room window to escape a fire burning in the kitchen, then there is a chance the pet owner will evade punishment. Ownership may or may not be enough to hold someone responsible for personal injury damages and costs. It all depends on the situation. Strict liability tort means a defendant is held fully liable for any injury sustained by another party regardless of whether the injury was intended. Strict liabilities due to the poor control of dangerous animals by their owners are one of the major strict liability categories. Another strict liability category is product liability. We will now deep dive into product liability next since it has a direct bearing on design for safety.

4.6 An Introduction to Product Liability Law

Product liability law remains of central importance to tort law. But the importance of this law—and the concept of strict liability that sits at its core—has long overshadowed its origin in food cases, where courts first worked out rules that allowed those injured by unsafe food to recover damages. Because of a consumer's complete vulnerability to the invisible danger of adulteration, the need for a rule of strict liability for food was not only justifiable on legal grounds but also necessary on practical grounds [12]. Then the focus shifted from the importance of food safety as something arguably unique to the importance for all consumers to have a remedy at law when injured by a defective product.

When a defective product that is unreasonably dangerous or unsafe injures a person, the injured person may have a claim or cause of action against the company that designed, manufactured, sold, distributed, leased, or furnished the product. Product liability of the company involved in a product means that company may be held responsible by a court of law under the law for a person's injuries. This company involved in a product may be required by the court to pay for the person's damages.

For purposes of this discussion, the terms "system," "product," and "goods" are synonymous. They may be used in different context, but are the same as related to consideration of product liabilities.

Many risks are assessed for products due to the potential of catastrophic hazards that result when a person is injured by a defective product. Many of these hazards are preventable, and yet, many organizations that develop products choose not to mitigate the risks of these hazards. The actual occurrence of a hazard that leads to personal injury may have previously been assessed as unreasonably dangerous or unsafe. These hazards could result in situations where an injured person has a claim or cause of action against the company, companies, or organizations that designed, manufactured, sold, distributed, leased, or furnished the product. Any number of companies or organizations involved in getting their product to the customer may be liable to a person for injuries for any number of reasons that led to dangerous hazards or unsafe conditions. As a result, companies or organizations may be required to pay for legal damages, and these legal damages, especially related to medical procedures and future complication possibilities, may be quite costly. Costs related to product liability may be high, but the costs to correct product designs may be higher. This is the reason it is best to find and fix problems early in the design stage. Companies or organizations weigh the advantages and disadvantages of correcting design weaknesses versus allocating funds in a reserve to pay out claims for product liability damages from court cases.

Many lawyers become specialists in handling legal cases related to product liability and understanding the law that governs product liability. Many books have been written related to the topic of product liability law. Product liability and the work of lawyers handling these types of cases deal with very complex subjects, no matter if they are arguing a case for a plaintiff or a defendant. This book does not mean to oversimplify a complex subject. Rather, we attempt to introduce the reader to a subject that has bearing in how safety engineers do their work.

Demonstrating negligence is one way to argue that an organization is liable for damages related to personal injury and unsafe conditions. Negligence as a legal argument to sue an organization over a product hazard that caused personal injury is an easy argument to win in a court of law. Proving negligence means the organization involved in a product design or an action related to getting a product to market was aware of an unsafe condition or high risk of personal injury, yet took no action or minimal actions, to correct the hazard or mitigate the risk of the hazard.

Privity is another way to argue that an organization is liable for damages related to personal injury and unsafe conditions. Basically, if an organization knew their product was unsafe and allowed it to be used by their customers regardless, this was cause for legal action against the organization.

4.7 Famous Legal Court Cases Involving Product Liability Law

Product liability law may have been derived in 1842 from a legal case called Winterbottom v. Wright decided in England [13]. In this famous case, Mr. Winterbottom was seriously injured while driving a horse-drawn mail coach. Mr. Winterbottom claimed the mail coach collapsed due to poor design and construction. Postmaster general bought the mail coach from Mr. Wright, the coach manufacturer. Mr. Wright provided the mail coach without horses. The postmaster contracted with a company, which Mr. Winterbottom worked for, to supply horses to pull the coach. This company then hired Mr. Winterbottom to drive the mail coach. Mr. Winterbottom could have sued his employer or the postmaster, but instead chose to sue Mr. Wright. This proved to be a bad decision. Mr. Winterbottom's case was dismissed based on the general rule that a product seller cannot be sued for negligence by someone without a direct business connection, such as a contracted worker. At this time, a party could not sue a company with whom the party is not "in privity." The law of contracts requires that there be "privity" if one party to a contract tries to enforce the contract by a lawsuit against the other party.

Read more: http://dictionary.law.com/Default.aspx?selected=1617#ixzz3if0AwQ3w

Professor William Prosser, expert on product liability law, known for his 1960 California Law Review article, titled "Privacy," stated that the history of product liability law is really the history of an assault on the "citadel of privity." "It is also the history of how injured people, like Mr. Winterbottom, were given back the keys to the courthouse and allowed a remedy at law for the injuries they suffered as a result of a defective and unsafe products" [13]. Professor Prosser was focused on tort privacy starting in the 1940s up to his death in 1972. He was a key person influencing how tort law conceptualized privacy.

With the publication of *Assault upon the Citadel* in 1960, Prosser heralded the increasing number of judicial decisions in which consumers were allowed to recover damages despite a lack of privity. To Prosser, this increase evidenced a kind of battle on behalf of consumers injured by defective products, allowing consumers to hold sellers strictly liable for product-related injury, which is to say, without the need to prove negligence. In telling the story of the battle, Prosser prominently featured food cases, discussing at length the consumer-protective rationales of these cases. Yet, six years later, when he announced the dawn of the strict liability era in a follow-up article, *The Fall of the Citadel*, Prosser seemed intent on deemphasizing the import of not only the food cases, but food safety as well. Suddenly, the cases that had been progenitors of consumer protection were mere footnotes—literally. The Washington Supreme Court's historic decision in Mazetti v. Armour is an excellent example of Prosser's strategic de-emphasis of food, with food safety sacrificed to the needs of the battle and to obtain victory." [12]

The assault upon the citadel of privity had begun with a battle "directed against a narrow segment of the wall, defended only by the sellers of food and drink" [14]. Having vanquished these product sellers, and then successfully expanded the battlefront to nearly all other products, it would have made no sense at all to pick a fight that might lead to greater losses just as near-total victory was so close at hand [12].

The assault on the citadel of privity most likely began in 1916. The change to the established general rule led to the creation of two exceptions. The first exception involved the situation where the seller knew that a product was dangerous, but failed to disclose the danger to the customer. The second exception included products that were deemed "inherently" or "imminently" dangerous. This category of dangerous products includes guns, ammunition, and explosives.

In 1916, the historic decision by Justice Benjamin Cardozo in the MacPherson v. Buick Motor Co case occurred. Justice Cardozo expanded the inherent danger exception and minimized the general rule of privity. Justice Cardozo wrote the following:

> We hold, then, that the principle of (inherent danger) is not limited to poisons, explosives, and things of like nature, to things which in their normal operation are implements of destruction. If the nature of a thing is such that it is reasonably certain to place life and limb in peril when negligently made, it is then a thing of danger. Its nature gives warning of the consequences to be expected. If to the element of danger there is added knowledge that the thing will be used by persons other than the purchaser, and used without new tests, then, irrespective of contract, the manufacturer of this thing of danger is under a duty to make it carefully.

In conclusion, he announced:

> We have put the source of the obligation where it ought to be. We have put its source in the law. [13]

Justice Cardozo explained that the MacPherson decision eliminated the notion that the duty to safeguard life and limb "grows out of contract and nothing else." The effect of the MacPherson decision was immediate and widespread. The doctrine of privity was changed forever. There were no longer disputes that a manufacturer could be held liable for its negligence in the poor design and construction of its products. The manufacturer's legal obligation extends to all product users and customers who suffer injuries or experience damages as a result of a defective product. This legal obligation means that any person injured by a defective product could sue the product manufacturer for negligence even if that person bought the product from another company, such as a wholesaler or retailer.

California Supreme Court became the first court in the United States to adopt the rule of strict product liability. Now it is the law in all 50 states. Strict product liability surfaced in 1963 when the California Supreme Court issued a decision known as Greenman v. Yuba Power Products. Chief Justice Roger Traynor issued the decision that created modern product liability law. This decision pointed out that an implied warranty that was associated with the product was a legal fiction created to achieve a desired result. Justice Traynor made it clear that the liability is not one governed by the law of contract warranties, but by the law of strict liability in tort. As explained by Justice Traynor:

> To establish the manufacturer's liability it was sufficient that the plaintiff proved he was injured while using the (product) in a way that it was intended to be used as a result of a defect in the design and manufacture of which the plaintiff was not aware that made the (product) unsafe for its intended use. [13]

In most jurisdictions a person injured by a product may base his or her recovery of damages on one basic legal theory—negligence. Breach of warranty and strict tort liability may also be the basis for a strong legal case. This section provides a brief introduction to the definitions and applications of negligence and warranties as legal theories. Tort liability was covered in the beginning of this chapter to establish the progress made in product liability laws.

4.8 Negligence

Definition of negligence: Negligence is a potential cause of a product liability. It is the failure to exercise care to avoid injuring someone to whom you owe the duty of care. Ignorance is no defense. There is the consideration of what a reasonable person could be aware of under the circumstances. This is called constructive knowledge in contrast to actual knowledge. For example, if you live in a city with a sidewalk between your property and the street and you know that it snowed all night, you have constructive knowledge that your sidewalk should be shoveled to prevent personal injuries to walkers from slipping and falling, regardless of whether you look out of the window and actually see how much snow has accumulated.

It is not enough for a buyer to prove that a seller acted in a negligent manner or that the seller neglected to take an action that is reasonable under similar circumstances. An injured person must demonstrate the existence of a seller's duty and that the duty was owed to the buyer. Many people assume the law requires everyone to act in the same reasonable way toward everyone; however, we do not owe everyone the same duty of care. For example, a person trespassing on my property in the dark that falls in my uncovered and unfilled pool and breaks his neck is the result of his own negligence. It would not be my fault that I did not replace the light bulb that might have alerted him to this hazard.

The failure to exercise ordinary care is referred to as a breach of the duty of care. This breach of duty must also have been the proximate cause of the injury or damages. The product customer, who was injured by the product or experience damages as a result of the product's normal use, must be able to demonstrate that the injury or damages would not have occurred without the breach of duty. It is most often an issue in "failure-to-warn" cases. In this type of legal case, the defendant (a product manufacturer) might argue that an injury or property damage would have occurred with or without a warning. The defendant might also argue that the cause of the injuries or damages was the affected person's own negligence and not that of the product.

In general terms, the duty to exercise ordinary care and to supply a safe and non-defective product, applies to everyone in the supply chain and distribution chain, starting with the product designer and manufacturer. The general name of the product designer and manufacturer is known as the Original Equipment Manufacturer (OEM). The OEM is responsible for every facet of the product's design including form, fit, and function and may also own the rights of the product through Intellectual Property (IP). With IP, an OEM could license the design to another company and receive royalties and monetary benefits. The OEM who carelessly makes a defective product, or another company other than the OEM that uses the product to integrate and assemble to something else without discovering the defect, or the retailer who sells the product to the general public who should exercise greater care in offering such products for sale may all be held liable for hazards resulting from the product. Under the law, these entities have a responsibility and duty of care to anyone who is likely to be injured by the product during normal use. This includes the product customer, the initial buyer, a family member to the product owner, an innocent bystander, and anyone who leases the item or holds it for the product owner.

The responsibility or duty to exercise care involves every phase of getting the product to the public. For example, the product must be designed in a way that it is safe when used as intended. The product must be inspected and tested at appropriate stages in the manufacturing, distribution, and selling process. The product must be made from appropriate (i.e., safe and non-defective) materials and assembled with appropriate care to avoid against its negligent manufacture. The product's container or packaging must be adequate (and not itself dangerous or defective) and contain appropriate warnings and directions for use. An otherwise non-defective product can be made unsafe by the failure to provide adequate instructions for its safe use. In many cases, depending on the type of product, there is a seller's duty to warn all users of obvious dangers. The criteria for what constitutes an obvious danger are far from obvious. Since the criteria that determines whether a danger or a hazard is obvious is unclear and ambiguous, there are many arguments related to obvious dangers that exist in a product's design or operation; this explains the reasons for many legal cases between sellers and buyers [13].

4.9 Warnings

Instructions and warnings are important in the prevention of accidents and injuries and in the reduction of risk of liability. However, they are not a substitute for good, safe product design. A warning cannot make a dangerous product safe. If a product can be made safer by design, that option should be implemented rather than just providing a warning. Understanding the trade-off between cost and time involved to redesign a product versus a product release to the marketplace for Time-to-Market (TTM) advantages, it is obvious that many companies will defer their decisions to redesign a product to make it safer until they know that a product will become popular and gain widespread acceptance. For this reason, as an example, a company may elect to use a product warning label to mitigate the risk of an accident or injury and schedule a redesign at a point of time in the future when feedback from customers are received, and a redesign will implement new features for customers as well as resolve or minimize safety risks. From the company perspective, it makes no business sense to redesign a product and hold up product release, missing their opportunity for TTM being the first to market a new idea, if their product only sits on a retailers' stores' shelves collecting dust. After all, there is no safety effect if no one uses an unsafe product. This is why there must be trade-offs for these types of decisions. However, the company must also reconcile their decision to forego a safety-related design change in favor of a warning, considering the amount of risk in this approach, if an unsafe product harms a user during this early release period.

If a company decides to minimize product liability by using product warnings, there is valuable guidance to help them move forward. It is recommended that the warnings should be designed to comply with the latest standards. The "best practices" for safety warnings can be found in the latest American National Standards Institute (ANSI) Z535 series, which comprises the following six individual standards:

- ANSI Z535.1 American National Standard for Safety Colors
- ANSI Z535.2 American National Standard for Environmental and Facility Safety Signs
- ANSI Z535.3 American National Standard for Criteria for Safety Symbols
- ANSI Z535.4 American National Standard for Product Safety Signs and Labels
- ANSI Z535.5 American National Standard for Safety Tags and Barricade Tapes (for Temporary Hazards)
- ANSI Z535.6 American National Standard for Product Safety Information in Product Manuals, Instructions, and Other Collateral Materials

The latest versions of these standards were published in 2011. The earlier 2007 publication made a substantial effort to harmonize the US standards in the ANSI Z535 series with the International Organization for Standardization (ISO) standard ISO 3864, *Graphical symbols: Safety colours and safety signs*, which is a four part standard that specifies design requirements, including shapes and colors of safety signs. Compliance with the ISO standard is compliance with the ANSI standard; therefore, a manufacturer designs warning options that can best fit their markets and customers.

4.10 The Rush to Market and the Risk of Unknown Hazards

In the previous section we mentioned the risk associated with rushing a product to market for a TTM advantage with a known hazard and using a warning label to alert the user to the danger. A larger risk of rushing a product to market occurs if there is an unknown hazard in the product and, because of insufficient analysis or testing, the hazard is discovered after the product is in the marketplace. This predicament is best illustrated by a very recent problem with Samsung Galaxy Note 7 smartphone. Samsung hurried their newest smart phone to the marketplace in an effort to beat the anticipated release of Apple's newest 5.5-inch iPhone 7 Plus [15]. "Samsung really wanted an innovative product in this case—they went from Note 5 directly to Note 7, it has slim battery and curved contours, etc.—and also prioritized speed-to-market to launch the product before iPhone 7 Plus, but overlooked robustness checks" [15].

Shortly after the release of the Samsung Galaxy Note 7 smartphone in August 2016, reports of exploding batteries in the phone started coming in from around the world. Within 2 weeks of the release and after some delays and confusing information being provided by Samsung, a recall was initiated. While there were several theories on the cause of the exploding batteries, Samsung blamed the problem on the battery manufacturer. Using a battery from a different manufacturer, Samsung released a new "safe" version of the smartphone as a replacement for the original phone. Within days of this replacement phone becoming available, batteries in this phone started to explode and/or catch fire including one on a Southwest Airline in Louisville, KY. Another recall had to be issued. This time, Samsung made the decision to discontinue production of the phone totally and offered refunds or replacement with existing models. There have been at least 92 reports of overheated batteries, 26 reported burns, and 55 reports with property damage [16]. It has been estimated that the recall will cost Samsung $5.3 billion [17]. This rush to market has cost Samsung not only the money for the recall but also a loss of reputation and no doubt customers. The damage to Samsung's market value has been substantial. Before the company announced discontinuing production of the phone, Samsung shares in the stock market fell more than 8% in the largest daily drop since 2008 and reducing its market value by $17 billion [16]. Sometimes the desire to rush a product to market to beat the competition may not be worth the risk involved.

4.11 Warranty

Definition of warranty: Warranty claims are governed by contract law. A warranty is a promise, claim, or representation made about the quality, type, number, or performance of a product [13]. Generally speaking, the law assumes that a seller always provides some kind of warranty concerning the product sold. The seller should be required to meet the obligation described in the warranty. Failure of the seller to meet the obligation of the warranty may be grounds for a breach of warranty claim.

The law that governs the sale of goods is Article 2 of the Uniform Commercial Code (UCC). Under the UCC, there are two kinds of warranties: express and implied [13].

Express warranty: An express warranty is created through an affirmation of fact made by the product seller or OEM to the product buyer or customer. It may stipulate the usage conditions, quality, safety, or reliability of the product at the time of the sale. This affirmation of the product becomes part of the "basis of the bargain" between seller and buyer. This affirmation or expressed warranty may be created in one of four ways:

1. Spoken: An express warranty may be made by spoken words during sales discussions or negotiations, or written on a receipt or invoice, or in a purchase order, sales agreement, or contract.
2. Silence: An express warranty may be created by silence in situations where omission of information has the effect of willfully hiding important information or creating a wrong impression about the quality of the product sold.
3. Samples: An express warranty may be created by samples shown to the buyer.
4. Written: An expressed warranty is written for customers to read and reference in the future. Marketing brochures may establish warranties as marketing claims. Warranties may be documented in published advertisements or in design specifications available on the Internet. Also, a warranty may be described on the label of the same type of product from an earlier purchase from the same seller or OEM. In this case, the buyer may reasonably expect the same levels of quality, reliability, and safety from a second purchase of the same product as the first purchase.

An express warranty can be about the quality of the product at the time of the sale, but it can also be about the quality of performance of the goods in the future [13]. Under the UCC, the time to file a lawsuit alleging a breach of warranty begins when a product is delivered. Typically, the time does not start when the product is put into use, and the defect is discovered. However, if the warranty concerns performance of a particular product, which will be free of defects for a specified period of time, then the clock continues for this specified period of time even after the warranty expires. For example, if the warranty period is one year but the OEM states the product is free of defects for three years, the OEM has an obligation to repair the product or provide a replacement product for two years after the warranty period expires. This additional two-year-defect-free period may be considered an extended warranty.

In the past, as previously discussed in this chapter, the law allowed protection of a product manufacturer or OEM, such that a product customer could not sue the OEM for a breach of an express warranty unless the customer was "in privity" with the OEM who made the warranty. This protection under the law is no longer possible. Many courts recognize that it is enough that the expressed warranty was made and that the product customer alleging breach of warranty most likely relied on it when making the decision to purchase the product [13].

Implied warranty: An implied warranty is presumed to exist unless the buyer clearly and unambiguously disclaims it in writing as a part of the sales agreement.

This is different from an express warranty, which is created by an affirmative act. There are two kinds of implied warranties in the UCC [13]:

1. The **implied warranty of merchantability** is a kind of minimum requirements warranty. Because the UCC is mostly concerned with commerce as conducted between merchants, in contrast to commerce between merchants and consumers, the definition of this implied warranty speaks in terms of goods that will pass without objection in the trade and that are fit for the ordinary purposes for which such goods are used. Typically, the implied warranty also includes a warranty of reasonable safety [13].
2. The **implied warranty of fitness for a particular purpose** imposes a similar requirement in cases where the seller knows or has reason to know of a particular purpose for which the goods are required. In such a case, where the buyer relies on the seller to select or furnish goods that are suitable for a particular purpose and the seller in fact has such expertise, an implied warranty of fitness for a particular purpose is created by law [13].

For example, assume an electronic component buyer tells the seller of a particular component that the buyer's company may be interested in using this particular component in a new electronic product design if certain product safety design features are added and production volume and capacity requirements are met. Assume the seller needs a new electromechanical machine that is able to efficiently add the new safety design features to the component and produce 25 electrical components per hour enabling the company to fulfill the production contract with the buyer. If the seller is known in the electronics industry as someone expert in the manufacture of this particular component and the seller recommends a particular electromechanical machine model to satisfy the production quantity requirements of the buyer, then the seller is making an implied warranty of fitness. If the seller proves to be unable to produce the component efficiently with the required safety design features that were agreed to by contract and only makes five electrical components per hour that is safety compliant, the buyer may file a lawsuit.

4.12 The Government Contractor Defense

While we focus mainly on product liability in commercial organizations doing business with the public, there is also the area of contractors providing products and systems to the government. In cases where there is a product or system that has caused accidents or injuries, under certain conditions the government "protects" the contractor from liability. The Supreme Court

Decision (*Boyle v. United Technology*) [18] and other later decisions provided an interpretation of the Federal Tort Claims Act that the contractor is protected, provided that three conditions are met:

1. The US government approved reasonably precise specifications.
2. The product conformed to those specifications.
3. The contractor warned the US government about the dangers in the use of the product that was known to the contractor, but not to the US government.

The Court stated,

> The first two of these conditions assure that the suit is within the area where the policy of the 'discretionary function' would be frustrated—i.e., they assure that the design feature in question was considered by a Government officer, and not merely by the contractor itself. The third condition is necessary because, in its absence, the displacement of state tort law would create some incentive for the manufacturer to withhold knowledge of risks, since conveying that knowledge might disrupt the contract, but withholding it would produce no liability. We adopt this provision lest our effort to protect discretionary functions perversely impede them by cutting off information highly relevant to the discretionary decision. [18]

Later cases have demonstrated that hazard analysis documentation provides useful evidence for satisfying the third condition. We will discuss various types of hazard analyses in Chapters 6, 7, 8, 9, 10, and 11.

4.13 Legal Conclusions Involving Defective and Unsafe Products

In general, a manufacturer is held liable for a product-related injury when their product is deemed to be unsafe. The unsafe condition of a product in normal use may result either by defective materials or defective design. A product may be defective, and still be safe, but a product that is unsafe is as a result of a defect of one type or another. The manufacturer is responsible for repairs to resolve defective products under a warranty contract, but may not be at risk of costly litigation for an unsafe product. If a defect causes an unsafe condition, then the manufacturer has reason to worry about costly litigation. The emphasis now is to focus on why a product is defective and avoid costly litigation. Follow the 10 paradigms within these pages and stay out of lengthy expensive legal battles.

References

[1] U.S. Department of Justice, Bureau of Justice Statistics, *Federal Tort Trials and Verdicts, 2002–03*, U.S. Department of Justice, Bureau of Justice Statistics, Washington, DC, 2005.

[2] State of Missouri, Department of Insurance, Financial Institutions & Professional Registration, 2014 Missouri Product Liability Insurance Report, Jefferson City, MO, November 2015.

[3] National Highway Traffic Safety Administration (NHTSA), Tracking Vehicle Recalls Over the Years, 2015, http://www.safercar.gov/Vehicle+Owners/vehicle-recalls-historic-recap (Accessed on July 31, 2015).

[4] NHTSA Press Release, U.S. DOT Imposes Largest Civil Penalty in NHTSA History on Takata for Violating Motor Vehicle Safety Act; Accelerates Recalls to Get Safe Air Bags into U.S. Vehicles, NHTSA 46–15 Tuesday, 3 November 2015.

[5] NHTSA Press Release, U.S. Department of Transportation Expands and Accelerates Takata Air Bag Inflator Recall to Protect American Drivers and Passengers, NHTSA 13–16, Wednesday, May 4, 2016.

[6] Orlando Sentinel, Toyota Adds 5.8M Vehicles to Global Takata Recall Total, October 27, 2016.

[7] Japanese Airbag Maker Takata Files for Bankruptcy, Gets U.S. Sponsor, Naomi Tajitsu, Reuters, June 26, 2017, from https://www.reuters.com/article/us-takata-bankruptcy-japan-idUSKBN19G0ZG (Accessed on July 31, 2015).

[8] Merriam-Webster Dictionary, Online, http://www.merriam-webster.com/dictionary (Accessed on July 31, 2017).

[9] Raheja, D. and Gullo, L.J., *Design for Reliability*, John Wiley & Sons, Inc., Hoboken, NJ, 2012.

[10] Hillson, D. and Murray-Webster, R., Understanding and Managing Risk Attitude, 2004, http://www.kent.ac.uk/scarr/events/finalpapers/Hillson%20%2B%20Murray-Webster.pdf (Accessed on July 31, 2017).

[11] United States Office of the Deputy Assistant Secretary of Defense for Systems Engineering, *Risk, Issue, and Opportunity Management Guide for Defense Acquisition Programs*, Office of the Deputy Assistant Secretary of Defense for Systems Engineering, Washington, DC, 2015.

[12] Stearns, D.W., Prosser's Bait-and-Switch: How Food Safety Was Sacrificed in the Battle for Tort's Empire, Marler Clark, LLP, 2014, http://scholars.law.unlv.edu/cgi/viewcontent.cgi?article=1574&context=nlj (Accessed on July 31, 2017).

[13] Stearns, D.W., An Introduction to Product Liability Law, Marler Clark, LLP, 2001, http://www.marlerclark.com/pdfs/intro-product-liability-law.pdf (Accessed on July 31, 2017).

[14] Prosser, W.L, The Assault upon the Citadel (Strict Liability to the Consumer), *The Yale Law Journal*, 69, 1099–1148, 1960.

[15] Howley, D., Why Samsung Exploded and How It Can Turn Itself Around, October 11, 2016, http://finance.yahoo.com/news/samsung-note-7-fire-recall-234416132.html (Accessed on July 31, 2017).

[16] Korszun, J., The Reason Behind Samsung Permanently Discontinuing the Galaxy Note 7, October 12, 2016, http://www.electronicproducts.com/Mobile/Devices/The_reason_behind_Samsung_permanently_discontinuing_the_Galaxy_Note_7.aspx (Accessed on July 31, 2017).

[17] Kenwell, B., Samsung to Lose More Than $5 Billion from Galaxy Note 7 Recall, October 14, 2016, https://www.thestreet.com/story/13854429/1/samsung-to-lose-more-than-5-billion-from-galaxy-note-7-recall-tech-roundup.html (Accessed on July 31, 2017).

[18] Supreme Court Decision, Boyle v United Technology (487 U.S. 500), 1988.

Suggestions for Additional Reading

Farrow, D.R., Speech at the Fifth International Workshop on Risk Analysis and Performance Measurement in Aviation Sponsored by FAA and NASA, Baltimore, August 19–21, 2003.

Raheja, D., *Preventing Medical Device Recalls*, Taylor & Francis, Boca Raton, FL, 2014.

Raheja, D. and Allocco, M., *Assurance Technologies Principles and Practices*, John Wiley & Sons, Inc., Hoboken, NJ, 2006.

5

Developing System Safety Requirements

Louis J. Gullo

5.1 Why Do We Need Safety Requirements?

Safety requirements ensure a safe design. Without safety requirements, buyers beware. System/product requirements must address the expectations of the customer in terms of safe system/product operation. To emphasize what was stated earlier in a previous chapter, inadequate system or product specifications and requirements plague the systems engineering discipline no matter which industry you work for. Generic specifications with vague requirements or incorrect requirements are provided to the hardware and software designers, leading to faulty designs or designs wrought with inherent defects, ambiguities, and weaknesses. From data collected through years of experience, the author has found that the vast majority of system failures and accidents from safety hazards were caused by either poorly written requirements or improper requirements decomposition or allocation flow down to hardware and software design specifications and detailed design description documentation.

System safety engineering provides the means to develop a risk management strategy for requirements based on potential or actual hazard identification and analysis. As a result of hazard analysis, there should be remedial

Design for Safety, First Edition. Edited by Louis J. Gullo and Jack Dixon.
© 2018 John Wiley & Sons Ltd. Published 2018 by John Wiley & Sons Ltd.

actions, process controls, design changes, risk mitigation plans, and preventive measures applied to a design using a systems-based approach [1, 2]. For any system or product, the most effective means of limiting product liability and accident risks is to implement an organized system safety effort, beginning in the conceptual design phase and continuing to its development, fabrication, testing, production, use, and ultimate disposal [1, 2]. Systems-based approach to safety requires the application of scientific, technical, and managerial skills to hazard identification, hazard analysis, and elimination, control, or management of hazards throughout the life cycle of a system, program, project, activity, or product [1, 2]. The concept of system safety is useful in demonstrating adequacy of technologies when difficulties are faced with probabilistic risk analysis [1, 3]. Communications regarding system risk is important in correcting risk perceptions by creating, analyzing, and understanding information models that show what factors create and control the hazardous process [1, 3].

System safety engineering provides requirements and methodologies on how to control nominal and worst-case conditions for safe system functionality. Besides safe system usage during normal system operation and periods of extreme conditions, safety engineering must provide requirements and methodologies on how to safely control a system or product during the occurrence of a system failure or during system maintenance. The system conditions during usage may be controlled through autonomous system behavior or through Human–Machine Interfaces (HMIs). The control of system conditions is implemented through system requirements that specify how the system's autonomous behavior should be performed or how the system operator or maintainer uses the HMI to work with the system. Requirements are flowed to the system users so they know the performance envelop and boundaries for system operation. These system requirements may be flowed to the system user through the HMI using help screens on a display or through alarms and warnings to prevent accidents. The requirements may also be provided through training manuals, technical courses to certify operators and maintainers, and operator or maintainer manuals to be used during the person's performance of any job working with the system. When operators and maintainers don't follow the establish procedures and work instructions provided to them, accidents occur through human error. System safety engineering determines causes of an accident either based on the recent hazard pattern analyses or as a result of investigation of past accident cases where root causes were determined. The means to prevent future initial occurrences of accidents or continuing accident occurrences are through sound and effective system safety requirements.

5.2 Design for Safety Paradigm 3 Revisited

Now let's revisit Paradigm 3 from Chapter 1.

Paradigm 3 from Chapter 1 States: "Spend Significant Effort on Systems Requirements Analysis"

Systems Requirements Analysis (SRA) is a significant task that deserves sufficient time and energy to make sure system requirements are good. System requirements must be written correctly so the customer's objectives are met and designers know how to interpret the words to develop the right system or product to ensure customer success. SRA must also include system safety requirements analysis. Where SRA ensures the system will perform as intended, based on the written requirements, the system safety requirements analysis makes sure the system's inability to perform as intended will not result in a system catastrophic failure effect or safety-critical hazard. Accidents and system failures will result from defective or poorly written requirements in system specifications. Defective requirements may result in safety-critical system failures with catastrophic effects.

The sources of most requirement failures are incomplete, ambiguous, and poorly defined device specifications. The bad requirements result in making expensive engineering changes in the later stages of the development or manufacturing process. When these engineering requirement changes occur consistently and continuously over a certain period of time, these changes are called "scope creep." Scope creep from constant requirement changes occurs at a time that represents an unstable period in the system development life cycle. It represents a fraction of the engineering design phase when immature requirements are prevalent; when requirements are being added, deleted, edited, improved, deferred, and replaced. Scope creep leads to requirements volatility and design instability. The longer it takes to resolve scope creep, the more cost is associated with engineering development. The longer it takes to make an engineering change during the System Development Process (SDP), and the longer it takes to detect a requirements defect, the more costly the change.

When a requirements defect is found during the requirements development phase of a program, this defect is considered to be an in-phase defect, which costs less to correct than an out-of-phase defect, which is detected later in the development process. A requirements defect found after the requirements development phase of a program is called an out-of-phase defect. A requirements defect found out of phase during the detailed design phase of a program could cost 10 times more to correct than a requirements defect found in phase during the requirements development phase of a program. A requirements defect found out of phase during the system integration and test phase of a program could cost 100 times more to correct with a design change than a requirements defect found during the requirements development phase and corrected with a design requirements change in the early stages of a program.

A single design change concerning one piece part in a design may not be a difficult change to make, but a robust integrated design change affecting multiple levels of a system hierarchy, or a design change involving intricate functional and operational design details integrated vertically from system level down to various piece part levels of hardware, is too hard to grasp and explain in a short design change review meeting and therefore too difficult to approve quickly. One reason for the difficulty of the change could be that it affects multiple technologies and part types in a complex mixed technology design and requires different types of Subject Matter Experts (SMEs) working together to explain how the design change works. The design change has far reaching effects on the system and may be too complex for a program to handle during later stages in system development when requirements are being verified and validated. Often robust changes cannot be made because the project is already delayed and there may not be design engineering resources at the later stages in development for implementing new features. After critical design reviews, designers move on to new development projects and programs. Unfortunately, most companies still discover design problems when the design is already in production. At this stage there are usually no resources and no time available for major design changes.

When conducting system safety requirements analysis, the system safety engineer should pay particular attention to missing functions in the specifications. When system requirements are volatile, requirements written for specific functions may be added, then deleted, and then added back again later. Traces between requirements may be broken so that orphan requirements exist, or childless parent requirements surface. This volatility could lead to mistakes in the system functional requirements, such as the incorrect number of functions, the misallocation of system functions to hardware and software, the poor integration of functions, or the inappropriate mix of certain types of hardware functions and software functions or software templates required for the design to operate as intended by the systems engineers and by the customers. The defects in system functional requirements may be found through functional dependency analysis using functional flow diagrams, functional process maps, activity diagrams, and sequence diagrams. Besides defects in system functional requirements that may be found through functional dependency analysis, there may also be defects related to timing dependencies. Functional requirements may be written for synchronous and asynchronous operations. Synchronous functions involve set frequencies of activities on specific timing cycles. Asynchronous functions do not follow timing cycles. They may be referred as interrupt-driven functions, which occur outside normal timing sequences. There are system requirements that may be necessary for only certain times in the life of a system, such as the "one-time event" when a system is installed. It is important to know when hazards may occur during the system development life cycle and the system's total life cycle. The engineer must assess the system requirements that are imperative during the various phases of the system life cycle, starting with development and ending at the End-of-Life (EOL) period or End-of-Service (EOS) period of the life cycle. System safety requirements should be written to prevent hazards from occurring:

- During development of all system components
- During manufacturing of all system components and assembly into the system
- During system installation at the customer site or when applied by a customer on their own
- During normal operation of a system
- During worst-case operation of a system
- During the occurrence of a failure that disrupts or halts normal system operation
- During maintenance actions to correct or prevent failures
- During system handling and transportation
- During system disassembly and destruction at the EOL or EOS period

Usually there are insufficient requirements for modularity to minimize interactions during multiple types of maintenance actions. These modularity requirements may be considered important for maintainability, reliability, safety, serviceability, logistics, human factors, testability, diagnostics capability, and prevention of old failures in current designs. Specifications need to address interoperability functions such as for the requirements for HMI interfaces, internal system interfaces, and external interfaces between systems. HMI system requirements interfaces involve requirements for user-hardware and user-software interfaces. Specifications should address how the product behaves when there is an occurrence of an unexpected system behavior such as a device shutting down from an external power surge or from an unexpected human error.

To identify missing functions in a specification, SRA is necessary and is always conducted by a cross-functional team. At least one member from each of these engineering disciplines should be present: R&D, design, quality, reliability, safety, manufacturing, and supportability or field service. If possible, a customer representative should be present as well. Everyone on the team should be aware of each other's requirements, with enough knowledge to be able to write about them, provide citations and references to them, peer review them, and edit them. Writing accurate and comprehensive performance specifications is imperative for a safe design.

People attending one particular safety training session revealed that trouble-shooting technicians on complex electronic products are unable to diagnose about 65% of the problems (i.e., cannot duplicate the fault), which can be blamed on requirements defect. Improved fault tolerance and test coverage requirements in the specifications are necessary for fool-proof product testing and troubleshooting, whether in development, manufacturing, or field support, but the cost to implement a fool-proof product test diagnostic and prognostic capability may be too cost prohibitive, and therefore not provided. In large complex systems, it is incredibly expensive to obtain near-100% test coverage and harder still to reach 100% fault coverage isolating to a single item that caused a failure.

To find system design requirement flaws early, a team has to view the system documentation and models from different angles. A cross-functional system team must be able to view the system requirements from at least the following perspectives:

- Functionality during nominal and worst-case system conditions
- Undesired functions the system or products within the system should never do
- Range of applications and uses
- Range of environments and operating conditions
- Active, semi-active, and passive safety controls
- Operating duty cycles during life (duty cycle is ratio of the on time period and total time)
- Design for reliability over the system life
- Design robustness for user/servicing mistakes over the system lifetime
- Sneak circuit paths in electrical designs
- Logistics requirements to avoid adverse events over the system lifetime
- Manufacturability requirements for defect-free production
- HMI interface requirements
- Human Factors Engineering (HFE) and human–machine engineering capabilities
- Installation requirements to assure safe functioning
- Shipping/handling capabilities to keep the device safe
- Serviceability/sustainability/testability/maintainability/diagnostics capabilities
- Prognostics and Health Monitoring (PHM) to warn users of imminent failures
- Interoperability with other systems or products
- Potential hazards, accidents, and abuses that may occur over the system lifetime

Most designers are likely to miss some of the aforementioned requirements. All of these requirements affect safety. This knowledge is not new to the various industries, but it may be new to the early career engineer. This knowledge can be shared and incorporated into the SDP by inviting system safety experts in these areas to join your development teams and brainstorm ways to conduct these SRAs. Requirements should include how the system should behave when a "sneak" failure occurs. A tool called Sneak Circuit Analysis (SCA) can be used for predicting undesirable sneak circuit occurrences or sneak circuit failures. A good specification will address what the system shall not do, such as "there shall be no sneak circuit paths" or "there shall be no events that cause sudden fires during normal system operation." If we do everything right the first time, we avoid pain and suffering to users.

5.3 How Do We Drive System Safety Requirements?

The primary focus of any system safety plan, hazard analysis, and safety assessment is to identify operational behaviors, anomalous conditions, faults, or human errors that could lead to a safety-critical failure, hazard, or potential mishap. This data is used to influence requirements to drive control strategies and safety attributes in the form of safety design features or safety devices to prevent, eliminate, control, or mitigation safety risk. Modern system safety is a comprehensive risk-based, requirements-based, functional-based, and criteria-based means to achieve structured goals, requirements, and objectives for the system, hardware, and software. The system safety efforts yield engineering evidence to verify system safety requirements and safe system/product functionality that is deterministic, or probabilistic, to show acceptable or unacceptable risk in the intended operating environment. Software-intensive systems that command, control, and monitor safety-critical functions require extensive software safety analyses to influence detail design requirements, especially in more autonomous or robotic systems with little or no operator intervention [1].

> A healthy sceptical attitude toward the system, when it is at the requirements definition and drawing-board stage, by conducting functional hazard analyses, would help in learning about the factors that create hazards and mitigations that control the hazards. A rigorous system safety analysis process is usually formally implemented as part of systems engineering to influence the design and improve the situation before the errors and faults weaken the system defenses and cause accidents. [1]

This rigorous system safety analysis process using a functional hazard analysis is one of the ways to drive system safety requirements.

It is only through design requirements and design changes that many hazards and faults may be eliminated or avoided. Therefore, writing system safety requirements should not be considered a trivial task. Lots of creative and innovative thought with design consideration of various options need to be applied in writing system safety requirements. The four key steps to writing system safety requirements are as follows:

1. Start by determining the safety-critical functions from all the system functions.
2. Analyze the overall system, hardware, and software design architecture.
3. Analyze the Human Systems Integration (HSI).
4. Derive system safety requirements from the system functions, architecture, and analysis results.

System safety requirements are specified during proven hazard analysis process to establish safeguards and to ensure essential functions do not cause catastrophic conditions that harm people or that damage equipment. System safety requirements are specified so that functions perform correctly in a predictable manner with predictable results. From hazard analyses, potential faults, failure conditions, and direct and indirect causal factors that can influence, contribute to, or cause hazards are determined. When each significant hazard is identified, an attempt must be made to understand how the absence or presence of something in the system design determines a life-or-death situation under normal system operating conditions, during performance of system corrective maintenance activities, or while trying to withstand harsh environments within the system (if the system is designed to enclose people) or within close proximity to the system. The system designers must ensure that operation of the system will not cause harm, prevent personal injury, avoid catastrophic damage, allow only safe conditions, and ensure that system operators collect data and monitor system health to predict performance in an environment where no one gets hurt. One simple example of a safety-related requirement that prevents hazards from occurring is to design all corners or edges of a mechanical assembly with smooth or rounded corners or edges, removing all burrs and sharp objects or surfaces, to avoid personal injuries due to cuts. System safety requirements must be an essential part of the systems engineering requirements process whether it is done through simple or complex SRAs. System safety requirements must be derived, developed, implemented, and verified with objective safety evidence and ample safety documentation, showing due diligence. Prevention of mishaps is the objective in writing system safety requirements.

5.4 What Is a System Requirement?

Developing and understanding requirements for new systems or products is a significant challenge. There are several reasons for this challenge.

First the customer typically has not thought through all of the product features and needs. There are functional requirements which jump out first. Other requirements seem to take time to discover. Then there are requirements of what the product should not do and how the product should behave when presented with off-nominal inputs (product robustness). Additionally there are requirements to meet business and regulatory needs. Then there are questions of infrastructure which often are not specified. They are just typically assumed to be present and only noticed in their absence. [4]

In 2006 C. J. Davis, Fuller, Tremblay, & Berndt found accurately capturing system requirements is the major factor in the failure of 90% of large software projects, echoing earlier work by Lindquist (2005) who concluded 'poor requirements management can be attributed to 71 percent of software projects that fail; greater than bad technology, missed deadlines, and change management issues'. This requirements challenge has been long recognized and cited. [4, citation Davy and Cope, 2008, 58]

A system requirement specifies what a system must do to satisfy a customer need or goal. A requirement documents anything that legally obligates a company to do something or deliver something. Requirements capture customer objectives, expectations, desires, limitations, and constraints. Requirements establish the direction for all work needed to ensure the customer's objectives are met and satisfied. A sentence containing a requirement will include one of two types of modal verbs in the future tense: "shall" or "will." This section introduces correct use of the terms "shall," "will," and "should." "Shall" designates a sentence as a requirement. It is a stronger term than "Will." "Will" denotes a goal or objective. It provides a statement of intent, or need, or prediction of the future and may also state a fact. It is a stronger term than "Should." "Should" reflects a desire that may or may not be possible to satisfy or comply with. It provides guidance or information. Any statement written with the intention to describe a requirement that includes the word "should" is an example of a bad requirement. Later on in this chapter, there will be more explanation on how to identify good and bad requirements and why requirements are bad, and the chapter recommends changes to fix bad requirements.

Requirements are a critical part of the systems engineering discipline. If the system requirements are wrong, there is a ripple effect that flows down to the hardware and software development phases of a program. Requirements must be correct at the beginning of the system development, or else costly rework will occur. The later in the system development phases that a requirements defect is found and corrected, the more costly it is to fix. This is why many companies use standard software tools and databases to document and analyze their requirements. Requirement generation and analysis tools prevent out-of-phase requirements defects and keep the costs to correct requirements defects down.

Use of requirements generation and analysis tools, such as IBM DOORS, ensures that the systems engineer will be able to find requirement gaps and requirements defects early in the SDP. Following the creation and documentation of requirements in specifications, the systems engineer will analyze the requirements using tools. Requirements analysis will include the tracing of requirements through the specification tree where every specification in the system is identified in a tree structure configuration or hierarchy. Tools such as IBM DOORS are capable of performing traceability analysis, forward requirements trace and backward requirements trace. This analysis is performed to fill any gaps in the requirements that may have been missed when requirements are flowed down from top-system specifications, such as in System-of-Systems Integration (SoSI), or in the decomposition of requirements that are apportioned or allocated to subsystem specifications. In performance of traceability analysis, an engineer will look for not only requirement gaps but also requirement linkages with forward and reverse traceability. The purpose is to ensure there are no lower-level orphan requirements, which are requirements with no parent requirement, or no childless parent requirements. In other words, each top-level requirement must be either flowed down or decomposed to lower-level requirements. The analysis also looks for requirement redundancies, ambiguities, vague requirements, unverifiable requirements, or requirements defects that contradict one another. New products usually experience requirement problems that account for approximately 40–60% of their total system problems. These requirement problems are caused by defective or poorly written requirements, which are related to missing or vague requirements' topology, ontology, terminology, or language.

5.4.1 Performance Specifications

Performance specifications define the functional requirements for the system or product, the environment that the system or product must operate within, the system or product interface characteristics, and the system or product interchangeability characteristics. A performance specification states these requirements in terms of the required results. However, a performance specification does not state the methods used for achieving these required results. It translates operational requirements into more technical language that tells the manufacturer what acceptable product performance is and how that product acceptability is determined. System safety professionals can make use of the performance specification process to include those requirements that will verify the elimination, the mitigation, and the control of hazards [5].

There are six types of requirements as described in the systems engineering fundamentals [5] and adapted here:

1. Operational
2. Functional
3. Performance
4. Design
5. Derived
6. Allocated

Operational requirements are statements that define the basic needs and expectations of the system in terms of mission objectives, environment, constraints, and measures of effectiveness and suitability. Functional requirements are the necessary task, action, or activity that must be accomplished. Performance requirements define how well the system must perform, generally measured in terms of reliability, quality, and other performance-related goals. Design requirements are the actual "how to build or buy" types of requirements. Derived requirements are implied requirements or requirements that must be followed due to higher-level requirements. Allocated requirements are established by dividing or apportioning a high-level requirement into multiple lower-level requirements.

5.4.1.1 System Specification Requirements (SSR)

System safety should provide Preliminary Hazard List (PHL) and Environment, Safety, and Occupational Health (ESOH) criteria and, from this, identifies ESOH requirements, constraints, and performance attributes for the system. Incorporate ESOH requirements into a system specification, as applicable. The identification of a hazard leads to a requirement to eliminate it or prevent its occurrence. The closure of a hazard does not eliminate the requirement to retain the hazard in the Hazard Tracking System (HTS). The hazard will be retained in the HTS and the SSR to provide future program visibility. The hazard and its disposition provide an audit trail of the actions taken to eliminate or mitigate the risk of occurrences. Also, the closed-out actions as documented in the HTS, including implementation status and accident data, are necessary to determine if further action is required later in the system life cycle.

5.4.1.2 Software Requirement Specification (SRS)

A collection of requirements for any particular Software Configuration Item (SCI) design is called a Software Requirements Specification (SRS). The SCI is the lowest level of code within a software design hierarchy to be managed with an SRS.

Requirements for functions, messages, interfaces, and processes are contained within the SCI's SRS. Text that contains "shall" statements in the SRS denotes the requirements of the SCI.

5.4.2 Safety Requirement Specification (SRS)

Safety functions requirements are collected and combined into a single document called a Safety Requirements Specification (SRS). This should not be confused with a Software Requirements Specification (SRS). The term and definition of SRS has ambiguities as you find applications of the acronym in different types of industries, such as military defense industry and medical industry. (Note: to avoid confusion, software requirements specifications may be called Computer Software Configuration Item Specifications (CSCISs) or software configuration item specifications).

The SRS should be comprehensive to include all hardware, software, and system functions that are attributed to safety engineering functionality. The SRS is the key system safety design document that specifies the detailed actions and processes that each safety function should execute: safety performance requirements for each safety function and the general and specific system environment factors to be considered in the design and development stages of the system, hardware, and software.

The Safety Requirements Specification (SRS) could be divided into four main sections, as a minimum:

1. Top-level system safety requirements
2. System safety functional requirements
3. Specific hardware and software functional requirements
4. System, software, and hardware safety Verification and Validation (V&V) requirements

5.5 Hazard Control Requirements

As you have seen, there are many types of requirements that are used in system and product development. Some performance requirements cause safety concerns, while others have no bearing on safety issues. Some requirements are written to prevent hazards from occurring. In essence, these performance requirements may have impacted safety in the past, but, due to sound design decisions in changing requirements, they have no effect on safety hazards now or eliminate the possibility of a hazard from occurring. Some requirements ensure safety risks are mitigated, but not eliminated. The severity of a hazard effect may have been lowered to an acceptable risk level due to the introduction of hazard controls. For instance, a mechanical system with two metal objects rubbing together causes friction and heat. If a flammable material comes into contact with the two metal objects, a flame may ignite and start a fire that could cause personal safety concerns. If the two metal objects are periodically cooled down with a liquid coolant, the possibility of a fire starting is reduced due to less time at high temperature, or the high temperature average may be lowered over a period of time to reduce the severe instance of a fire starting. Adding a liquid coolant to the mechanical system process involving the two metal objects is a means of hazard control.

Hazard control determination is an output from a hazard analysis and risk assessment [6]. When hazard controls are planned, the need for hazard controls is transformed into a set of design or process requirements forming a safety standard. In developing hazard control requirements, there are several areas to consider:

- Conformance to existing safety-related standards is required to meet minimal levels of protection or risk mitigation; however, acceptable levels of risk may not be assured. These acceptable levels of risk may not be possible due to many reasons, such as inappropriate decisions, poor consensus, biases, limited proactive safety assumptions, over-generalization, or poor investigative and analysis work.
- When deriving requirements, the specific safety-related issue or the output of an unbiased accident analysis, system safety analysis, safety study, safety assessment or review, survey, observation, test, simulation, or inspection must be considered and addressed.
- Design requirements must be validated and verified. Provide specific means to ensure that the requirement is translated effectively to the system or product design and does work as intended and as needed. The requirements' V&V must be performed by formal documented approaches, including tests, modeling, simulations, analysis, inspection, or demonstrations.
- System dynamics must be understood by an expert of the system and conveyed to others by writing requirements that accommodate dynamic system changes to assure continued acceptable levels of risk over diverse operating profiles.
- Standardized language, simple common language, lexicons, taxonomies, and usage criteria such as ontologies must be applied in writing requirement documentation.
- Requirements development tools must be evaluated and certified.
- The real-world application and customer use environment must be considered when developing requirements. When addressing functions or operations, know what drives the function or operation, such as the combinations of human, hardware, and software actions, and the interfaces between them (e.g., HMI, hardware interfaces, software interfaces, etc.).

- Requirement abstraction, semantics, context, terms, and written tense should be well understood. Not every statement in a requirements document needs to be a "shall" statement. Some statements are normative and others are informative.
- A picture is worth a thousand words, so use pictures, depictions, illustrations, and diagrams to support the writing of requirements with logical expressions.
- After writing specifications, analyze the requirements language and tone, paying attention to the naming conventions, stereotypes, and intended user. The language should be analyzed to ensure it does not offend anyone. Jargon that only a certain demographic representing a small percentage of the potential user population would understand should be minimized or replaced with more common expressions.
- The intent of a questionable requirement should be independently analyzed, defined, and verified using peer reviews.
- The requirements should be defined to show consistency between high-level, mid-level, and lower-level abstractions and provide tractability or traceability, forward and backward.
- System or product requirements development processes and reviews must be documented.
- Configuration control must be provided during requirements development.
- System and product requirements must be consistently applied, leveraging typical specialty engineering discipline requirements.
- Risk-based and contractual criteria should be defined for ranking, validating, and verifying requirements.

5.6 Developing Good Requirements

According to MIL-STD-961E [7], a good specification should do four things: (1) identify minimum requirements, (2) list reproducible test methods to be used in testing for compliance with specifications, (3) allow for a competitive bid, and (4) provide for an equitable award at the lowest possible cost. These four things may seem more relevant to a PM perspective, rather than a system safety engineering perspective. Items 1 and 2 would apply, however, to both perspectives. Let's focus on writing requirements strictly from an engineer's perspective.

Engineers want clear and concise requirements. The requirements should be correct and consistent. Good requirements avoid ambiguities and redundancies, are verifiable and testable, and avoid imprecise language and terminology. For instance, a good requirement uses numbers and concise metrics to signify when or how something must occur. The term periodic in a requirement is not precise if used alone, but if the term periodic is combined with a frequency metric (e.g., transmit

Table 5.1 Qualities of a well-written performance specification

Attribute	Quality
Clear	Easily understood, unambiguous
Complete	Contains everything that is pertinent
Consistent	Free of conflicts with other requirements
Correct	Specifies what is actually required
Feasible	Technologically possible
Objective	Leaves no room for subjective interpretation
Need-oriented	States problem only, not solutions
Singular	Focuses on only one subject
Succinct	Free of superfluous material, avoids over specification
Verifiable	Can be measured to show that the need is satisfied

periodically one (1) status message every minute, ±1 second), then it is a good requirement. Also, the requirement should include an error tolerance for each metric, such as ±1 second, to know how accurate the measurement must be to verify a requirement metric.

Table 5.1 defines qualities of well-written requirements to summarize what was just mentioned.

5.6.1 Recognize Bad Requirements

Bad requirements negatively impact system performance, development cost, and program schedule. They are not clear and concise and are prone to errors in designing system hardware or software. They cause conflicts, such as a requirement stating that function xyz will occur after another event, in a sequence of events, while another requirement states the function xyz occurs at a certain time interval independent of events in a sequence.

The following are words that contribute to badly written requirements: about, adequate, appropriate, approximately, as necessary, avoid, bad, big, close, fast, few, frequent, good, high, if necessary, immediate, intermittent, intuitive, large, long, low, many, maximize, minimize, most, optimize, periodic, quick, rapid, regular, robust, short, slow, small, sufficient, and timely, to name a few examples.

Types of bad requirements include

- Ambiguous
- Vague
- Contradictory
- Inconsistent
- Unnecessary

- Untraceable
- Not rational
- Unverifiable
- Untestable
- Incomplete
- Missing or omitted
- Without tolerances or limits

5.6.2 Requirements at the Top of the Issues List

Referencing data from the Office of the Secretary of Defense (OSD) [8], we cite a chart on the top 10 emerging systemic issues to show where requirements issues rank in terms of top DoD concerns. This data was extracted from 52 program reviews from the OSD since March 2004. Figure 5.1 contains these top 10 issues. These are the major contributors of poor program performance. Notice that bad requirements are the second largest issue.

Top 10 Emerging Systemic Issues

DEVELOPMENTAL TEST & EVALUATION	
1. Management	• IPT roles, responsibilities, authority, poor communication
	• Inexperienced staff, lack of technical expertise
2. Requirements	• Creep/stability
	• Tangible, measurable, testable
3. Systems Engineering	• Lack of a rigorous approach, technical expertise
	• Process compliance
4. Staffing	• Inadequate Government program office staff
5. Reliability	• Ambitious growth curves, unrealistic requirements
	• Inadequate "test time" for statistical calculations
6. Acquisition Strategy	• Competing budget priorities, schedule-driven
	• Contracting issues, poor technical assumptions
7. Schedule	• Realism, compression
8. Test Planning	• Breadth, depth, resources
9. Software	• Architecture, design/development discipline
	• Staffing/skill levels, organizational competency (process)
10. Maintainability/Logistics	• Sustainment costs not fully considered (short-sighted)
	• Supportability considerations traded

Major contributors to poor program performance

DT&E – From Concept to Combat

Figure 5.1 Top 10 emerging systemic issues [8]

Correcting bad requirements in the early stage of system development has little cost impact. Corrections to requirements later in development cause increased cost impact so the problem is found and corrected. A bad requirement introduced in the requirements stage of development has no cost impact, while a bad requirement detected in the system detailed design phase can have a 10× cost increase compared with the cost of initially developing a requirement. A bad requirement detected in the system integration and test phase of development can have a 10× to 100× cost impact compared with the cost to correct the bad requirement during the detailed design phase.

Requirement issues are may be categorized as requirements creep and volatility, unstable and immature requirements, and requirements written that are not tangible, measurable, and testable.

5.6.3 Examples Good Requirements for System Safety

The following list includes several examples of well-written safety requirements that could benefit any design program:

- Require the system design to define safe states for each operation, function, or process.
- For instance, the system shall power up system initialization functions with a predefined power up sequence ending in a predetermined safe state and indicate a "system ready" status.
- Require timing constraints and limits in all safety design features. For instance, a safety requirement with a timing constraint is "safety interlock circuit shall engage 10 seconds (±1 second) after safety interlock switch is actuated."
- Require the system design to notify the operator when an operational transition is about to occur between safe and unsafe states or transition between safe and unsafe modes.
- The system element controlling hardware subsystems shall return the hardware to a designated safe state when unsafe hardware conditions are detected. These hardware conditions may be detected by environmental stress sensors, such as temperature or vibration sensors like thermocouples and accelerometers, which are used to predict and prevent catastrophic hardware failures.
- The system shall detect high severity failures in external safety-critical hardware input/output hardware devices and interfaces and shall revert to a safe state upon their occurrence.
- The system shall be designed to failover and recover to a designed safe state of reduced functionality in the event of a failure of system components.

- A failure of the system shall not cause system damage or personal injury during performance of a certain function at a specific period of time under certain stresses or conditions.
- Normal operation shall not cause system damage or personal injury during performance of a certain function at a specific period of time under certain stresses or conditions.
- Accuracy of data flowing into and out of safety-critical processes shall be checked by other processes, using redundancy and backup systems in a fault-tolerant architecture configuration.
- The system processes shall provide slack in message traffic timing so that message traffic errors may be eliminated or reduced through retransmission of messages at least three times before flagging an error.
- Watchdog timers or similar devices shall provide an allowable error rate to assure that the computer is operating properly and not hung up or frozen.
- Fault detection and isolation routines shall be written with specified probabilities of fault detection (Pfd) and probabilities of fault isolation (Pfi) for testability of safety-critical computing subsystems.
- False alarms will be minimized to less than one per one million test results, where test results report a pass or fail decision. (Note: A false alarm may be a false positive or a false negative. A false negative means a fail decision is the result that is found to be incorrect. A false positive means a pass decision is the result that is found to be incorrect. Many times, false positives are not considered false alarms, when they should be considered a type of false alarm due to the similarity of the failure mode to a false negative (an alarm which was false).)
- Interlocking mechanism shall be implemented to ensure safe entry into dangerous access ports or through doors.
- Fail-safe circuitry shall be designed for any component that has a failure effect probability greater that 0.00001 of causing a fire or explosion.

5.6.4 Negative versus Positive Requirements

Positive requirements are requirements written with "shall" statements. Negative requirements are requirements written with "shall not" statements.

Many of the system level requirements involve specification of system functions. A typical system requirement may be stated as "The system shall perform the functions necessary to store and back-up key system parametric data every 24 hrs." Systems engineering requirements are usually written in a positive form, such as

"The system shall perform…"

System safety requirements are usually written by SSEs in the negative form, such as

"The system shall not…"

A typical system safety requirement may be stated as

"System Operation shall not cause fatalities and bodily harm."

It may be too costly and time-consuming to verify these types of safety requirements written in a negative form. How do you verify a device will not operate as required? How much testing is required to demonstrate safe performance? These are difficult questions to answer. Since systems engineers want requirements that are verifiable, system requirements are written in a positive form. System safety requirements may be written in a positive form, such as

"The system operation shall prevent personal injury, avoid catastrophic damage, and allow only safe conditions for operation," or "the system shall perform in an environment where no one is hurt."

Any negative requirement shall have a timing constraint and actions associated with the condition it is designed to prevent. For instance, an example of a negative requirement is "the system shall not cause an operator injury." This requirement may be considered a bad negative requirement since it lacks details that would be useful in verifying the requirement is met. Therefore, this requirement could be rewritten to become a good negative requirement, such as "the system shall not cause an operator injury within two hours following the initialization of the function 'xyz', while the operator is seated at the operator console." To prove this requirement is met, the operator must be seated at the console for two hours following the start of the "xyz" function. Also, the operator must be in a sitting position at the console to verify this requirement is met. If the operator is standing or another position, this requirement is invalid.

5.7 Example of Certification and Validation Requirements for a PSDI

This section provides an example of certification and validation of requirements for a system safety device called a Presence Sensing Device Initiation (PSDI) used in mechanical power presses. The certification and validation requirements for a PSDI are written within an Occupational Safety and Health Administration

(OSHA) document [9]. There are many requirements and standards under OSHA jurisdiction. The US Department of Labor is responsible for Title 29 of the Code of Federal Regulations. Part 1910 of Title 29 is concerned with OSHA standards. An example of an OSHA standard in Title 29 of the Code of Federal Regulations (29CFR) is Standard Number 1910.217, which is the mandatory requirement for certification and validation of safety systems for PSDI. The purpose of this standard is to ensure that the safety systems and PSDI used in mechanical power presses are designed, installed, and maintained in accordance with all applicable requirements 29 CFR 1910.217 (a)–(h) clauses and Appendix A of the standard. This OSHA standard makes sure that the PSDI is safe to operate as a safety system for mechanical power presses [9].

General Design Certification and Validation Requirements Using 29 CFR 1910.217 [9]

Certification and Validation Program Requirements

General design certification requirements for an Original Equipment Manufacturer (OEM) involve two main criteria. The OEM manufacturer shall certify that

- The design of hardware and software components, integrated, and tested within subsystems and assemblies of a system design meet OSHA performance requirements and is ready for the intended use of the system.
- The system performance of integrated subsystems including hardware and software meets OSHA operational requirements.

The general design certification validation of the requirements is usually performed by an OSHA-recognized third-party validation organization that validates that the OEM completed the certification of the two main criteria.

Level of Risk Evaluation Requirements

The manufacturer shall evaluate and certify that the design and operation of the safety system by determining conformance with the following.

Design Features

The safety system shall have design features with the ability to sustain a single failure or a single operating error and not cause injury to personnel as an operational hazard. Acceptable design features shall demonstrate, in the following order of precedence, that

- No single point failures may cause injury.
- Redundancy, data comparison checks, and diagnostic checks exist for the critical items that may cause injury.

- Electrical, electromechanical, and mechanical piece parts and components are selected for their particular design applications so that they can withstand operational and external environmental stress conditions expected for that system.
- The key performance parameters of each critical component shall be derated. Component derating reduces the electrical stress of a parameter by a percentage lower than their full rated parametric values as specified in the manufacturers' data sheets. These derating percentages will be safety factors documented and specifically noted in design guidelines for each company's design program.

Electrical design parameters, such as power, voltage, or current, are specified for maximum values in the component manufacturers' data sheets. These specified values are called ratings. When an electrical component is used in a system at the full rated value as specified by the component manufacturer, the component is 100% stressed. If the component is used above the full rated value as specified by the component manufacturer, the component is considered to be overstressed. Overstressing a component usually voids any manufacturer's warranty of that component. If the component is used in a system at a level below the full rated value as specified by the component manufacturer, the component is derated. The derating of a particular component's parameter is calculated by measuring or calculating the stress of the component and subtracting from 100%. The stress is calculated by dividing the actual value of a parameter by the rated value of the parameter as defined in the manufacturer's data sheet. The amount of derating has a direct correlation to the stress on the component and the probability of failure of that component. The reliability of the component is directly proportional to the derating percentage of the component's rated parameter. The amount of component derating varies from one derating guideline to another. Derating guidelines are available from multiple sources. Each company should develop their own derating design guides based on design guides readily available in the public domain.

The manufacturer shall design, evaluate, test, and certify that the PSDI safety system meets the following requirements:

- Environmental limits, such as
  - Temperature
  - Relative humidity
  - Vibration
  - Fluid compatibility with other materials
- Design limits, such as
  - Power (e.g., power input, power output, and power dissipation)
  - Power transient tolerances

o Material stability to long term power fluctuations
o Material compatibility with other materials within close proximity
o Material electrical, mechanical, and electromagnetic stress limits
o Material electrical mechanical and electromagnetic tolerance limits
  – Electromagnetic tolerance limits includes conducted and radiated emissions.
o Sensitivity of receivers to signal strength for signal and data acquisition
o Repeatability of reading and recording consistent parameter measurements without errors
o Adequate space allowance for consistent parameter measurements without blocking view or causing inadvertent initiation of a function or actuation of a switch
o Reliability and operational life of components in terms of Cycles to Failure (CTF) or Time to Failure (TTF) or both

The general design certification program level of risk evaluation requirements is usually validated by an OSHA-recognized third-party validation organization that validates that the OEM completed the certification. This third-party validation organization is the honest broker for OSHA, ensures candid discussions between OSHA and the manufacturer, and keeps the manufacturer honest.

Detailed Design Certification and Validation Requirements

The manufacturer or the manufacturer's representative shall certify that the documentation necessary to demonstrate that the PSDI safety system design is in full compliance with the requirements of 29 CFR 1910.217(a)–(h) and Appendix A. The manufacturer or the manufacturer's representative shall certify by means of analysis, tests, or combination of both, establishing that the following requirements are fulfilled:

- Reaction time
- Function complete
- Full stop
- Test signal generation

For the purpose of demonstrating compliance with reaction time requirements, the test documentation shall use the following definitions and requirements:

- "Reaction time" means the time measured in seconds for a signal required to activate or deactivate the system, to travel through the system, measured from the time of signal initiation to the time the function being measured is completed. Reaction time could be measured for any function of a system with adequate interface and communication commands. For particular digital simulations and microprocessor-based command and control (C2) systems, this reaction time begins counting when the command to "start" a function is transmitted from a command processor, up to the time when the "started" function command is received by the command processor from the digital processor initializing and executing the function.

- "Function complete" means the system function performed by a digital processor in the system that began processing a command initiated by a command processor with a "start" command has now completed its processing and produced a change of state in its output element. When the change of state is motion (e.g., a rotating electromechanical device), the measurement shall be made at the completion of motion before a "function complete" command is acknowledged back to the command processor.
- "Full stop" means that there is no movement in any mechanical or electromechanical devices, such as gears, bearings, motors, or shafts. Full stop is a command that affects multiple system functions simultaneously or sequentially. When a single full stop command is given, and multiple functions react, some functions react to the command immediately, while other functions must wait for their position to stop in a sequential pattern. Because of this sequential order of events that must occur, the reaction time to execute a full stop could be considerably longer than the time to start a single function. The reaction time for full stop begins counting when the command to "full stop" is transmitted from a command processor, up to the time when the "full stop" function command is received from the digital processor executing the function and acknowledged by the command processor.
- "Test signal" generation is required to ensure reliable functionality of the system. Test signal generation is usually an essential function within a Built-In-Test (BIT) or embedded diagnostic test system. The generation of the test signal introduced into the system for measuring reaction time shall be such that the initiation time can be established with an error of less than 0.5% of the reaction time measured. The instrument or device used to measure reaction time shall be calibrated to demonstrate consistent accuracy. The instrument or device used to measure reaction time may be designed as part of the BIT functions embedded in the system or may be external support test equipment that is connected during a period of system downtime to conduct preventive or corrective maintenance. The instrument or device used to measure reaction time shall be calibrated to be accurate to within 0.001 second (±1 ms tolerance).

For certain types of machinery, PSDI for mechanical power presses and machine guarding equipment, timing is very critical so the requirements are measuring timing is taken very seriously. The average value of stopping time (T_s) shall be calculated as the arithmetic mean of at least 25 stops for each stop angle initiation measured for three types of criteria: (1) with the brake and/or clutch unused, (2) each 50% worn, and (3) 90% worn. The recommendations of the brake system manufacturer shall be used to simulate or estimate the brake wear. The manufacturer's recommended minimum lining depth shall be identified and documented, and an evaluation performed that the minimum depth will not be exceeded before the next periodic (e.g., annual) recertification (recert) or revalidation.

Each reaction time required to calculate the safety distance, including the brake monitor setting, shall be documented in separate reaction time tests. These tests shall specify the acceptable tolerance band sufficient to assure that the tolerance buildup will not render the safety distances unsafe. Tests that are conducted by the manufacturer of electrical components to establish stress, life, temperature, and loading limits must be tested in compliance with the provisions of the National Electric Code. Electrical and/or electronic cards or boards assembled with discrete components shall be considered a subsystem and shall require separate testing that the subsystems do not degrade in any of the following conditions:

- Ambient temperature variation from −20°C to +50°C
- Ambient relative humidity of 99%
- Vibration of 45G for one millisecond per stroke (1 ms/stroke) when the item is to be mounted on the press frame
- Electromagnetic interference at the same wavelengths used for the radiation sensing field, at the fundamental power line frequency and its harmonic frequencies
- Electrical power variations of ±15%

Installation Certification and Validation Requirements

The employer shall evaluate and test the PSDI system installation, shall submit supporting documentation to the third-party OSHA-recognized validation organization, and shall certify that the requirements of 1910.217 (a)–(h) and Appendix A have been met and the installation is proper.

Recertification and Revalidation Requirements

A PSDI safety system that has received installation certification and validation shall undergo recertification and revalidation every year or sooner if any of the following occurred:

- Each time the system's hardware is significantly changed, modified, or refurbished
- Whenever the operational conditions are significantly changed including environmental, application or facility changes
- Upon failure occurrence of a significant component, which may affect safety, and a corrective action has been made to resolve the failure occurrence

5.8 Examples of Requirements from STANAG 4404

Standard NATO Agreement 4404 (STANAG 4404) [10] is the safety design requirements and guidelines for munition-related safety-critical computing systems. These requirements and guidelines are normally tailored to a particular system in development before inclusion in any contractual documents. STANAG 4404 Annex A provides guidelines for tailoring the requirements and guidelines to different system types and sizes. Annex A also provides guidelines for the implementation of the requirements and guidelines in various munition system development phases. The analyst must ensure that the selected requirements and guidelines are tailored to the munition system under consideration and that the final design implements the safety design requirements and achieves the overall system safety goal, subject to review by the appropriate safety authority. The developer cannot claim compliance with this STANAG without providing documentation supporting the tailoring and implementation of the design requirements and guidelines contained herein.

The following are examples of well-written system safety design requirements and guidelines from STANAG 4404 [10]:

Design Safe States
The system shall have at least one safe state identified for each logistic and operational phase.

Standalone Computer
Where practical, safety-critical functions should be performed on a standalone computer. If this is not practical, safety-critical functions shall be isolated from noncritical functions to the maximum extent possible.

Ease of Maintenance
The system and its software shall be designed for ease of maintenance by future personnel not associated with the original design team. Documentation specified for the computing system shall be developed to facilitate maintenance of the software.

Safe State Return
The software shall return hardware subsystems under the control of software to a designed safe state when unsafe conditions are detected.

Restoration of Interlocks
Upon completion of tests and/or training wherein safety interlocks are removed, disabled, or bypassed, restoration of those interlocks shall be verified by the software prior to being able to resume normal operation. While overridden, a display shall be made on the operator's or test conductor's console of the status of the interlocks, if applicable.

I/O Registers

Input/Output (I/O) registers and ports shall not be used for both safety-critical and noncritical functions unless the same safety design criteria are applied to the noncritical functions.

External Hardware Failures

The software shall be designed to detect failures in external hardware input or output hardware devices and revert to a safe state upon their occurrence. The design shall consider potential failure modes of the hardware involved.

Safety Kernel Failure

The system shall be designed such that a failure of the safety kernel (when implemented) will be detected and the system returned to a designed safe state.

Circumvent Unsafe Conditions

The system design shall not permit detected unsafe conditions to be circumvented. If a "battleshort" or "safety arc" condition is required in the system, it shall be designed such that it cannot be activated either inadvertently or without authorization.

Fallback and Recovery

The system shall be designed to include fallback and recovery to a designed safe state of reduced system functional capability in the event of a failure of system components.

Simulators

If simulated items, simulators, and test sets are required, the system shall be designed such that the identification of the devices is fail-safe and that operational hardware cannot be inadvertently identified as a simulated item, simulator, or test set.

System Errors Log

The software shall make provisions for logging all system errors detected. The operator shall have the capability to review logged system errors. Errors in safety-critical routines shall be highlighted and shall be brought to the operator's attention as soon as practical after their occurrence.

Positive Feedback Mechanisms

Software control of critical functions shall have feedback mechanisms that give positive indications of the function's occurrence.

Peak Load Conditions

The system and software shall be designed to ensure that design safety requirements are not violated under peak load conditions.

Power-Up Initialization

The system shall be designed to power up in a safe state. An initialization test shall be incorporated in the design that verifies that the system is in a safe state and that safety-critical circuits and components are tested to ensure their safe operation. The test shall also verify memory integrity and program load.

Power Faults

The system and computing system shall be designed to ensure that the system is in a safe state during power-up, intermittent faults, or fluctuations in the power that could adversely affect the system or in the event of power loss. The system and/or software shall be designed to provide for a safe, orderly shutdown of the system due to either a fault or power-down, such that potentially unsafe states are not created.

5.9 Summary

In conclusion, when system safety engineering is properly integrated in the early phases of the SDP, many hazards can be eliminated or mitigated as development teams generate and analyze requirements for performance specifications. When potential hazards are tied to performance results and metrics, verification of a hazard's elimination or mitigation is given high priority. System safety engineers integrated with system, hardware, and software development engineers, and Test and Evaluation (T&E) personnel in a diverse development team writing good requirements in a tight and cohesive partnership, an appropriate balance between safety, cost, schedule, and performance, is ensured. Hazard controls should be developed and converted into safety requirements that form an appropriate system safety specification or product safety specification. System safety engineering and other members of a cross-functional team should apply reasoning and logic based on risk management approaches and data-driven decisions to influence and convince Program Managers (PMs) and customers why requirements are bad and need to be changed to improve the design for safety of their systems and products. Obviously, system safety engineering with backup from their teams should recommend changes to requirements that are within the range of acceptability to PMs and customers considering cost and schedule constraints.

The power to influence decisions to spend time and money to fix requirements to ensure risk of unsafe conditions are eliminated or minimized is not easy to develop or apply for many system safety engineers. People skills to influence people are usually a tough challenge, but are worth the effort. Help with skill building to influence people is contained throughout this book, especially in Chapters 13 and 14. The benefits from the challenges to enact requirement changes are realized and proven worthy of the effort when demonstrations and tests show that only safe conditions occur during operation and maintenance of the systems or products over their life cycles.

References

[1] System Safety, Wikipedia, https://en.wikipedia.org/wiki/System_safety (Accessed on 1 August 2107).

[2] Roland, H.E. and Moriarty, B. (1990). *System Safety Engineering and Management*. John Wiley & Sons, Inc., New York.

[3] Fischhoff, B. (1995). Risk Perception and Communication Unplugged: Twenty Years of Process. *Risk Analysis*, Vol. 15, No. 2, pp 137–145.

[4] Raheja, D. and Gullo, L.J. (2012). *Design for Reliability*, John Wiley & Sons, Inc., Hoboken, NJ.

[5] Wilkinson, P.K. (2014). Using the Performance Specification Process in Hazard Elimination and Control, presented at 30th International System Safety Conference, Atlanta, GA, August 6, 2012.

[6] Allocco, M. (2014). Eliminating or Controlling System Risks via Effective System Safety Requirements and Standards. *Journal of System Safety*, Vol. 50, No. 1, 30–31.

[7] MIL-STD-961E, Military Standard: Defence and Program-Unique Specifications Format and Content, U.S. Department of Defense, Washington, DC.

[8] United States Office of Secretary of Defense (OSD), Top 10 Emerging Systemic Issues, http://www.dtic.mil/ndia/2007systems/Thursday/AM/Track2/5675.pdf (Accessed on August 1, 2017).

[9] Occupational Safety and Health Administration (OSHA), Law and Regulations (e.g., Standards: 29 CFR)

[10] STANAG 4404—Standard NATO Agreement 4404, Safety Design Requirements and Guidelines for Munition Related Safety Critical Computing Systems.

6

System Safety Design Checklists

Jack Dixon

6.1 Background

Checklists are very important tools for system safety engineers. There are simply too many things that a system safety engineer is required to think about and consider when evaluating a design. In general, checklists are valuable tools for all of us to use in our daily lives. They are an easy way to remind us of either something to consider or something that requires action. Checklists are memory aids. Many of us use checklists in our daily lives without even thinking about them. Checklists serve as job aids or training aids to increase our learning curve on using a particular process. For a simple example, everyone has a "to-do" list to keep track of the things that must get done during an event or a period of time. The shopping list or the "start of the working day to-do list" are checklists. Your grocery shopping list, which contains the food and beverage items you need for preparing meals at your home, is a checklist. Your holiday shopping list containing the names of gifts you want to buy for your family and friends during the year and give to them on the next holiday or birthday is a checklist. Your daily calendar is another example.

As you will see in this chapter, checklists play an important part in the world of system safety engineering. They help to ensure safety in processes, in production, in maintenance operations, and in design of new products and systems.

Design for Safety, First Edition. Edited by Louis J. Gullo and Jack Dixon.
© 2018 John Wiley & Sons Ltd. Published 2018 by John Wiley & Sons Ltd.

6.2 Types of Checklists

There are many different types of checklists and numerous ways to categorize them, but for purposes of system safety, we divide them into three categories:

1. Procedural
2. Observational
3. Design

   All three types of checklists have applications in safety.

6.2.1 Procedural Checklists

Procedural checklists are used to attempt to mistake-proof a complicated process or procedure that contains many steps. It provides a guide to a user so that they don't miss important, critical steps in a complex operation. While the procedural checklist can be used for many processes, they can be a key factor in safety and accident prevention. The most common and most familiar procedural checklists that come to mind are the checklists that pilots use for flying an airplane.

6.2.1.1 Flying Planes

The airplane checklist had its origin in 1935 after a deadly accident occurred during the final evaluation by the US Army of three manufacturers' aircraft. The three airplanes were Martin's Model 146, Douglas's Model DB-1, and Boeing's Model 299. In the earlier phase of the Army's evaluation, Boeing's Model 299 was the leader; it was faster, could fly further, and could carry five times more bombs that the Army had requested. However, during the final test flight, it taxied, took off, and climbed smoothly, but stalled suddenly turned on one wing and crashed in a fiery explosion, killing two of the five people on board including the pilot. Pilot error was blamed for the accident. The pilot had forgotten to release the elevator lock prior to taking off, realized the mistake later, and tried to unlock the handle, but it was too late. Newspaper reports dubbed the airplane, "too much plane for one man to fly." The Army gave contracts to Douglas. However, because of the obvious superiority of the Boeing Model 299, the Army purchased several more for further testing, knowing that if any more accidents occurred, it would be the end of the Model 299. So, a group of test pilots put their heads together to come up with a way to make sure that everything required to fly this more complex aircraft was done properly. They came up with a pilot's checklist. They actually came up with four checklists—one for takeoff, one for flight, one for pre-landing, and one for post landing. As they say, "the rest is history." The Army approved the plane, ordered thousands of them, and renamed it the B-17. B-17 "Flying Fortress" flew 1.8 million miles without an accident and helped the United States win the Second

World War. Checklists became a common tool in the airplane flying business (Adapted from Refs. [1] and [2]).

A short excerpt from a typical preflight checklist for an MD-80 is shown in Figure 6.1 [3].

BEFORE STARTING ENGINES

LOG BOOKS AND SEL .... CHECKED
* RUDDER PEDALS AND
     SEATS .... ADJUSTED AND LOCKED
* WINDOWS .... CLOSED AND LOCKED
   O_2 PANELS/MASKS/INTERPHONE/
     GOGGLES ....SET AND CHECKED
   EMERGENCY LIGHTS ....ARMED
* PROBE HEAT .... CAPT
* WINDSHIELD ANTI-ICE .... ON
   ANTI-SKID .... OFF
   PRESSURIZATION .... AUTO (UP) AND SET
* AIR COND SHUTOFF .... AUTO
* FLIGHT GUIDANCE PANEL .... SET AND CHECKED
* FLT INSTR/SWITCHES/BUGS .... SET AND
                               CROSSCHECKED
* FUEL PANEL/QUANTITY AND
     DISTRIBUTION .... SET/__LBS AND CHECKED
   GEAR HANDLE AND
     LIGHTS .... DOWN AND GREEN
* TRANSPONDER.. SET
* STABILIZER TRIM .... SET
   SPOILER LEVER .... RET
   THROTTLES .... CLOSED
   FUEL LEVERS .... OFF
   FLAPS/SLATS .... UP/RETRACTED
* AILERON/RUDDER TRIM ....ZERO/ZERO
* PARKING BRAKE/PRESSURE ....PARKED/NORMAL
* SHOULDER HARNESSES (If Operative) .... ON
* FLIGHT FORMS .... CHECKED
* NO SMOKING SIGNS ....ON
* SEAT BELT SIGNS (5 Minutes Prior To Departure)......ON

PRIOR TO ENG START OR PUSH-OUT

GALLEY POWER .... OFF
ENGINE IGNITION .... CONTIN
FUEL PUMPS .... ON
AUX HYDRAULIC PUMP .... ON
ANTI-COLLISION/EXTERIOR LIGHTS.....ON/AS REQUIRED
DOOR ANNUNCIATORS .... OUT
AIR CONDITIONING SUPPLY SWITCHES .... OFF

Figure 6.1 MD-80 preflight checklist

6.2.1.2 Operating Room

While there are procedural checklists for many functions, another example of a procedural checklist that has only recently been developed and has shown tremendous potential to save lives is a checklist for surgical operating rooms.

According to the World Health Organization (WHO) [4],

- The mortality rate after major surgery is between 0.5 and 5%.
- Complications after operations occur in up to 25% of patients.
- In industrialized countries, nearly half of all adverse events in hospitalized patients are related to surgery.
- At least half of the cases in which surgery led to harm are considered to be preventable.

Complication rates from surgery are estimated to be between 3 and 17%. Given that in 2004, there were 230 million operations performed worldwide, a substantial number of people are actually harmed by the surgery that is supposed to be helping them. For example, in the United States, over 300,000 operations result in a surgical site infection, and more than 8,000 deaths are associated with these infections [1].

In the mid-2000s, Dr. Atul Gawande teamed up with WHO to develop a global program to reduce avoidable deaths and harm from surgery. This effort resulted in the development of a one-page, 19-item checklist. The checklist was tested in eight hospitals around the world, and the reductions in complications and deaths arising from surgical procedures were astounding. Major complications were reduced by 36% and deaths were reduced by 47% [1, 5]!

A few of the questions on the checklist include the following:

- Has the patient confirmed his/her identity, site, procedure, and consent?
- Is the site of surgery marked?
- Does the patient have a known allergy?
- Has antibiotic prophylaxis been given within the last 60 minutes?
- Is essential imaging displayed?
- Nurse verbally confirms completion of instrument, sponge, and needle counts.

So, even a simple checklist asking relatively simple questions applied to something as complicated as surgery can yield substantial benefits. The complete surgical checklist can be found at the WHO website [6].

 Paradigm 10: If You Stop Using Wrong Practices, You Are Likely to Discover the Right Practices

6.2.2 Observational Checklists

The checklists used by quality auditors fall into the category of observational checklists. These checklists can be used to observe a process and, by tracking certain data or characteristics, determine if the process in under control and within specification limits. These processes might include production, chemical processes, control room processes, and so on. Tally sheets, a form of checklist, can be used by quality personnel to count and track defects of various types to identify a need for process improvement.

Audits are another variant of the observational checklist. While audits are usually associated with quality or financial functions, they can also be used by safety personnel to ensure that safety requirements are met, to evaluate a safety program's effectiveness, or to improve a safety-related process.

Some typical questions that might appear in a system safety program audit include, but are not limited to, the following:

- Is there a system safety program policy in place?
- Does the system safety engineer report directly to, or have direct access to, the program manager?
- For each program, is there a system safety program plan?
- Is there a process in place to identify and track hazards?

6.2.3 Design Checklists

While the procedural and observational checklists have an important place in safety, the design checklist is most important in designing for safety. The design checklist is typically used as an aid to jog the analyst's memory and can be used to support any number of safety analyses. Design checklists provide a source material for identifying hazards in the early stages of design. There is no one design checklist that will be sufficient to identify all hazards. The safety engineer should not rely exclusively on checklists to help to recognize hazards. Other methods should be used to supplement the use of checklists. These methods may include brainstorming with a group of experts knowledgeable of the system being developed, analysis of similar systems, lessons learned databases, and so on. However high-quality design checklists provide a good starting point.

It is always good to use more than one checklist. Design checklists can have different orientations, which may help the safety engineer to recognize potential hazards when developing a new product or system. For example, some common orientations include

- Requirements from contracts, specifications, and/or standards
- Energy sources
- Generic hazards
- Hazards from similar, specific systems
- Hazards related to generally hazardous operations

Examples of each of these types of checklists are provided in the following paragraphs. They are only samples to stimulate the reader's imagination and are not meant to be exhaustive checklists. Many sources of checklists exist and some are provided in the suggested reading at the end of this chapter.

6.2.3.1 Requirements Checklist

An example of a requirement-type checklist is shown in Table 6.1. It shows some typical excerpts from MIL-HDBK-454, a common standard that might be required by a government contract [7]. This type of checklist not only provides a check on contract, specification, and/or standards compliance but can also stimulate recognition of hazards that apply to the product or system at hand.

Table 6.1 Example of requirement-type checklist

Paragraph number	Title	Requirement
4.1	COTS equipment	COTS equipment that has been listed or certified to an appropriate commercial standard by a Nationally Recognized Test Laboratory (NRTL) (e.g., Underwriters Laboratories (UL), Canadian Standards Association (CSA), or TUV Rheinland (TUV))
4.2	Fail-safe	The design and development of all military electronic equipment should provide fail-safe features for safety of personnel during the installation, operation, maintenance, and repair or interchanging of a complete equipment assembly or component parts thereof
4.3	Bonding in hazardous areas	Electronic equipment to be installed in areas where explosive or fire hazards exist should be bonded in accordance with MIL-STD-464 for aerospace systems, MIL-STD-1310 for shipboard systems, and NFPA 70, for facilities, or as otherwise specified in the equipment specification
4.5	Electrical	The design should incorporate methods to protect personnel from inadvertent contact with voltages capable of producing shock hazards
4.5.3	Accidental contact	The design should incorporate methods to protect personnel from accidental contact with voltages in excess of 30 V rms or dc during normal operation of a complete equipment

6.2.3.2 Energy Sources Checklist

Using a checklist of energy sources can help the analyst to identify hazards that result from having these energy sources in a product or system. Some typical energy sources that are common in systems are shown in Table 6.2.

6.2.3.3 Generic Hazard Checklist

Just as with energy sources, a list of general types of hazards can provide a starting point for the safety engineer to identify hazards that may be in the system being developed. Table 6.3 illustrates some of these types of hazardous sources.

6.2.3.4 Similar System Checklist

Many times the product being developed is similar to a product that has previously been developed. Hazards that were identified with the earlier system can serve as a useful checklist for the engineer to use while working on the new system. An example of a previously developed laser system is shown in Table 6.4 and can be used as a guide for development of a new laser system.

Table 6.2 Example of energy source-type checklist

No.	Energy source
1.	Batteries
2.	Fuel
3.	Explosives
4.	Pressurized components
5.	Springs
6.	Electricity
7.	Rotating devices

Table 6.3 Example of generic hazard-type checklist

No.	Generic hazard
1.	Electricity
2.	Radiation
3.	Chemical reaction
4.	Fire
5.	Pressure release
6.	Noise
7.	Vibration

Table 6.4 Example of similar system-type checklist

No.	Hazard	Effect
1.	Cadmium plating	Cadmium causes problems in disposal
2.	Residual laser radiation escapes from laser	Eye injury
3.	Weight exceeds lifting limits	Back injury
4.	Inadvertently placing mode switch in incorrect mode	Inadvertent firing
5.	Unclear ARM mode	Inadvertent firing in wrong mode
6.	Confusing firing controls	Inadvertent firing results in eye injury
7.	Inadvertent firing of laser	Eye damage or burns from laser
8.	Laser radiation	Eye damage or burns from laser
9.	Hot surfaces are personnel hazard	Burns to personnel
10.	High voltage	Shock hazard
11.	Shock from capacitor discharge	Possible electrocution from high voltage discharge

6.2.3.5 Generally Hazardous Operations Checklist

Considering the operations that may be required while using the product or system being developed is another way to stimulate the thought processes to identify potential hazards that may be presented by the new product. Table 6.5 provides a short list of example operational hazards.

6.3 Use of Checklists

As can be seen from the earlier, there are numerous types of checklists with many different applications. They can be used to help ensure that all the proper steps are taken in a process to make sure it remains safe. They can be used to double-check

Table 6.5 Example of hazardous operation-type checklist

No.	Operational hazard
1.	Testing
2.	Lifting
3.	Cleaning
4.	Welding
5.	Handling hazardous material
6.	Confined space
7.	Heavy equipment operation
8.	Painting

a production operation. The design checklists can be used to help identify hazards in a new design or for creating requirements or verifying that requirements are met.

Design checklists should be tailored to support the effort at hand. They may be tailored in numerous ways to help the analyst. They should be tailored to reflect the type of equipment being developed, to reflect the subsystems involved, and/or to reflect the specifications and standards being used.

The most common use of the design checklist is early in the development of a product or system to help the safety engineer create the preliminary hazard list and preliminary hazard analysis. When using the design checklist to identify hazards in a new system or product, the analyst must remember that no one checklist can be all encompassing. More than one checklist should be used to help guide the engineer through the process of identifying hazards. Design checklists are not the end-all for identifying hazards; they are only meant to be a job aid to help stimulate the engineers' thinking.

Two more complete sample checklists are provided in the Appendices. Appendix A contains a checklist that is a combination of energy sources and generic hazards [8]. Appendix B is a more detailed design checklist that can be used to help identify hazards or as a design guide to provide to design engineers at the beginning of a project, or it can be used to verify a design is safe during first article inspection/testing [9].

In summary, the checklist is a valuable tool that can be used for many purposes, but they must be used cautiously and must not be relied on as the only tool.

References

[1] Gawande, A. (2010) *The Checklist Manifesto*, Picador, New York.
[2] Schamel J. How the Pilot's Checklist Came About, http://www.atchistory.org/History/checklst.htm (Accessed on August 1, 2017).
[3] Turner, J. and Huntley, S. (1991) *The Use and Design of Flightcrew Checklists and Manuals*, DOT/FAA/AM-91/7, U.S. Federal Aviation Administration, Office of Aviation Medicine, Washington, DC.
[4] World Health Organization (WHO), http://www.who.int/patientsafety/safesurgery/en (Accessed on August 1, 2017).
[5] Haynes, A., et al. (2009), A Surgical Safety Checklist to Reduce Morbidity and Mortality in a Global Population, *New England Journal of Medicine*, 360, 491–499.
[6] WHO, Surgical Safety Checklist, http://www.who.int/patientsafety/safesurgery/en (Accessed on August 1, 2017).
[7] MIL-HDBK-454B, Department of Defense Handbook General Guidelines form Electronic Equipment, April 15, 2007.
[8] Goldberg, B.E., Everhart, K., Stevens, R., Babbitt III, N., Clemens, P., and Stout, L. (1994) *System Engineering "Toolbox" for Design-Oriented Engineers*, NASA Reference

Publication, 1358, National Aeronautics and Space Administration, Marshall Space Flight Center, Huntsville, AL.

[9] U.S. Army Communications-Electronics Command (CECOM), SEL Form 1183, System Safety Design Verification Checklist, February 2001.

Suggestions for Additional Reading

Raheja, D. and Allocco, M. (2006) *Assurance Technologies Principles and Practices*, John Wiley & Sons, Inc., Hoboken, NJ.

Raheja, D. and Gullo, L.J. (2012) *Design for Reliability*, John Wiley & Sons, Inc., Hoboken, NJ.

Additional Sources of Checklists

Space and Missile System Organization, SAMSO-STD-79-1 SAMSO Standard Integrated System Safety Program for the MX Weapon System

Air Force Systems Command Design Handbook, DH 1-6, *System Safety*

Air Force System Safety Handbook, July 2000

7

System Safety Hazard Analysis

Jack Dixon

7.1 Introduction to Hazard Analyses

Hazard analysis is the cornerstone of safe systems development. Design for safety requires the elimination or mitigation of all hazards. Therefore, the first, and the most important, step in eliminating hazards is to identify all the hazards. Once hazards are recognized, they can be evaluated and then eliminated or controlled to an acceptable level. The evaluation of hazards includes determining their causes and effects. Hazard analysis is also used to determine the risk presented by the hazard, which, in turn, provides a way to prioritize the hazards. This information is then used in the analysis to help determine design options that will eliminate or mitigate the risk of having an accident or mishap.

In this chapter after introducing some terminology and discussing risk in greater detail that we did in Chapter 4, we will cover several of the most widely used hazard analysis techniques including

- Preliminary Hazard List (PHL)
- Preliminary Hazard Analysis (PHA)
- Subsystem Hazard Analysis (SSHA)
- System Hazard Analysis (SHA)

Design for Safety, First Edition. Edited by Louis J. Gullo and Jack Dixon.
© 2018 John Wiley & Sons Ltd. Published 2018 by John Wiley & Sons Ltd.

- Operating & Support Hazard Analysis (O&SHA)
- Health Hazard Analysis (HHA)

While there are over a hundred different techniques that have been described in the literature and used for various projects, we focus on the most common in this book. Later chapters will cover numerous additional hazard analysis techniques. The "Suggestions for Additional Reading" at the end of each chapter should be consulted for discussions of additional but less often used analysis techniques.

7.1.1 Definition of Terms

It is important that we define some basic terms commonly used in hazard analysis.

Accident: A sudden event that is not planned or intended and that causes damage or injury [1].
Mishap: An unlucky accident or mistake [1]. Therefore, an accident and a mishap are synonymous. A better, more complete definition is an event or series of events, resulting in unintentional death, injury, occupational illness, damage to or loss of equipment or property, or damage to the environment [2].
Hazard: A real or potential condition that could lead to an unplanned event or series of events (i.e., mishap), resulting in death, injury, occupational illness, damage to or loss of equipment or property, or damage to the environment [2].
Risk: A combination of the severity of the mishap and the probability that the mishap will occur [2].

7.2 Risk

We briefly discussed risk in Chapter 4 in broad terms. In this chapter we will focus on design risk as it applies to system safety and hazard analyses, in particular. "System safety contributes to mishap prevention by minimizing system risks due to hazards consistent with other cost, schedule, and design requirements" [3]. System safety requires that risk be evaluated and the level of risk accepted or rejected. "All risk management is about decision-making under uncertainty. It is a process wherein the risks are identified, ranked, assessed, documented, monitored, and mitigated" [4].

The following sections on risk are adapted from *Design for Reliability* [4] with updates included to reflect the latest standards.

7.3 Design Risk

There are numerous types of risk. Program risk is the overall risk to a project or program. Technical risk is a subset of program risk, and design risk can be considered a subset of technical risk. However, since this book is focused on safe product design, we will expand the topic of design risk in this section.

Design risk can encompass several types of risk—engineering risk, safety risk, risk of failure to perform the intended function, and so on. We have chosen to use the term "design risk" to encompass all these terms. Furthermore, since design risk is most often associated with safety, we will approach design risk largely from the safety point of view. Of course, the same principles apply whether we are talking about a hazardous situation or just one where a product fails to perform its function for the customer.

7.3.1 Current State of the Art of Design Risk Management

Risk is a concept that is common to both hardware and software products and products containing both. Concern over failures in products is a long-standing problem. The continuing growth of complexity in today's systems and products heightens the concern of risk. Because of this increasing complexity, the risks are increasing, and the risks need to be identified, assessed, and managed in a more formal way than they have in the past.

The engineer's job entails making technical decisions. Many of these decisions are made with limited or incomplete information. Incomplete information leads to uncertainty, and therefore risk is inherent in the engineering decision-making process.

The traditional approach to reducing risk has been to design and regulate products or systems very conservatively. This can entail designing in large safety margins, including multiple safety barriers in the design, excessive quality control, and regular inspections. This conservative approach can lead to very expensive systems that still do not guarantee safety or success. The current trend is to use more formal methods of assessing risk through analysis.

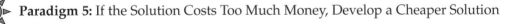 **Paradigm 5:** If the Solution Costs Too Much Money, Develop a Cheaper Solution

7.3.2 Expression of Risk

Risk is an expression of the possibility/impact of a mishap in terms of hazard severity and hazard probability. From the beginning of the design process, the goal is to design to eliminate hazards and minimize the risks. If an identified hazard cannot be eliminated, the risk associated with it must be reduced to an acceptable level.

Risk is the product of consequences of a particular outcome (or range of outcomes) and the probability of its (their) occurrence. The most common way of quantifying risk is as the product of consequences of a particular outcome (or range of outcomes) and the probability of its (their) occurrence. This is expressed as

$$R = C \times P \text{ (consequence severity} \times \text{probability of occurrence)}$$

The severity of the consequences of an undesirable event or outcome must be estimated. Severity is an assessment of the seriousness of the effects of the potential event if it does occur.

The probability of occurrence is the likelihood that a particular cause or failure occurs.

7.3.2.1 Qualitative Risk Analysis

Qualitative risk analysis is the most widely used type of risk analysis, largely because it is quickest and simplest to perform. This approach uses terms such as high, medium, and low to characterize the risk. These levels of risk are determined by the combination of likelihood of occurrence and severity of loss in relative terms described by words rather than numerically. This form of risk analysis does not use actual hard data in the risk assessment process and as a result is subjective and relies heavily on the experience of the analyst.

7.3.2.2 Quantitative Risk Analysis

Quantitative risk analysis is used to estimate the probability of an undesirable event happening and to assess, in quantitative terms, the magnitude of the consequences in order to evaluate the risk. This approach is obviously the preferred approach, but it requires significant amounts of data, either historical or test results, to be able to accurately estimate the probability of occurrence and magnitude of the losses that may occur. This approach is complicated, time-consuming, and costly. There is also uncertainty associated with the estimates that are being made. When data is limited or questionable, the risk assessment can be disputed. Communication of the risk becomes more difficult.

7.3.3 Risk Management

Managing risk entails the identification of the risk, the assessment of the risk including the estimation of the likelihood of its occurrence and the consequences it presents, the cost-benefit trade-off, and the mitigation, or acceptance, of the risk.

Paradigm 9: Taking No Action Is Usually Not an Acceptable Option

7.3.3.1 Risk Assessment

Risk assessment is the quantification of possible failures. In order to do a risk assessment, we need to know what can go wrong in the system, how likely is the failure to happen, and what will be the consequence of the failure if it occurs. All engineering design begins with the consideration of what will work and what might go wrong. In a bit of irony, the design is only as successful if the designer can foresee how the system might fail.

Risk combines the probability of failure with the consequences of failure. An essential factor in risk assessment is uncertainty, which adds some of the unknown to the mix because the designer doesn't know exactly which failures may occur nor when or where they may occur. There is uncertainty in each of the three factors mentioned earlier—uncertainty in what can go wrong, uncertainty in the likelihood of a particular failure occurring, and uncertainty in defining the consequences of a failure if it does occur. All of this is due to a certain amount of ignorance about the risk. Our goal in reliable design is to minimize this ignorance, thus minimizing the uncertainty in our decisions and minimizing the risk inherent in our designs.

Everyone is familiar with Murphy's law that states, "If something can go wrong, it will." Risk assessment provides a means to determine how likely something is to go wrong and what will happen if it does. Risk assessment provides the designer with a quantitative input to the design process and helps the designer to focus resources on the most significant problems (failures) in order to reduce our exposure to risk.

Risk Identification

The first step to controlling risk is to identify those areas that present the risks. Usually a designer will attempt to figure out what can go wrong with the product or system being designed. This can be done using numerous tools such as Failure Modes, Effects, and Criticality Analysis (FMECA), in which all failure modes are identified or the designer may choose to use a PHA, which is used to determine the hazards that may be present in a product or system. The PHA is used at the earliest stages of concept design to achieve "preliminary" results. As the design matures, the PHA is expanded into other forms of hazard analyses, each being more detailed.

Key to the success of the product design is the tracking of these failures/hazards throughout the design process to ensure that they are mitigated to acceptable levels. Both the FMECA and the PHA are usually recorded in some form of matrix that allows updates and tracking as more information about the design becomes available. The FMECA is described in Chapter 8 and PHA is described later in this chapter.

Risk Estimation

As discussed previously, risk estimations are made by combining the probability of occurrence of the undesired event with a measure of the severity of the consequences. This combination of severity and probability is the Risk Assessment Code (RAC).

At the early stages of design, it is typical to assess the risk using an Initial Risk Assessment Code (IRAC). This assessment is usually a worst-case assessment made based on very preliminary assessments made during the concept development stage of product design. Later, a "current" RAC can be used to reflect the latest risk assessments. At the end of the development program, the "current" RAC becomes the Final Risk Assessment Code (FRAC).

Probability

Initially a qualitative statement of estimated probability of occurrence as shown in Table 7.1 is used to make the initial risk estimate. As the design progresses, and time and budget permits, the probability of occurrence can be refined to become more quantitative. This can be accomplished by using reliability analysis techniques to estimate product or system failure rates or other specialized analysis techniques such as Fault Tree Analysis (FTA), event tree analysis, and so on.

Consequences

Usually the first step in assessing hazardous consequences is to make a qualitative ranking of the severity of the hazard is shown in Table 7.2. Much like estimating

Table 7.1 Hazard probability levels

Probability level	Qualitative probability of occurrence	Applicability to item
A	Frequent	Likely to occur often in the life of an item
B	Probable	Will occur several times in the life of an item
C	Occasional	Likely to occur sometime in the life of an item
D	Remote	Unlikely, but possible to occur in the life of an item
E	Improbable	So unlikely that it can be assumed occurrence may not be experienced
F	Eliminated	Incapable of occurrence. This level is used when potential hazards are identified and later eliminated

Table 7.2 Hazard severity levels

Severity	Category	Mishap definition
Catastrophic	1	Death, system loss, or severe environmental damage
Critical	2	Permanent partial disability, injuries, or occupational illness that may result in hospitalization of at least three personnel, reversible significant environmental impact
Marginal	3	Injury or occupational illness resulting in one or more lost work days, reversible moderate environmental impact
Negligible	4	Injury or occupational illness not resulting in a lost work day, minimal environmental impact

probability of occurrence, severity of consequences can be analyzed in greater depth as the design progress using more quantitative techniques, such as FMECA or FTA.

Risk Evaluation

The next step is the evaluation of the risks to determine if further action is warranted or if the risk is acceptable.

Significance
Risks must be evaluated to determine their significance and the urgency of their mitigation.
The RAC is used to determine if further corrective action is required. It is used to make this determination as shown in Table 7.3. The determination of RAC allows the risks to be grouped into typically five groups that represent their risk priority—high, serious, medium, low, or eliminated. The ranking of the risk into these groups determines if further action is necessary and by what authority the hazard can be accepted and closed. Further action can then be taken on the most important risks first.

Risk Acceptability
The RAC are divided into four groups that represent their risk priority. The ranking of the hazard into these groups determines if further action is necessary and by what authority the hazard can be accepted and closed (see Table 7.4 for risk levels).

Cost
In addition to the probability of occurrence and the severity of the consequences, it is often necessary to consider the cost of mitigation. There are almost always budget limitations and these must be taken into account during the process of deciding how far to go in mitigating the risks and which ones to concentrate on. Obviously, the risks with the highest rank need the most attention to preferably totally eliminate them or to reduce the consequences if they do occur. The lower ranking risks may have to be accepted.

Table 7.3 Risk assessment matrix

Frequency of occurrence	Hazard severity categories			
	1 Catastrophic	2 Critical	3 Marginal	4 Negligible
A—Frequent	High	High	Serious	Medium
B—Probable	High	High	Serious	Medium
C—Occasional	High	Serious	Medium	Low
D—Remote	Serious	Medium	Medium	Low
E—Improbable	Medium	Medium	Medium	Low
F—Eliminated	Eliminated			

Table 7.4 Risk levels

Risk level	Risk assessment code	Guidance	Decision authority
High	1A, 1B, 1C, 2A, 2B	Unacceptable	Management
Serious	1D, 2C, 3A, 3B	Undesirable	Program manager
Medium	1E, 2D, 2E, 3C, 3D, 3E, 4A, 4B	Acceptable	Program manager or safety manager
Low	4C, 4D, 4E	Acceptable (without higher level review)	Safety team

Risk Mitigation

Risk mitigation uses design measures to reduce the probability of occurrence of a failure or undesired event and/or by reducing the consequences that result if the failure or undesired event occurs.

The best solution is always to eliminate the risk by designing it out of the system or product.

The first step in mitigation of risk is the identification of risk mitigation measures. The designer must identify potential risk mitigation alternatives and their expected effectiveness. Risk mitigation is an iterative process that culminates either in the complete elimination of the risk or in a residual risk when the risk has been reduced to a level acceptable to the decision-making authority.

The order of precedence for mitigating identified risks is provided in MIL-STD-882 [2]:

- Eliminate hazards through design selection. Ideally, the hazard should be eliminated by selecting a design or material alternative that removes the hazard altogether.

- Reduce risk through design alteration. If adopting an alternative design change or material to eliminate the hazard is not feasible, consider design changes that reduce the severity and/or the probability of the mishap potential caused by the hazard(s).
- Incorporate engineered features or devices. If mitigation of the risk through design alteration is not feasible, reduce the severity or the probability of the mishap potential caused by the hazard(s) using engineered features or devices. In general, engineered features actively interrupt the mishap sequence and devices reduce the risk of a mishap.
- Provide warning devices. If engineered features and devices are not feasible or do not adequately lower the severity or probability of the mishap potential caused by the hazard, include detection and warning systems to alert personnel to the presence of a hazardous condition or occurrence of a hazardous event.
- Incorporate signage, procedures, training, and Personal Protective Equipment (PPE). Where design alternatives, design changes, and engineered features and devices are not feasible and warning devices cannot adequately mitigate the severity or probability of the mishap potential caused by the hazard, incorporate signage, procedures, training, and PPE. Signage includes placards, labels, signs, and other visual graphics. Procedures and training should include appropriate warnings and cautions. Procedures may prescribe the use of PPE. For hazards assigned catastrophic or critical mishap severity categories, the use of signage, procedures, training, and PPE as the only risk reduction method should be avoided.

The hazard reduction precedence is shown in Figure 7.1.
Robust design is always the best way to mitigate product or system risk.

7.3.3.2 Risk Communication

Risk communication is the exchange of information between interested parties concerning the nature, scale, importance, disposition, and control of risk. Risk communication can, and should, occur at all stages of the risk assessment process. On a macroscale, interested parties could be government agencies, companies, unions, individuals, communities, and the media. Although ultimately these external parties may need to be informed of the risk associated with our product, for our purposes we are more interested in the internal risk communication process that must take place during the design process.

For any product design effort, the program/project manager is responsible for the communication, acceptance, and tracking of hazards and residual risk. The program manager must communicate known hazards and associated risks of the system to all interested parties. As changes are introduced into the system, the program manager must update the risk assessments and is also responsible for

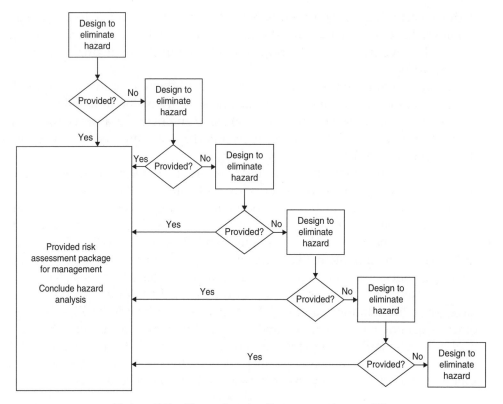

Figure 7.1 Hazard reduction precedence [3]

informing system designers about the program manager's expectations for handling of newly discovered hazards. The program manager will evaluate new hazards and the resulting residual risk and either recommend further action to mitigate the hazards or formally document the acceptance of these hazards and residual risk. He or she must evaluate the hazards and associated residual risk in close consultation and coordination with the ultimate end user to ensure that the context of the user requirements, potential mission capability, and operational environment are adequately addressed. Providing the documentation of the hazard and risk acceptance to the interested parties facilitates communication. Any residual risks and hazards must be communicated to system test efforts for verification.

The designer is responsible for communicating information to the program manager on system hazards and risk, including any unusual consequences and costs associated with hazard mitigation. After attempting to eliminate or mitigate system hazards, the designer should formally document and notify the program manager of all hazards breaching thresholds set in the safety design criteria.

Key to the successful risk communication process is the system safety organization and the lines of communication within the program organization and with associated organizations. Interfaces must be established between system safety and other functional elements of the program. Also important to the risk communication process is the establishment of the authority for resolution of identified hazards.

Another critical functional area within the project organization that plays a major role in risk resolution and communication is the test organization. Risk mitigations that have been implemented in the design must be tested for effectiveness during the system test process. Residual risk and associated hazards must be communicated to the system test efforts for verification.

Communication with a test organization is a two-way street. Newly identified hazards, safety discrepancies, and product failures found during testing must be communicated by the testing organization to design engineering and program management.

7.4 Design Risk Management Methods and Hazard Analyses

Over the last half century or so, many different techniques have been developed that can contribute to the analysis of hazards and risks. Each technique has its best use, and it is imperative that the analyst considers very carefully the suitability of a particular analysis technique prior to its application.

Some factors that should be considered when assessing these techniques to determine the most appropriate one to use for a given situation include

- Cost
- Manpower requirements
- Simplicity
- Availability of data
- Development phase
- Level of risk
- Flexibility to be updated
- Traceability of results [4]

7.4.1 Role of Hazard Analysis

Hazard analyses are performed to identify and define hazardous conditions/ risks for the purpose of their elimination or control. Analyses should examine the system, subsystems, components, and their interrelationships, as well as logistic support, training, maintenance, operational environments, and system/component disposal. Such analyses should

- Identify hazards and recommend appropriate corrective action.
- Assist the individual(s) actually performing the analyses in better evaluating the safety aspects of a given system or element.
- Provide managers, designers, test planners, and other affected decision makers with the information and data needed to permit effective trade-offs.
- Demonstrate compliance with given safety-related technical specifications, operational requirements, and design objectives [3].

Detailed description of the various techniques used to identify and evaluate hazards and the risk associated with them are provided in the following sections and in several chapters that follow.

7.5 Hazard Analysis Tools

There are numerous hazard analysis tools to choose from—over 100. The following sections will discuss the most common ones. We have included a typical approach to each of these techniques. There are many variations of each type and the analyst should feel free to modify them as needed to fulfill the demands of the particular problem at hand.

7.5.1 Preliminary Hazard List

The PHL is the next level up from a design checklist and is typically the first analysis conducted on a system. A design checklist is often used as the starting point for the PHL.

7.5.1.1 Purpose

The purpose of the PHL is to compile a list of potential hazards very early in development or the product or system and provide an initial assessment of the hazards. It helps to identify the hazards that may require special safety emphasis and hazardous areas where more in-depth analyses may be required. The results of the PHL may be used to determine the extent of needed follow-on hazard analyses. The best time to prepare a PHL is in the concept phase of the development. The PHL is typically used to develop the PHA.

7.5.1.2 Process

The PHL process usually begins with the safety engineer reviewing the preliminary design including initial drawings, concept plans, preliminary product descriptions, similar products, and historical data. After identifying potential energy sources, environments, functions, hardware/software involved, and so on, the analyst would produce a list of potential hazards that might be presented by the system. Several design checklists might be used to help the safety engineer to recognize potential hazards. A brainstorming session with the design engineers, systems engineer, logisticians, and so on could be used to help create the initial PHL. The goal at this point is to include as many possible hazards that the system might present. Later techniques will refine these hazards and dismiss the ones that end up not applying.

7.5.1.3 Worksheet

The best way to document this, and many other hazard analyses, is to use a matrix format. A typical format for the PHL is shown in Figure 7.2.

Preliminary Hazard List					
System: _____					
Preparer: _____					
Date: _____					
Identifier	**Hazard**	**Causal Factors**	**Effects**	**IRAC**	**Comments**

Figure 7.2   Preliminary hazard list format

Column Descriptions
Header Information: Self-explanatory. Any other generic information desired or required by the program could be included.
Identifier: This can simply be a number (e.g., 1, 2, 3, … as hazards are listed). Or it could be an identifier that identifies a particular subsystem or piece of hardware that will be in the system with a series of numbers as multiple hazards are identified with that particular item (e.g., motor 1, motor 2, etc.).
Hazard: This is the specific potential hazard that has been identified.
Causal Factors: This should be a brief description of how the potential hazard could result in a mishap.
Effects: This is a short description of the possible resulting mishap. What is the effect(s) resulting from the potential hazard—death, injury, damage, environmental damage, and so on?
IRAC: This is a preliminary qualitative assessment of the risk.
Comments: This column should include any assumptions made, recommended control, requirements, applicable standards, actions needed, and so on.

7.5.1.4 Advantages/Disadvantages

The main advantage of the PHL is that it is relatively quick, easy, and inexpensive. It results in a list of potential hazards that need to be further assessed as the system development process continues. It helps to focus management and engineering attention on the major hazards that might be present in the system and have them addressed early in the development. It can also help management allocate resources to where they will be most productive in developing a safe system.

7.5.2 Preliminary Hazard Analysis

PHA is typically an expansion of the PHL. While it is not necessary to have a PHL in order to create a PHA, it helps to give the safety engineer a start. It is probably the most commonly performed hazard analysis technique and its use is highly recommended for every program.

7.5.2.1 Purpose

The main purpose of the PHA is to expand the PHL by refining the analysis of the hazards in the PHL and identifying any new hazards that arise as the system design is developed further and more design details emerge in the preliminary design phase. It is also started in the early stages of development. The PHA identifies the causal factors, the consequences, and the risks associated with the preliminary design concept(s). Recommended hazard controls are also presented. In addition, the results of the PHA help to determine the level of effort needed for additional hazard analysis. With the PHA, we want to positively influence the design for safety as early as possible in the development program.

7.5.2.2 Process

As the preliminary design continues, the PHL is expanded by adding more details to the various hazards that have already been identified, adding more hazards as new ones are found and elaborating on the recommended controls to mitigate the risk. Again, checklists and brainstorming with people knowledgeable of the system will help to flesh out the PHA.

Things to consider when performing a PHA that might help to stimulate the identification of hazards include

- System components
- Energy sources
- Explosives
- Hazardous materials
- Interfaces and controls

- Interface considerations to other systems when in a network or in a System-of-Systems (SoS)
- Material compatibilities
- Inadvertent activation
- Software
- Operating environments
- Procedures for operating, test, maintenance, Built-In-Test, diagnostics, emergencies, and disposal
- Modes of operation
- Health hazards
- Environmental impacts
- Human factors and human errors
- Life support requirements
- Safety in manned systems (e.g., crash safety, egress, rescue, survival, and salvage)
- Support equipment

For each hazard identified, an initial risk assessment should be included and risk mitigation actions should be recommended.

7.5.2.3 Worksheet

A typical format for the PHA is shown in Figure 7.3.

Preliminary Hazard Analysis											
System: _____											
Preparer: _____											
Date: _____											
Identifier	Operational Mode	Hazard	Causal Factors	Effects	IRAC	Risk	Recommended Action	FRAC	Risk	Cmts	Status

Figure 7.3 Preliminary hazard analysis format

Column Descriptions
Header Information: Self-explanatory. Any other generic information desired or required by the program could be included.

Identifier: This can simply be a number (e.g., 1, 2, 3, … as hazards are listed). Or it could be an identifier that identifies a particular subsystem or piece of hardware that will be in the system with a series of numbers as multiple hazards are identified with that particular item (e.g., motor 1, motor 2, etc.).

Operational Mode: This is a description of what mode the system is in when the hazard is experienced (e.g., operational, training, maintenance, etc.).

Hazard: This is the specific potential hazard that has been identified.

Causal Factors: This should be a brief description of how the potential hazard could result in a mishap.

Effects: This is a short description of the possible resulting mishap. What is the effect(s) resulting from the potential hazard—death, injury, damage, environmental damage, and so on?

IRAC: This is a preliminary qualitative assessment of the risk.

Risk: This is a qualitative measure of the risk for the identified hazard, given that no mitigation techniques are applied.

Recommended Action: Provide recommended preventive measures to eliminate or mitigate the identified hazards. Hazard mitigation methods should follow the preferred order of precedence provided earlier in this chapter.

FRAC: This is a final qualitative assessment of the risk, given that mitigation techniques and safety requirements identified in "Recommended Action" are applied to the hazard.

Comments: This column should include any assumptions made, recommended control, requirements, applicable standards, actions needed, and so on.

Status: States the current status of the hazard (open, monitored, closed).

7.5.2.4 Advantages/Disadvantages

As with the PHL, the main advantage of the PHA is that it is relatively quick, easy, and inexpensive. It provides an early assessment of potential hazards, their risk, and recommendations for mitigation. It identifies hazards that need to be further assessed as the system development process continues. It helps to focus management and engineering attention on the major hazards that might be present in the system and have them addressed early in the development. It also helps management allocate resources to where they will be most productive in developing a safe system.

7.5.3 Subsystem Hazard Analysis (SSHA)

As the design progresses, the next analysis typically conducted is the SSHA.

7.5.3.1 Purpose

The purpose of the SSHA is to provide a more detailed analysis than the PHA as the develop proceeds. It helps to derive detailed safety requirements to be incorporated into the system design to help ensure a safe design. The SSHA may also identify new hazards that were overlooked previously. The SSHA only looks at a particular subsystem to identify hazards within that subsystem and provides detailed analysis of each subsystem to determine hazards created by the subsystems to themselves, to related or nearby equipment, or to personnel.

7.5.3.2 Process

The SSHA is performed when detailed design information becomes available, but as early as possible. The SSHA process involves performing a detailed analysis of each subsystem in the system being developed.

Consideration should be given to the following when conducting an SSHA:

- Modes of failure
- Component failure modes
- Human errors
- Human system interfaces
- Single point failures
- Common mode failures
- Effects of failures that occur in subsystem components
- Functional relationships between components and equipment comprising each subsystem
- Subsystem hardware events
- Software events
- Faults
- Improper timing
- Erroneous inputs to the subsystem

7.5.3.3 Worksheet

A worksheet similar to the PHA can be used with a slightly different emphasis on cause and effect as shown in Figure 7.4.

Subsystem Hazard Analysis										
System: _____										
Preparer: _____										
Date: _____										
Identifier	Operational Mode	Hazard	Cause	Effects	IRAC	Risk	Recommended Action	FRAC	Risk	Cmts	Status

Figure 7.4 Subsystem hazard analysis format

Column Descriptions

Header Information: Self-explanatory. Any other generic information desired or required by the program could be included.

Identifier: This can simply be a number (e.g., 1, 2, 3, … as hazards are listed). Or it could be an identifier that identifies a particular subsystem or piece of hardware that will be in the system with a series of numbers as multiple hazards are identified with that particular item (e.g., motor 1, motor 2, etc.).

Operational Mode: This is a description of what mode the system is in when the hazard is experienced (e.g., operational, training, maintenance, etc.).

Hazard: This is the specific potential hazard that has been identified.

Cause: This column identifies conditions, events, or faults that could cause the hazard to exist and the events that can lead to a mishap.

Effects: This is a short description of the possible resulting mishap. What is the effect(s) resulting from the potential hazard—death, injury, damage, environmental damage, and so on? This would generally be the worst-case result.

IRAC: This is a preliminary qualitative assessment of the risk.

Risk: This is a qualitative measure of the risk for the identified hazard, given that no mitigation techniques are applied.

Recommended Action: Provide recommended preventive measures to eliminate or mitigate the identified hazards. Hazard mitigation methods should follow the preferred order of precedence provided earlier in this chapter.

FRAC: This is a final qualitative assessment of the risk, given that mitigation techniques and safety requirements identified in "Recommended Action" are applied to the hazard.

Comments: This column should include any assumptions made, recommended control, requirements, applicable standards, actions needed, and so on.

Status: States the current status of the hazard (open, monitored, closed).

Several other techniques are sometimes used as SSHA's in lieu of the matrix type worksheet. These include the Failure Modes and Effects Analysis (FMEA) and the FTA. These techniques are covered in separate chapters of this book.

7.5.3.4 Advantages/Disadvantages

The SSHA adds additional rigor to the analysis by focusing on hazards, causes, and effects. It takes more effort than the PHL and PHA, but is still cost-effective.

7.5.4 System Hazard Analysis (SHA)

The SHA is a detailed analysis that focuses on the hazards at the system level. This helps to develop the overall system risk. The SHA evaluates hazards associated with system integration and involves evaluating all hazards across subsystem interfaces. The SHA can be viewed as an expansion of the SSHA in that it includes analysis of all subsystems and their interfaces, but does not include all the subsystem hazards identified in the SSHA.

7.5.4.1 Purpose

The main purpose of the SHA is to ensure safety at the system level. Its major focus is on the internal and external interfaces.

7.5.4.2 Process

The SHA is performed using detailed design information. The SHA process involves performing a detailed analysis of possible hazards created by the system operating as a whole or created by interfaces between subsystems or with other systems, including humans. The SHA should identify hazards associated with the integrated system design, including software and subsystem interfaces. It should also provide recommended actions necessary to eliminate the identified hazards or mitigate their risks.

Consideration should be given to the following when conducting an SHA:

- Possible hazard resulting from integration
- Hazardous procedures
- Software interfaces
- Subsystem interfaces
- Human interfaces
- Performance degradation
- Functional failures
- Timing errors
- Inadvertent functioning
- Subsystems interrelationships
- Possible independent, dependent, and simultaneous hazardous events
- System failures
- Failures of safety devices
- Common cause failures
- Degradation of a subsystem or the total system

- Design changes that affect subsystems
- Human errors

7.5.4.3 Worksheet

The same worksheet used for the SSHA can be used for the SHA. This time, the emphasis is on the system level rather than the individual subsystems. The worksheet is shown in Figure 7.5.

System Hazard Analysis											
System: _____											
Preparer: _____											
Date: _____											
Identifier	Operational Mode	Hazard	Cause	Effects	IRAC	Risk	Recommended Action	FRAC	Risk	Cmts	Status

Figure 7.5 System hazard analysis format

Column Descriptions

Header Information: Self-explanatory. Any other generic information desired or required by the program could be included.

Identifier: This can simply be a number (e.g., 1, 2, 3, … as hazards are listed). Or it could be an identifier that identifies a particular subsystem or piece of hardware that will be in the system with a series of numbers as multiple hazards are identified with that particular item (e.g., motor 1, motor 2, etc.).

Operational Mode: This is a description of what mode the system is in when the hazard is experienced (e.g., operational, training, maintenance, etc.).

Hazard: This is the specific potential hazard that has been identified.

Cause: This column identifies conditions, events, or faults that could cause the hazard to exist and the events that can lead to a mishap.

Effects: This is a short description of the possible resulting mishap. What is the effect(s) resulting from the potential hazard—death, injury, damage, environmental damage, and so on? This would generally be the worst-case result.

IRAC: This is a preliminary qualitative assessment of the risk.

Risk: This is a qualitative measure of the risk for the identified hazard, given that no mitigation techniques are applied.

Recommended Action: Provide recommended preventive measures to eliminate or mitigate the identified hazards. Hazard mitigation methods should follow the preferred order of precedence provided earlier in this chapter.

FRAC: This is a final qualitative assessment of the risk, given that mitigation techniques and safety requirements identified in "Recommended Action" are applied to the hazard.

Comments: This column should include any assumptions made, recommended control, requirements, applicable standards, actions needed, and so on.

Status: States the current status of the hazard (open, monitored, closed).

Again, several other techniques are sometimes used as SHA's in lieu of the matrix type worksheet or to supplement the SHA by providing further in-depth analysis of a particular hazard. These include the FMEA and the FTA. These techniques are covered in separate chapters of this book.

7.5.4.4 Advantages/Disadvantages

The SHA identifies system interface-type hazards and provides the basis for making an assessment of overall system risk. It identifies system level hazards not found by earlier analyses, particularly those propagated by the incompatibility of subsystem interfaces.

7.5.5 Operating & Support Hazard Analysis (O&SHA)

The O&SHA is used to identify hazards that may occur during various modes of system operations. Like other analyses, the O&SHA should be started as early as possible to allow for problems to be fixed as early as possible. However, since the system design needs to be substantially complete and operating and maintenance procedures must be available, at least as drafts, the O&SHA is typically started late in the design phase or early in production, which is much later than most other analyses.

7.5.5.1 Purpose

O&SHA is used to identify and evaluate the hazards associated with the operation, support, and maintenance of the system. Its emphasis is on procedures, training, human factors, and human–machine interface. O&SHA will also typically identify additional cautions and warnings that may be required.

7.5.5.2 Process

The process of conducting an O&SHA will use design and operational and support procedure information to identify hazards associated with all operational modes. The hazards are identified through the thorough analysis of every detailed procedure that is to be performed during system operation and support. Input to the O&SHA will include design and operating information, user and maintenance manuals, and hazards identified during the conduct of other hazard analyses.

Consideration should be given to the following when conducting an O&SHA:

- Facility/installation interfaces to the system
- Operation and support environments
- Tools or other equipment
- Tooling
- Support/test equipment
- Operating procedures
- Maintenance procedures
- Task sequence, concurrent task effects, and limitations
- Human factors
- Personnel requirements
- Workload
- Human errors
- Testing
- Installation
- Repair procedures
- Training
- Packaging
- Storage
- Handling
- Transportation
- Disposal
- Emergency operations
- PPE
- Hazardous materials
- Hazardous system modes under operator control
- Chemicals and materials used in production, operation, and maintenance

7.5.5.3 Worksheet

A worksheet similar to that used for the SSHA and SHA can be used for the O&SHA. However, this time, the emphasis is tasks involved in the operation and support of the system. The worksheet is shown in Figure 7.6.

	Operating & Support Hazard Analysis										
	System: _____ Preparer: _____ Date: _____										
Identifier	Task	Operational Mode	Hazard	Cause	Effects	IRAC	Risk	Recommended Action	FRAC	Risk	Cmts	Status

Figure 7.6 Operating and support hazard analysis format

Column Descriptions

Header Information: Self-explanatory. Any other generic information desired or required by the program could be included.

Identifier: This can simply be a number (e.g., 1, 2, 3, ... as hazards are listed). Or it could be an identifier that identifies a particular subsystem or piece of hardware that will be in the system with a series of numbers as multiple hazards are identified with that particular item (e.g., motor 1, motor 2, etc.).

Task: This column is used to identify the task being evaluated.

Operational Mode: This is a description of what mode the system is in when the hazard is experienced (e.g., operational, training, maintenance, etc.).

Hazard: This is the specific potential hazard that has been identified.

Cause: This column identifies conditions, events, or faults that could cause the hazard to exist and the events that can lead to a mishap.

Effects: This is a short description of the possible resulting mishap. What is the effect(s) resulting from the potential hazard—death, injury, damage, environmental damage, and so on? This would generally be the worst-case result.

IRAC: This is a preliminary qualitative assessment of the risk.

Risk: This is a qualitative measure of the risk for the identified hazard, given that no mitigation techniques are applied.

Recommended Action: Provide recommended preventive measures to eliminate or mitigate the identified hazards. Hazard mitigation methods should follow the preferred order of precedence provided earlier in this chapter.

FRAC: This is a final qualitative assessment of the risk, given that mitigation techniques and safety requirements identified in "Recommended Action" are applied to the hazard.

Comments: This column should include any assumptions made, recommended control, requirements, applicable standards, actions needed, and so on.
Status: States the current status of the hazard (open, monitored, closed).

7.5.5.4 Advantages/Disadvantages

O&SHA provides a focus on operational and procedural hazards. However, this analysis can become quite tedious if there are many tasks to be analyzed in a large system.

7.5.6 Health Hazard Analysis (HHA)

HHA is used to evaluate a system to identify hazards to human health.

7.5.6.1 Purpose

The purpose of an HHA is to identify human health hazards and to evaluate proposed hazardous materials and processes that use hazardous materials. HHA also is used to propose measures to eliminate or control the risk presented by the hazards.

7.5.6.2 Process

As with other analysis techniques, HHA uses design and operating information and knowledge of health hazards to identify the hazards involved with the system under development. The system is evaluated to identify potential sources of human health-related hazards such as noise, chemical, radiation, and so on. An assessment must then be made of the quantity of the hazardous source and the exposure levels that may be involved. Risk is assessed and mitigations are recommended for implementation. Other analysis, such as PHL, PHA, and O&SHA, can be sources of information for health hazards that may be present. Health hazard-related checklists can also be used to stimulate the identification of health hazards.

Health hazard considerations include

- Noise
- Chemicals
- Radiation (both ionizing and nonionizing)
- Toxic materials
- Shock
- Heat
- Cold
- Vibration
- Human–machine interfaces
- Accidental exposures

- PPE
- Environments
- Stressors of the human operator or maintainer including their synergistic effects
- Carcinogens
- Biological hazards (e.g., bacteria, viruses, fungi, and mold)
- Ergonomic hazards (e.g., lifting, cognitive demands, long duration activity, etc.)
- Blast overpressure

7.5.6.3 Worksheet

A worksheet similar to that used for the O&SHA can be used for the HHA. Actually, the two techniques are similar, but O&SHA is focused on tasks performed during operation and maintenance; HHA is focused entirely on hazards to human health. The worksheet is shown in Figure 7.7.

Health Hazard Analysis											
System: _____ Preparer: _____ Date: _____											
Identifier	Hazard Type	Hazard	Cause	Effects	IRAC	Risk	Recommended Action	FRAC	Risk	Cmts	Status

Figure 7.7 Health hazard analysis format

Column Descriptions

Header Information: Self-explanatory. Any other generic information desired or required by the program could be included.

Identifier: This can simply be a number (e.g., 1, 2, 3, ... as hazards are listed). Or it could be an identifier that identifies a particular subsystem or piece of hardware that will be in the system with a series of numbers as multiple hazards are identified with that particular item (e.g., motor 1, motor 2, etc.).

Hazard Type: This column shows the type of health concern presented (e.g., noise, chemical, radiation)

Hazard: This is the specific potential health hazard that has been identified.

Cause: This column identifies conditions, events, or faults that could cause the hazard to exist and the events that can lead to a mishap.

Effects: This is a short description of the possible resulting mishap. What is the effect(s) resulting from the potential hazard—death, injury, damage, environmental damage, and so on? This would generally be the worst-case result.

IRAC: This is a preliminary qualitative assessment of the risk.

Risk: This is a qualitative measure of the risk for the identified hazard, given that no mitigation techniques are applied.

Recommended Action: Provide recommended preventive measures to eliminate or mitigate the identified hazards. Hazard mitigation methods should follow the preferred order of precedence provided earlier in this chapter.

FRAC: This is a final qualitative assessment of the risk, given that mitigation techniques and safety requirements identified in "Recommended Action" are applied to the hazard.

Comments: This column should include any assumptions made, recommended control, requirements, applicable standards, actions needed, and so on.

Status: States the current status of the hazard (open, monitored, closed).

While the matrix approach can provide for a thorough analysis and identification of health hazards presented by the system, often more detailed assessments of particular hazards may be required. These assessments are usually conducted by medical or health personnel and result in a detailed report being prepared.

7.5.6.4 Advantages/Disadvantages

HHA quickly identifies the human health hazards inherent in the system and the ways to eliminate or mitigate the risk. However, exposures to more exotic substances or hazards may require experienced medical or health personnel.

7.6 Hazard Tracking

Hazards should always be tracked throughout the product life cycle using a closed-loop Hazard Tracking System (HTS). The data elements in the HTS should include, as a minimum, the following:

- System identification information
  - System
  - Subsystem (if applicable)
  - Applicability (version of specific hardware designs or software releases)
  - Requirements references
- Hazard information
  - Hazard
  - System mode
  - Causal factor (e.g., hardware, software, human, operational environment)

- o Effects
- o Associated mishaps
- o IRAC
- o Target or final risk assessment code
- o Mitigation measures (with traceability to version specific hardware designs or software releases)
- o Verification and validation method
- o Hazard status
- Miscellaneous information
  - o Action person(s) and organizational element
  - o Record of risk acceptance(s)
  - o Hazard management log (record of hazard entry and changes to the hazard record during the system's life cycle)
  - o Hazardous Material (HAZMAT)

Figure 7.8 illustrates a typical closed-loop hazard tracking process where once a hazard is identified, it is categorized and entered into the HTS, and as solutions

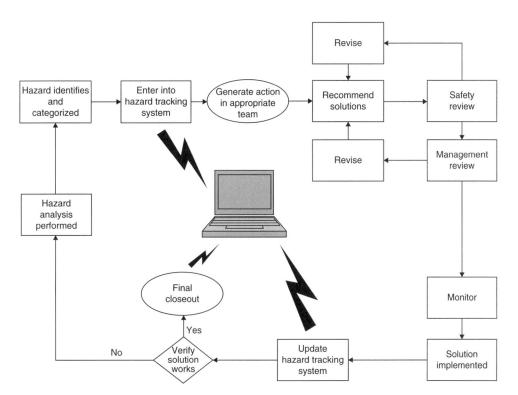

Figure 7.8 Typical closed-loop hazard tracking process

to resolve the hazard are developed, they must be agreed to by the design team and the management and are tracked until they are closed out.

7.7 Summary

As can be seen from the matrices for the different analyses, there is a lot of similarity. They can be tailored to fit the needs of the project. The important thing is to do the analysis and to include enough data to understand the problem and to provide adequate documentation to allow acceptance of the results.

References

[1] Merriam-Webster Dictionary, Online, http://www.merriam-webster.com/dictionary (Accessed on August 1, 2017).
[2] MIL-STD-882E (2012) System Safety, U.S. Department of Defense, Washington, DC.
[3] NM 87117-5670 (2000) *Air Force System Safety Handbook*, Air Force Safety Agency, Kirtland Air Force Base, Albuquerque, NM.
[4] Raheja, D. and Gullo, L.J. (2012) *Design for Reliability*, John Wiley & Sons, Inc., Hoboken, NJ.

Suggestions for Additional Reading

Raheja, D. and Allocco, M. (2006) *Assurance Technologies Principles and Practices*, John Wiley & Sons, Inc., Hoboken, NJ.
Raheja, D. and Gullo, L.J. (2012) *Design for Reliability*, John Wiley & Sons, Inc., Hoboken, NJ.
Roland, H.E. and Moriarty, B. (1990) *System Safety Engineering and Management*, John Wiley & Sons, Inc., New York.

8

Failure Modes, Effects, and Criticality Analysis for System Safety

Louis J. Gullo

8.1 Introduction

To develop a safe system, numerous types of system safety analyses may be required to get the job done. The goals of all safety analyses are to identify potential hazards associated with the product or system being developed, to assess the risk of those hazards, and to eliminate the risk or control the risk to an acceptable level. There are many types of system safety analyses as summarized in Chapter 3. This chapter describes how the Failure Modes and Effects Analysis (FMEA) and Failure Modes, Effects, and Criticality Analysis (FMECA) are useful for system safety analysis.

FMEA and FMECA were originally developed for use as an engineering analysis tool in the 1950s within the aerospace industry. Soon afterward, FMEA was adopted for use within the military and nuclear industries for system safety and reliability engineering applications. Today, FMEA and FMECA methodologies are widely recognized tools in the reliability analysis toolbox and are especially effective when applied in system failure risk assessments. FMEA and FMECA methodologies should be considered a critical analysis practice for any system safety

Design for Safety, First Edition. Edited by Louis J. Gullo and Jack Dixon.
© 2018 John Wiley & Sons Ltd. Published 2018 by John Wiley & Sons Ltd.

engineering program. These are structured processes for assessing hazard risks from failure mechanisms with associated failure conditions and for mitigating the effects of those failure mechanisms through compensating provisions and corrective actions for design improvement.

This chapter discusses what is necessary to perform an appropriate FMEA or FMECA related to system safety engineering. Cooperative teamwork with hardware, software, and system designers is imperative in conducting an effective FMEA or FMECA. System safety engineering varies widely from one system design to another, due to the types of simple and complex systems, system-of-systems, and families-of-systems that may exist. Care should be taken to assess the similarities of the new designed systems to predecessor systems to determine how much of the new designs are leveraged from legacy systems. The historical hazard and failure data from these legacy systems are valuable information to use in preparing an FMEA or FMECA for your new design.

8.1.1 What Is an FMEA?

FMEA is a complex engineering analysis methodology used to identify potential hazards that could occur during normal or worst-case system operation. Worst-case system operation may be system missions performed under harsh environmental conditions or during high electrical stress loading during peak usage scenarios. An FMEA is used to identify potential hazards from failure symptoms that could develop from time degradation, fatigue accumulation, or physical wear-out mechanisms. These mechanisms may evolve into high severity safety hazards that may occur over continuous operation during long periods of time. An FMEA is also used to identify failure symptoms such as fires or explosions that manifest instantaneously as hazards. These types of hazards may seem to manifest from obvious physical damage causes that detrimentally affect personal health and well-being, or the hazards may be caused by slow material or chemical breakdowns that take time to manifest into harmful effects. Besides failure symptoms, FMEAs are used to document and study failure modes, failure causes, and failure effects. This engineering analysis methodology isolates potential problem areas affecting system mission success and reliability, maintainability, and safety of both hardware and software.

8.1.2 What Is an FMECA?

The FMECA is an analysis similar to FMEA, with the inclusion of quantitative Criticality Analysis (CA), which uses equations to calculate the criticality of each hazard based on related failure modes and failure causes. Criticality allows an analyst to assess severity of potential hazards caused by failures in order to objectively rank hazard priorities for follow-up studies, possibly, leading to design improvements to ensure the elimination of safety risks or to minimize the likelihood of safety hazards. There are many ways to assess criticality, besides hazard severity. Criticality assessment may also include probability of occurrence, detection probability, failure mode distribution, and failure effect distribution to name a few. Criticality is useful for calculating the Risk Priority Number (RPN) for ranking the failures and hazards in terms of priorities for taking actions to mitigate the risks of hazards and failures. Usually, RPN is calculated based on Severity (SEV), probability of Occurrence (OCC), and Detection probability (DET). RPN is discussed in more detail later in this chapter.

8.1.3 What Is a Single Point Failure?

FMEA and FMECA are especially valuable to uncover and resolve single point failure modes. These single point failures occur at discrete levels of the system or product design. They relate to one specific independent failure or hazard event that leads to a mission critical or safety critical effect, which involves catastrophic personal injuries or equipment damage. Instead of multiple hazard events that must occur either sequentially or simultaneously to cause personal injury or equipment damage, a single point failure occurs by itself without other dependencies. Single point failure modes are system or product design weaknesses or latent manufacturing defects that have an unacceptably high probability of occurrence or a critically severe failure effect that causes personnel injury or high system repair cost due to loss of system functionality. Elimination of these single point failures is one of the primary concerns of the analyst performing FMEAs. For key findings, design changes should be incorporated to add more robust components, add redundant circuits, reduce the severity of the failure effects, or minimize the probability of occurrence of the particular failure mode, or increase the detectability of the failure mode in order to reduce risk. The toughest thing to do in either an FMEA or FMECA is to know the latent failure modes and hazards prior to the design release and prevent those failures and hazards by design changes.

Many FMEAs assess designs to determine the possibility of accidents that can happen from a single point failure cause, such as a component failure or a human error, or from multiple failure causes. According to the system safety theory, at least two events happen in any accident. There has to be a latent hazard in the system and a trigger event, such as human error, to activate an accident. From the Space Shuttle Challenger accident example in Chapter 1, we know the failure mode that caused a catastrophic hazard was caused by the rubber o-ring seal that was overstressed at cold temperature during launch. The trigger event was the decision by NASA management to lift off (human error) when the specification warned against launching when the outside temperature was below 40°F. The catastrophic event could have been prevented if either the hazard (o-ring seal

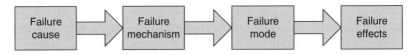

Figure 8.1 Failure modes and effects relationship diagram [1]

weakness) was redesigned or the decision-making process by management was improved to prevent errors in deciding to launch.

Paradigm 4 from Chapter 1 states: "Prevent Accidents from Single as well as Multiple Causes." The Fault Tree Analysis (FTA) is the system safety analysis methodology used to assess the effects of multiple failure causes and hazard events that occur either sequentially or simultaneously. Ample attention is given to multiple causes in numerous places in the book. The most attention to multiple causes are discussed with FTAs discussed in Chapter 9.

8.1.4 Definitions

Figure 8.1 illustrates the relationship between failure cause, failure mechanism, failure mode, and failure effect.

The definitions of failure cause, failure mechanism, failure mode, and failure effect are the following:

- Failure cause: The conditions or situation during design, production, or application of a device, product or system, which explain the underlying root of a problem that led to a failure mechanism
- Failure mechanism: The physical results of a root cause failure of a device, product, or system that led to a failure mode
- Failure mode: The unwanted state or function of a device, product, or system that are either inherent to the product or induced and explains the inability of a desired function or demonstrates a gradual degradation in performance that could lead to a failure effect
- Failure effects: The results of a failure mode on the higher levels of assembly up to the end item level or integrated functions adjacent to the device, product, or system, which has failed to perform its function, has reached an unacceptably high probability of failure, or has an increasing probability of failure

8.2 The Design FMECA (D-FMECA)

One type of FMECA is a Design FMECA (D-FMECA), which is an analysis of system or product performance considering what happens when or if a failure occurs. This type of FMECA is performed by examining assembly drawings, part datasheets, electrical schematics, and specifications. The D-FMECA does not include analysis of manufacturing-related failures, workmanship defects, and random isolated incidences related to variations in assembly or component supplier build processes.

D-FMECA is a living document during the development of the product or system hardware and software design. The value of FMECA is determined by the early identification of all critical and catastrophic subsystem or system failure modes so they can be eliminated or minimized through design improvements early in development prior to production and delivery to the customer. It is important to continually update the data contained in D-FMECA with actual failure modes and effects data from testing and actual field applications to keep pace with the evolving design so it can be used effectively throughout development and sustainment phases of the product or system life cycle. Furthermore, the results of a D-FMECA are valuable for Logistics Support Analysis Reports (LSARs) and other tasks that may be part of an Integrated Logistics Support (ILS) plan [1].

The purpose of the D-FMECA is to analyze a system/product design to determine the results or effects of failure modes on system/product operation and to classify each potential system failure according to its severity, failure occurrence, detection method, and RPN, which is a value that summarizes the effects of the severity, occurrence, and detectability. There are several references as cited at the end of this chapter that provide approaches to calculate the failure criticality or RPN. Each identified failure mode will be classified with an RPN, which will be used in the design process to assign priority for design corrective actions. For many times, an RPN limit is set to a predefined threshold, such as 6, on a scale of 0–9. RPNs are calculated for each failure mode and compared to the RPN limit, and a design change decision is made based on the scoring of the failure modes. That is, a failure mode with an RPN > 6, on a scale of 0–9, must be eliminated, or, as a minimum, its effects reduced. An alternative method is to rank all the listed entries by RPN and select the higher-value RPN entries for follow-up [1].

Most sources and references have previously defined RPN as a multiplicative calculation method. This method may result in large numbers within a wide range of values that are separated by several orders of magnitude. This method may be deemed too complicated by the casual analyst for rapid analysis, to quickly differentiate failure modes that designers should be concerned about from failure modes that designers may defer fixing until later in the development cycle. The approach for calculating RPN in this chapter is to provide an additive model, which is easier for mainstream usage and simpler for the casual analyst to grasp.

RPNs assist the designer in prioritizing which failure modes to fix immediately. Risk mitigation techniques are developed to correct the high-risk single point failures first, to offset the risks, to reduce the risks, or to eliminate the risks (also known as risk avoidance). The design improvements could be planned as scheduled system/product enhancements incorporated at a later date or incorporated immediately, if severity warrants it.

The goal of an FMECA is zero design-related single point failures that may become safety hazards. Where single point design-related failures and hazards cannot be eliminated, then minimize the effects of the single point failures or hazards and minimize the likelihood of their occurrence.

Design-related failures or hazards are due to

- Incorrect or ambiguous requirements.
- Deficiency in accounting for human use, lack of "error proofing."
- Incorrect implementation of the design in meeting requirements.
- Unspecified parameters in the design that should have been specified to ensure the design works correctly.
- Inherent design flaws that should have been found during design analyses, such as FMECAs, or verification testing.
- High electrical or mechanical stress conditions, which are beyond the strength of the design (exceed design derating guidelines and manufacturer ratings).
- Design process weaknesses—lack of robust review sub-process.
- Probabilistic pattern defect occurrences and systemic defects that are random isolated incidences. These pattern defects and systemic defects are related to design weaknesses, or poor manufacturing processes, or workmanship defects.

Probabilistic pattern failures and systematic failures that are random isolated incidences related to design weaknesses might need further discussion. Examples of these types of design failures are intermittent random failure events, such as race conditions related to static or dynamic hardware or software timing and incorrect usage of shared data or global variables. These design-related failures include defects in the specification and requirements. One type of timing failure mode, the race condition, may be prevented with properly worded requirements, such as synchronizing timed events and controlling the application of processor interrupts, which occur asynchronously. If functional timing requirements and interface requirements are properly specified, timing and race conditions can be eliminated.

8.3 How Are Single Point Failures Eliminated or Avoided in the Design?

Single point failures are addressed by assessing the corresponding criticality or RPN parameters, severity (SEV), probability of occurrence (OCC), and probability of detection (DET). Once the RPN or criticality is calculated for the particular single point failure, the consideration to implement a design change is made based on the relative ranking of the failure. If the failure is ranked high, a design change may be warranted. The design change may take several forms. There are instances when the results of an FMECA conclude the need for design changes to improve reliability, in which the improvement in the design involves changes in the system architecture. An example of a design improvement related to the system architecture is the enhanced capability to allow for fault tolerant operation. Fault tolerance is the ability of a product or system to subdue the effects of a failure without causing a mission critical effect and is usually achieved by ensuring the system architecture includes redundant elements such as parallel redundant channels or signal paths with active or passive data replication. For redundant hardware elements in a subsystem that is part of fault tolerant system architecture, a number of elements may fail, while a minimum number of elements are required to operate to meet system performance requirements. This minimum number of elements compared with the total number of elements installed and configured in the system equates to a redundancy ratio used in system modeling approaches. This type of system model may be referred to as an "m out of n" system. The "m out of n" system is one type of a fault tolerant system, where "m" is the required number of elements that must be operating for the system to function, while "n" represents the total number of system elements.

When redundant elements involve multiple microprocessors, the architecture will include master–master or master–slave configurations for certain critical functions. The difference between master–master and master–slave configurations is the preset functional assignments of the processors. In a master–master configuration, both master processors are performing the functions together on the critical path. In a master–slave configuration, the master performs the preset assignment, while the slave is held in reserve or backup and usually not performing a critical path function until it is promoted from slave to master when the master fails. Based on the system assets and the functional priorities, as a processor fails in a critical mission path involving a master–master or master–slave configuration, the processor assignments outside the critical path may be reassigned and a processor with a low priority function may be promoted to serve in a master–master or master–slave configuration. Processor promotions and demotions occur continuously depending on critical function failures, with noncritical functions taken offline, critical functions recovery and failed processors brought back online to perform the noncritical functions. The fault tolerant architecture offers flexibility to system designers who depend on certain critical functions to be available 100% of the time.

During the execution of D-FMECA, the analyst must identify all the causes of failure for a particular failure mode including one or many failure symptoms, which are the characteristics of the failure that can define the failure in different levels, such as physical, electrical, mechanical, molecular, or atomic. Failure symptoms are failure effects at higher levels in the configuration of the system or product. There is a multiple one-to-many relationship in this analysis. For each failure identified (failure to meet spec or process), there are one or more failure modes. For each failure mode, there are one or multiple failure causes.

The following are key support considerations to performing FMECA. Accounting for these points is needed to assure a successful analysis—one that points out key risk, how such risks can be mitigated, and why they are recommended:

• Management support
• Cross-functional team
• Support in planning and completing all tasks

Since FMECA is performed by a team, management support is vital to the success of any FMECA. This provides an optimal environment for uncovering possible concerns and improvements. Management must support the time, money, and resources it takes to perform this effort. For more complex, critical processes, the process is longer, but the results are also more valuable.

There are three approaches to D-FMECA: a functional approach, hardware approach, and a software approach.

• In the functional approach, each item is analyzed by the required functions, operating modes, or outputs that it generates. The failure modes of a system are analyzed for specification and requirements ambiguities and for hardware or software defects that allow for a high potential for system faults due to a lack of fault tolerant design architecture. A functional block diagram is created to illustrate the operation and interrelationships between functional entities of the system and lower level blocks as defined in engineering data, specifications, or schematics. Since this FMECA approach is highly dependent on complete and accurate product or system level requirements, the functional FMECA may also be called a requirement FMECA. This type of FMECA may also be called a system D-FMECA, or an architecture FMECA, or a top-level FMECA.
• In the hardware approach, all predictable potential failure modes of assemblies are identified and described. Each of the part/component level failure modes and failure mechanisms is analyzed to determine the effects at next higher indenture levels and product/system levels. Actual failure analysis data that identifies the physics of failure mechanisms are useful in providing empirical data to replace the analytical data in FMECA in terms of failure causes, failure modes, and failure effects. The hardware FMECA may also be called an electrical D-FMECA or a mechanical D-FMECA. These types of FMECAs are piece

part or component bottom-up FMECAs. The hardware FMECA for assemblies may be performed at the board level or box level.

- In the software approach, software design functions are analyzed. The software D-FMECA includes analysis of software components, configuration items, modules, and blocks that are analyzed during code walk-throughs and code reviews to determine potential of failure modes such as static and dynamic timing issues and race conditions caused by probabilistic failure mechanisms that could lead to system/product effects. A Software FMECA is very similar to a functional FMECA and is performed using the top-down approach. Further details of the software D-FMECA process are described later in this chapter.

Start a D-FMECA early in the design process, when the design specifications are written, before drawings, schematics, and parts lists are created [1]. This is a type of D-FMECA known as the functional D-FMECA, which is done from the top-level requirements down to ensure the design requirements will incorporate features to handle mission-critical failure modes and mitigate their effects. Fault-tolerant design capabilities and system sparing are the most common architecture approaches to handle mission critical failure modes and mitigate their effects. The concept of redundancy is the easiest fault tolerance implementation, but maybe the most costly. The cost of redundancy depends on how much of the redundant capability is applied in spare mode. In spare mode, the redundant capability may only be exercised when a failure occurs or during peak operation, but not during normal operation. This type of spare concept is called cold spare. By contrast, the concept of warm or hot spare may be applied for normal operating conditions, where the spare capability shares the load and balances the stresses of the system, and is not part of the redundant capability that is initiated upon occurrence of a failure during normal operation. Also, the cost is dependent on how much parallel capability is needed at the various hardware and software configuration levels. When employing redundancy, consider if the redundancy should be implemented using hot spares, warm spares, or cold spares. In this case, the term "spares" refers to redundant elements of the design that are configured as powered-on elements operating in backup mode (hot spares), powered-on in standby mode (warm spares), or powered-off until needed or not connected to power (cold spare). Also, consider the redundancy in a master–master configuration, or master–slave configuration, and if active replication or passive replication is needed. Other costs associated with redundancy are added weight, space, and power usage. Besides redundancy, other considerations for design improvements during FMECA are part quality, new design/application risks, and design margin.

After the initial top-down functional D-FMECA, the next D-FMECA that may be performed is the hardware D-FMECA, which is performed when the drawings, schematics, and parts lists are created, but prior to building production hardware. This phase in design is the prototype phase, when a hardware D-FMECA is performed. In this D-FMECA, all parts are analyzed, looking at part failure modes involving functions and bus interfaces, failure modes on pins such as opens, shorts, and low impedances for linear or power devices, and "Stuck-at-One" (SA1) or "Stuck-at-Zero" (SA0) states for digital or logic devices. A hardware D-FMECA uses a "bottom-up" approach, which begins at the lowest level of the system hierarchy, such as a component or piece part device level. The hardware D-FMECA is typically implemented by systems engineers at the assembly or circuit card level, considering all causes of system failures or hazards originating at the circuit card or its interconnects [1].

The software D-FMECA is typically conducted at the software unit or module level. This software level may also be called the Computer Software Configuration Item (CSCI) level or the SCI level. Software D-FMECAs are very similar to the functional D-FMECA using a top-down approach to software design-related failure mode or software safety hazard analysis. The software D-FMECA traces the effects of software design failure modes up the system software functional hierarchy to determine the next higher effects and the end effect on system performance. Software D-FMECAs are described later in this chapter and in greater detail in Ref. [1].

Example of a D-FMECA (simple step-by-step) process using in a hardware design analysis application:

- Start at the circuit card or assembly level to document all failure modes related to the component or piece part level, such as a digital IC on a Circuit Card Assembly (CCA).
- List and number all failure modes.
- Consider the part type failure causes, focusing on the pins of the IC, and the functions of the device. Each pin could have an open or short, high or low impedance, low voltage or high current leakage.
- List the failure causes for each failure mode. Each failure mode could have several causes, such as a high temperature failure mode caused by power dissipation, open bond wire (in a dual wire bond application for current handling), or delamination of the die substrate from the die paddle.
- List the failure effects for each failure cause. Separate failure effects by the hardware indenture levels starting with the CCA and working up the configuration levels to the system level for the end item or end user effects.
- Assess the probability of occurrence, detection method, and severity for each failure effect.
- Calculate the criticality for each failure effect in terms of RPN.
- Determine compensating provisions, safety issues, or maintenance actions for each that could change the criticality, increasing or decreasing the criticality calculation results or the RPN.

- Enter all FMECA data into a database for easy data retrieval and reporting.
- Sort by RPN or criticality.

After failure modes for a particular design are found, steps should be taken to mitigate the failure effects and manage the failure correction process. The first choice is to verify failure modes and failure causes in FMECA using Failure Reporting, Analysis, and Corrective Action System (FRACAS) data. When test or field data is not available, the second choice is engineering analyses, failure mechanism modeling, durability analysis and models, or physics of failure models. Failure mechanism models, and the like, exist for many failure modes and failure causes. The failure mechanisms that are possible for electronic assemblies over its life are associated with fatigue, corrosion, diffusion, wear out, fracture, and many other types of failure mechanisms. These can be estimated through engineering physics modeling and analysis.

Four of the most important reasons for performing an FMECA are

- To improve the design reliability through design changes
- To learn about the design for documenting why a failure might occur and how the design detects and reacts to a failure
- To conduct a system safety analysis
- To perform risk assessment

Performing an FMECA alone does not improve the reliability of the design. The reliability is improved when design changes are implemented that avoid failure modes, that minimize the probability of failure mode occurrences, and that lessen the severity of failure or hazard effects. The reliability is also improved when design changes are implemented that alter the system architecture to incorporate fault tolerance features, which may include functionality and circuit redundancy, or an increase in the capability and the efficiency of the design to detect, isolate, and recover from failure modes.

As defined in IEEE 1413.1, IEEE Guideline for Reliability Prediction [2], the time-related characteristics of a failure mode falls into one of the four following categories:

- Items, including electronic and electrical components, that exhibit constant rates of failure
- Items, including electromechanical and mechanical components and pyrotechnic devices, that exhibit degradation over time or "wear-out" failures
- Items, including low quality parts and systems that are subject to a reliability growth program, that exhibit decreasing rates of failure over time or "early life" failures
- Items that exhibit combinations of the previous three types of failures

An inherent failure mode is due to characteristics of the product that cannot be eliminated or controlled to avoid failure. An example of an inherent failure mode is a dead flashlight battery, because a battery has a finite life. On the other hand, an induced failure mode is due to characteristics of the product or external factors that can be either eliminated or controlled at some point in the life cycle of the product to avoid failure during operation. If the opportunity to eliminate or control a failure mode is not exercised, whether voluntarily or unknowingly, then the failure mode is classified as process induced. Process-induced failure modes can be identified and mitigated by means of verifiable compensating provisions, such as design changes, special inspections, tests, or controls, or by operational corrective actions, such as replacement, reconfiguration, repair, or preventive maintenance.

Failure mode data are often represented in the form of a Pareto chart or histogram, which plots the failure modes by decreasing frequency of the occurrence of failures over a certain time period or decreasing RPNs. For a graphic example of a Pareto diagram, see Figure 8.2 [1].

In addition to a failure mode being either inherent or induced, it is also either potential or actual. A potential failure mode is the capability of a product to exhibit

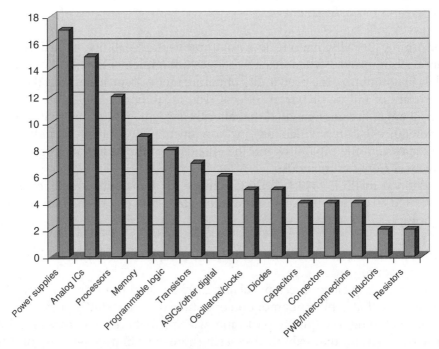

Figure 8.2 Failure cause Pareto diagram [1]

a particular mode of failure. An actual failure mode is the affectivity of a product to exhibit a particular mode of failure. In order for a potential failure mode to become an actual failure mode, a failure cause must act as the catalyst through evidence of failure mechanisms. In general, the exhibited failure mode at one level of the system becomes the failure cause for the next higher level. This bottom-up failure flow process applies all the way to system-level. The same process flow can be used in the opposite direction, with the top-down approach. For every observable failure mode, there is an initiating failure mechanism or process that is initiated by a root cause. A failure flow diagram could be developed for an FMEA to describe the steps that could be followed in a D-FMECA analysis using available diagnostics to analyze a failure using the top-down approach. The hazard/failure flow diagram should illustrate all failure modes at one functional level that are failure effects at lower functional levels of the hardware design.

The top-down system hazard/failure analysis flow example for a computer monitor hazard is shown in Figure 8.3.

FMECA will prove to be value added during the hardware and software design development process when its outputs are useful to design engineers and other engineering disciplines that support the development process.

8.4 Software Design FMECA

FMEA and FMECA methodologies are not just for system analysis or hardware analysis. FMECA is a tool to structure thought about anticipated software hazards or software failures and their root causes. There are 15 steps to follow in performing an FMECA on a software design.

Step 1: Software Elements Identification

The first step of the Software FMECA is focused on the identification of the software "elements" to be included in the analysis. The software "elements" must be identified in context of the abstraction level of the Software FMECA. The first step involves identifying its business and customer requirements that form the basis for determining the proper scope. The scope and motivation for the Software FMECA also depend on the given situation and relationship of the software with the entire product (assuming the software is embedded in a product). For that reason, a system block diagram or a context diagram may be useful to understand the scoping of the Software FMECA. Although depicted as a top-down, cascading approach, stand-alone FMECAs may be performed at any level of the software abstraction. This may often be the case when a stakeholder desires immediate feedback on the risks to the software without waiting for higher level FMECAs to be conducted.

System hazard mode: Computer may cause an unsafe condition during normal operation which could lead to equipment damage or personal injury.
- **System failure mechanism:** Operator sees images flash intermittently on computer monitor
- **System hazard cause:** Subsystem failure mode

Subsystem failure mode: Computer monitor not functioning properly at times, but still works
- **Subsystem failure mechanism:** Flat panel Thin Film Transistor (TFT) glass receiving higher than normal voltage peaks during execution of commands to illuminate screen
- **Subsystem failure cause:** Assembly failure mode

Assembly failure mode: TFT controller board not functioning
- **Assembly failure mechanism:** OP AMP hybrid output signals are generating high power outputs due to an interconnect defect on the Printed Wiring Board (PWB) that routes amplified signals on the TFT controller board to the flat panel monitor
- **Assembly failure cause:** PWB interconnect failure mode

Component failure mode: OP AMP is electrically overstressed by interconnect defect on PWB which can cause OP AMP package to ignite and burn up and potentially start a fire in the flat panel monitor.
- **Component mechanism:** PNP transistor output signals in the OP AMP are exceeding electrical specification parameters
- **Component failure cause:** Manufacturing process failure mode

Manufacturing process failure mode: Solder connection to OP AMP output on PWB is shorting to high power trace paths used for power distribution to other components used for TFT operation
- **Manufacturing process failure mechanism:** Too much solder volume applied to PWB pads
- **Manufacturing process failure cause:** Inadequate testing, defective soldering equipment, faulty soldering procedures, or human error

Figure 8.3 Top-down system hazard/failure analysis flow example

Table 8.1 depicts examples of possible software elements to be considered in FMECA for the different software abstraction levels.

Step 2: Potential Failure Modes

The second step of the Software FMECA is to identify the failure modes of each of the software elements from the previous step [1].

Table 8.1   Possible software elements for FMECA

Software abstraction artifact	Possible software elements
Requirements documentation	Tagged requirements; use cases; features and/or sub-features; functions; events and transitions between states in state transition diagrams
Architecture documentation	Components from four views of software architecture [3] are 1. Logical view: modules or objects 2. Physical view: allocation of the software to the hardware processing or communication nodes; 3. Process view: concurrency and distribution of functionality 4. Development view: files, libraries, directories mapped to the software development environment
High level design documentation	Calls in a software system call tree; sub-trees derived from a software system call tree; predominant interactions between major software objects in an object oriented development; major threads of execution in real-time, interrupt-driven embedded software; relations in object relationship diagrams or entity relationship diagrams; message sequences in message sequence charts; specific action times in timing diagrams
Low level design documentation	Low level calls among detailed software components, subroutines, libraries; Interfaces between low level software components; control and data flows between low level software components
Interconnection documentation	The Open System Interconnection (OSI) model is a standard conceptual model that characterizes communication functions within a computing system to encourage interoperability using standard protocols. The OSI model has seven layers: 1. Physical layer 2. Data link layer 3. Network layer 4. Transport layer 5. Session layer 6. Presentation layer 7. Application layer
Source code (e.g., programmed language)	Variables; constants; pointers; arrays; other data storage or memory items; key decisions in code; key calculations in code; key inputs or outputs of a given software component

Step 3: Potential Effect(s) of Failure

The third step of the Software FMECA is to analyze the effect of each of the failure modes identified in the previous step. As such, the Software FMECA team should explore and postulate what the effect would be on the product/system output(s), given that a software element is presumed to be misbehaving and experiencing a failure or corrupted.

To achieve this postulation, a multi-perspective Software FMECA team is required to enable a comprehensive product/system review of the potential effect of the failure mode [1].

Step 4: Severity Rating

The fourth step of the Software FMECA is to specify the criticality/severity for each of the failure modes in the previous step. The criticality/severity measure is normally represented on a 1–10 scale with higher values representing greater criticality/severity. From the Software FMECA standpoint, this is little difference here from a traditional hardware FMECA, but with a reminder that the scope of the Software FMECA will heavily influence the effect and this severity rating. For many reasons, it remains prudent to adopt a severity scheme that is consistent with the severity scheme used in the hardware and product/system FMECAs [1].

Step 5: Potential Cause(s) of Failure Modes

The fifth step is to identify the potential root or underlying cause(s) of each previously identified failure mode. The nature of the potential root or underlying cause(s) varies greatly by the nature of the failure mode and the level of abstraction of the Software FMECA [1].

Step 6: Failure Occurrence Rating

The sixth step is to establish the probability of occurrence of the previously identified root or underlying cause. Within traditional FMECA operation, the FMECA team usually assesses historical failure data to assess the probability or likelihood of occurrence of the cause. Within Software FMECAs, the team may use a number of different ways to establish the occurrence rating. To begin with, a combination of historical test and/or field failure data may be used to objectively establish the rating. However, for most Software FMECAs, objective occurrence data usually does not exist especially at the different levels of abstraction. In these cases, the Software FMECA team must use subjective expert opinion or surrogates for the occurrence ratings. For example, a team establishing occurrence ratings at the code level of abstraction for a Software FMECA may use a software reliability prediction model based on the Capability Maturity Model Integration (CMMI) assessment results as a valuable surrogate for assessing the failure potential of individual software units, prior to accumulation of actual empirical data from test environments or customer application environments. Software reliability prediction models, like the CMMI-based model, are described in IEEE-Std-1633-2004 [4].

After the Software FMECA team identifies the occurrence rating for the software units associated with the causes in Step 5, the team may then identify a software unit with a known field failure experience or an expert consensus of an occurrence rating, to be the benchmark. However, a word of caution deserves attention by Software FMECA teams utilizing surrogate measures to establish the occurrence rating. The surrogate values are meant to be applied early in the development process and should be updated with higher confidence data later in the development process. The value of the RPN rating system is not in the absolute ratings assessed, but rather the relative ratings assessed. The RPN scores are meant to distinguish high from medium from low risk so that management can take action on the priority high-value RPN values. As such, a surrogate measure should provide a reasonable differentiation of occurrence scores. The surrogate measure adds little value to the RPN exercise if all the software units receive essentially the same occurrence scores.

Step 7: Detection Rating

The seventh step is to assess the probability that a given root or underlying cause would escape detection and ultimately affect an output (based on the scope of the Software FMECA). A number of development and test activities, as well as design and code conventions, are most critical in preventing defective or misbehaving software "elements" from impacting output(s) of the software or product/system (depending on the scope of the Software FMECA).

In traditional FMECA operation, the FMECA team would normally access historical data related to the detection of root or underlying causes and base the detection rating on that hard data. However, most Software FMECA teams find themselves with little, if any, software data of that nature. As a result, Software FMECA teams have found it useful to follow a structured approach of subjectively assessing the detection rating using a quantitative scheme leveraged from hardware reliability block diagrams.

To begin with, the software team first delineates the complete list of applicable detection activities and conventions. This list should encompass all the abstraction levels applicable to the organization's software development, test, and maintenance lifecycle activities and practices, as well as conventions with regard to fault tolerance and recovery.

Next, for each individual root or underlying cause, the Software FMECA team then subjectively assesses the independent probability that each detection activity or convention will detect or stop the root or underlying cause from impacting an output (in terms of the scope of the Software FMECA).

Lastly, the Software FMECA team calculates the detection score for each root or underlying cause by treating the list of applicable detection activities and conventions as a parallel reliability block diagram. In this manner, the overall probability of a root or underlying cause escaping complete detection is computed [1].

Step 8: Calculating the Risk Priority Number

The eighth step is to calculate the RPN. This is a standard convention within the FMECA process in which the three scores (severity, occurrence rating, and detection rating) are multiplied together to get an overall risk score (RPN). The Software FMECA teams most often implement the standard industry convention of 1–10 value range for each of the scores, which then produce RPN scores in the range of 1–1000. The FMECA method usually then requires an FMECA team to establish an RPN threshold, such that any RPN score above the threshold requires additional treatment, as discussed in the subsequent prevention and mitigation actions. The industry norm for an RPN threshold, using a 1–10 value range on each of the individual scores, is 200. However, the Software FMECA team may decide to select a lower threshold to identify a larger portion of items requiring prevention and mitigation action planning in the upcoming FMECA steps.

Step 9: Improvements Related to Mitigation

The ninth step involves critical Software FMECA team brainstorming on how to mitigate the effects of each cause and/or failure mode, for example, lessen the severity of each effect. Depending on the nature of the cause and failure mode, different possible mitigation actions may be taken. Generally, the team will identify ways in which the software and/or product/system may be designed to be more robust and insensitive to the software cause and/or failure mode. Standard thinking surrounding design margins and redundant mechanisms equally apply in the software domain, thereby reducing the severity when the root or underlying cause occurs.

Step 10: Updated Severity Rating

The tenth step involves reassessing the severity rating by assuming that the mitigation actions just identified will be implemented successfully. The updated severity rating now depicts the reduced risk resulting from the mitigation actions.

Step 11: Improvements Related to Prevention

The eleventh step involves critical Software FMECA team brainstorming on how to prevent the root or underlying causes from occurring. Prevention actions may include process and training changes related to the software development, test, and maintenance activities, as well as software design and code implementation that immediately identifies and stop causes at the point of origin. There may be a fuzzy line of demarcation between prevention improvement and the upcoming discussion of detection improvement. The reader may choose to include the immediate catch of a cause as prevention, while any subsequent activity falls in the realm of detection.

Step 12: Updated Occurrence Rating

The twelfth step involves reassessing the occurrence rating by assuming that the prevention actions just identified will be implemented successfully. The updated occurrence rating now depicts the reduced risk resulting from the prevention actions.

Step 13: Improvements Related to Detection

The thirteenth step involves critical Software FMECA team brainstorming on how to detect and stop the hypothetical root or underlying causes from escaping through the development, test, and maintenance process and the subsequent operation of the software to impact an output of concern (based on the scope of the Software FMECA). At this point, the Software FMECA team will need help from software process experts to identify what new or modified software processes would detect the causes earlier and more assuredly. Additionally, software architects and designers will need to advise the Software FMECA team on what new or modified software architecture components and design structure or code is needed to detect causes during software runtime. Often, the recommendations from this step form the needed improvement plan for enhanced fault tolerance and recovery including the classes of exceptions and the expected exception handling.

Step 14: Updated Detection Rating

The fourteenth step of the Software FMECA involves reassessing the detection rating by assuming that the detection improvement actions just identified will be implemented successfully. The updated detection rating now depicts the reduced risk resulting from the detection improvement actions.

Step 15: Updated RPN Calculation

The fifteenth step is to update the calculated RPN numbers for each of FMECA line items that were above the established RPN threshold and for which additional prevention and mitigation actions were identified. For this step, Software FMECA team reviews all of the recommended prevention, mitigation, and detection actions for a given FMECA line item and then computes the updated RPN value based on the updated three ratings. The updated RPN score now reflects the overall revised risk assessment. In most organizations, the initial and revised RPN numbers form the risk profiles that management finds convenient to manage during development and fielding of products. Often management will monitor the volume and age of outstanding RPN values above the acceptable threshold so that they can ensure proper engineering resources are applied to prevention, mitigation, and detection activities and conventions. A number of organizations also use the results of the initial and updated RPN scores to feed more advanced models that seek to predict project and product success outcomes, such as schedule and cost performance, customer satisfaction and even early product returns.

A Software FMECA example is provided in Chapter 16 of this book, as well as an example of software failure causes for potential software failure modes. We provide potential software failure mode effects for software elements of a navigation system.

8.5 What Is a PFMECA?

FMECA is a tool used for addressing risk of hazards or failures. In the case of the D-FMECA it is extremely valuable in reviewing the design aspects of a product—from the standpoint of the ability to investigate possible issues for the product meeting its design requirements. Just as the D-FMECA focuses on possible design failure modes at multiple levels of hardware or software structure, Process Failure Modes, Effects, and Criticality Analysis (PFMECA) is a detailed study, focused on manufacturing and test processes and steps, required to build reliable products. It includes review of materials, parts, manufacturing processes, tools and equipment, inspection methods, human errors, and documentation. It reviews the build, inspection effectiveness, and test aspects of the product for possible risks of process step failures at multiple levels of processes, including the severity of problems after the product is in the customer's hands.

8.5.1 What Is the Difference Between a Process FMECA and a Design FMECA?

Process FMECA (PFMECA) is an FMECA that assesses the failure modes and failure causes related to the manufacturing process or the design process. The design process (D-FMECA) is an FMECA that assesses the failure modes and failure causes related to the system or product design.

The PFMECA process is broken down into sub-processes and lower level steps. For each step, possible problems, their likelihood of occurring and their ability to be detected before extensive consequences are explored. The impacts to the customer, as well as to subsequent sub-processes, are considered. Throughout this process, a matrix is used to help define the scope of effort, track the progress of the analysis, document the rationale of each entry, and facilitate quantification of the results. Figure 8.4 is an example of one type of matrix format used for such an analysis. This quantification process, along with the rationale, is extremely useful, when working to prioritize possible improvement recommendations. The details of how to use this form as a tool to complete the analysis are discussed in the subsequent paragraphs.

| Product or Product Line | | | **PFMECA** | | | Team Members | | | |
| Process or Sub-Process | | | | | | Revision Date | | | |

Item #	Process Name and Description	Process Step	Failure Mode	Failure Effect	SEV	Potential Causes	OCC	Verification Method	DET	RPN	Recommended Actions

Figure 8.4 Example PFMECA form [1]

8.5.2 Why PFMECAs?

Review of failure data in many programs indicates that many of the causes of product failures were related to processes, especially the manufacturing and test processes. These tie directly to the design function responsibilities, especially as they relate to reliability and safety. The following considerations should be taken into account:

- The design engineer knows more than anyone how the system or equipment works—what is important to its operation and what is not. Often, the design engineer plays a significant role in making decisions that drive the manufacturing process.
- The designer selects parts, defines layout and materials, and makes a myriad of decisions that define how the design is to be built and tested. For instance, if a new material or component is required to meet requirements, this could impact the thermal profile used to attach the part. Often, issues with manufacturing or test engineering are avoided or resolved by changing and improving the design. Design decisions must be selected with cost, schedule, and performance considered.
- The PFMECA is best performed with a team of process stakeholders—including manufacturing, test engineering or technicians, system safety engineering, maintenance (when that is a key issue), reliability engineering, quality engineering, and design engineering. This diversity of viewpoints provides a variety of viewpoints to consider for various issues.

An example of a PFMECA as used in a safety analysis is to consider causes of a fire in a building. One question is "How would this affect us?" Many different and interesting answers might ensue. "What would cause this to occur?" Almost without exception, the design engineers would provide issues with assembly, installation, maintenance, and inspection, while the maintenance engineers and technicians would bring up likely design problems.

The point is the fidelity of FMECA is improved by having multiple viewpoints. This is not without cost: The more personnel involved in the analysis, the more discussion takes place, and this takes time. However, if properly guided, such discussion is enlightening, and the effort yields more complete and useful results. For instance, a design may include an elegant means of alignment of piece in an assembly. However, if the operator must perform the alignment by "touch" or "eyeballing," the results almost certainly vary with operator experience, state of mind, and physical ability. Adding a fixture to assure less variability in results could be an invaluable improvement to the process with little cost. Upon completion of a successful PFMECA, some or all of the following benefits are realized:

- Improved reliability
- Improved quality, less variability (higher yields, diminished schedule and cost risks)
- Improved safety in the manufacturing processes
- Enhanced communication between the process owners
- Improved understanding of processes and interfaces by participants

8.5.3 Performing PFMECA, Step by Step

Step 1: Team Preparation

In selecting specific team members, it is important to assure that the team participants can understand the specific processes and steps to be performed, the tooling and fixtures involved, the design of the equipment undergoing the processes, and how the unit or system is to be used by the customer.

The first step in performing the PFMECA is to list the processes and steps to be considered. In effect this task actually translates the defined scope into the specific areas that will be covered. A process flow diagram, as well as the specific procedures and assembly drawings to be analyzed are needed to perform the PFMECA. An example of a process flow diagram is contained in Chapter 5 of this book. They should be kept on hand during the analysis as reference material. In fact, in many cases (where possible), examples of the actual hardware to be built assembled or otherwise processed should be available as well. Also, trips to the process lines to view the actual activities are extremely useful to assure the participants are aware of the details involved. Ground rules of the effort should be established up front.

Each team member may have their own point of view in how to conduct the PFMECA. For instance, team members may disagree on the scoring for each failure mode and the criteria used to calculate RPNs. Therefore, it is valuable to settle on the criteria for ranking of specific factors and their corresponding descriptions. These factors are used as inputs into the RPN equation. This section describes tables for sample scoring of the three factors (SEV, OCC, and DET) used in the RPN.

Step 2: Defining the Processes and Sub-processes

Figure 8.4 shows the two columns "process name" and "process step" being completed. This is the first part of the PFMECA effort. By completing the first two columns, the team actually defines the scope of the analysis. That is, if the process steps are analyzed and there are no other additions or changes, then this step actually defines the level of detail and which steps are covered in the analysis. It is sometimes necessary to include additional columns. This may be necessary, based on the way the process owner has defined the process to be studied. Also, in the course of the PFMECA, other processes may be identified as worthwhile to analyze, or another change may be considered necessary. These changes are expected and can be implemented, as long as the team recognizes that this is a change in the analysis scope.

Step 2a: Failure Modes and Effects: SEV Factor

When the analysis scope is defined, then the steps are reviewed for how they might fail. In this context, a "failure" is when a step is missed or done incorrectly or a defect can result. The idea is to delineate the list of possible ways in which mishaps can occur (failure modes) and look at the rest of the process to see how the whole process could be impacted (failure effects). Any failure could result in increased cost, loss of quality, schedule impact, and even harm to equipment or personnel. More specific ways a failure could impact the process are

- Interruption of subsequent process steps or processes
- Defects that could result in wasted material, rework labor, or schedule time to repair
- Unreliability or poor quality in the shipped product
- Unsafe situation on the line or in the field

Table 8.2 is reused from Table 16.1. Table 8.2 is provided to show the correlation between severity and priority. Severity 1 and 2 codes refer to safety critical failures and hazards and align with priority 1 and 2 codes from Table 8.2. Furthermore, the severity and priority 1 and 2 codes align with the category 1 and 2 codes used in risk assessments and some FMEA/FMECA standards and guidelines.

Table 8.2 Priority factors

Priority	Description of the problem
Critical (priority 1)	Safety critical with catastrophic effects involving loss of life
	Prevents the accomplishment of an operational or mission essential capability
	Jeopardizes security or other requirement designated critical
Major (priority 2)	Safety critical with damage to equipment without loss of life or personal injury
	Adversely affects the accomplishment of an operation or mission essential capability and no work-around solution is known
	Adversely affects technical, cost, or schedule risks to the project or the life cycle support of the system, and no work-around solution is known
Minor (priority 3)	Adversely affects the accomplishment of an operational or mission essential capability, but a work-around solution is known
	Adversely affects technical, cost, or schedule risks to the project or the life cycle support of the system, but a work-around solution is known
Annoyance (priority 4)	Results in user/operator inconvenience or annoyance and does not affect a required operational or mission essential capability
	Results in inconvenience or annoyance for development or support personnel, but does not prevent the accomplishment of those responsibilities
Other (priority 5)	Any other effect, such as cosmetic or enhancement change

Priority refers to the important of the impact of an issue or problem. Priority means the level of urgency related to correcting an issue. Severity 1 and 2 hazards or failures has the corresponding priority to enable focus on the top priority issues that need to be resolved before any others. The importance of the impacts of these issues is the basis for each of the severity (SEV) factors defined. This is why each failure mode has a separate line, and all the major possible effects are listed on that line. The SEV factor (1–10) is selected from the choices in Table 8.3, based on the most severe possible effect. For purposes of calculating an RPN, the SEV/ priority 1 and 2 codes correlate to the SEV factors 9 and 10.

Tables 8.4 and 8.5 provide the codes for the likelihood of occurrence (OCC) of failures or hazards and detectability of the failure or hazard occurrence (DET), Step 2b and Step 2c, respectively, which are used with the SEV code in the RPN calculations.

Step 2b: Possible Causes: OCC Factor

This step is intended to consider the possible ways the failure mode could occur. It is not uncommon for there to be several causes. An "immediate cause" is often defined as the step done incorrectly that would result in the problem.

Table 8.3 Severity (SEV) factors

SEV	Description of severity
10	Safety critical with catastrophic effects involving loss of life System, product, or plant safety at risk; risk of non-compliance to government regulations Jeopardizes security or other requirement designated critical
9	Significant interference with subsequent process steps or significant damage to equipment Could result in safety-critical or mission-critical failure with no work-around solutions Major impact on ability to produce safe and quality system or product on time
7–8	System or product test failure during storage and operational specified environments that could adversely affect the accomplishment of an operational or mission essential capability System or product defect causes in-process rejection or disruption to subsequent process steps with some impact to safety, quality, and on-time performance
4–6	Results in user/operator inconvenience to conduct additional operations and steps to stay on mission Customer dissatisfaction, some degradation in performance, loss of margins, or delays in process that result in inconvenience for development or support personnel
2–3	Slight customer annoyance, slight deterioration in performance or margin, minor rework action or in-line process delays
1	Little or no effect on product or subsequent steps; may include additional steps, such as offline cosmetic or process enhancement change

Root cause is found by asking the question "What would have caused that?" This is asked multiple times until an actionable cause is determined. The point is to get to a reason for a possible problem that can ultimately be addressed and corrected. For instance, it is not enough to state the operator did something incorrectly. It is more important to look at what kinds of pressures are on the operator, how well trained are the operators, and whether there are adequate instructions and pictures for the intended result, and there are design possibilities for the tooling/fixturing or the product itself that could lower the likelihood of failure.

Each of the causes listed is assigned an "OCC" factor. As with all the three factors, the values vary from 1 to 10. The following table is an example of how the values could be made assigned to failure modes and provide consistency in scoring risks between different FMEAs/FMECAs.

This particular example is especially useful, if the process being analyzed is tracking defect data for Statistical Process Control (SPC). In that case the "sigma" values represent a statistical method for estimating process issues in a

Table 8.4 Occurrence (OCC) factors

OCC	Failure Rate (FR)	Defect rate	Sigma	Description of likelihood of failure or hazard occurrence
10	FR>5/hour	1 in 2	$< +2\sigma$	Failure or hazard is certain
9	1/hour<FR<5/hour	1 in 8	$< +2\sigma$	Failure or hazard is almost inevitable
8	0.1/hour<FR<1/hour	1 in 20	$< +2\sigma$	Failure trend likely; process step not in control or SPC not used
7	0.01/hour<FR<0.1/hour	1 in 40	$< +2\sigma$	Failure trend very likely; similar steps consistently repeating known problems; little or no experience with new tool or step
6	1.0 E-03<FR<1.0 E-02	1 in 80	$< +2\sigma$	Possible failure trend; process step varies in and out of SPC control (CPK<1.00); out of control greater than 50% of the time
5	1.0 E-04<FR<1.0 E-03	1 in 400	$\sim+2.5\sigma$	Possible failure trend; process step is in control (CPK<1.00) ~50% of the time; similar steps with intermittent problems
4	1.0 E-05<FR<1.0 E-04	1 in 1000	$\sim+3\sigma$	Possible failure trend; process step is in control (CPK<1.00); similar steps with occasional problems
3	1.0 E-06<FR<1.0 E-05	1 in 4000	$\sim+3.5\sigma$	Low likelihood of failure trend; process step in control (CPK>1.00); similar steps with isolated occurrences
2	1.0 E-09<FR<1.0 E-06	1 in 20,000	$\sim+4\sigma$	Very low likelihood of failure; process step in control (CPK>1.33); rare occurrence of dissimilar problems in similar steps
1	FR<1.0 E-09/hour	1 in 1,000,000	$\sim+5\sigma$	Remote failure trend or no failure data after many months (e.g., three months) of testing or operational hours accumulated in similar steps (CPK>1.67)

Note: CPK is a statistical measurement of design or manufacturing process capability, which stands for process capability index. FR is a reliability measurement of failures over time, such as Failures per Million Hours (FPMH).

Table 8.5 Detection (DET) factors

DET	Description of detectability
10	No means of detection; no process or equipment to find problem in time to affect an outcome
9	Controls would probably not detect a problem; operator/maintainer intervention is necessary
7–8	Controls have poor chance of detecting problems; periodic operator inspection and maintenance checks are required to detect problems on a regular basis
5–6	Controls have high test coverage for detection of some problems and low coverage for other problems; operator/maintainer inspection and tests required to detect most problems
3–4	Controls have a good chance to detect problem with high test coverage; process and test equipment detects presence of problem under most situations with minimal downtime
1–2	Controls will always or almost always detect a problem to avoid system or product downtime

quantifiable way. If this data is not available, then other historical information or the judgment of the team must be used to arrive at a reasonable value for OCC.

Step 2c: Verification Method: DET Factor

The final factor to be determined is the detection (or "DET") factor. The main underlying idea is to gain insight into how well the process is setup to detect a failure, defect other flaws in a manner that would prevent a worsening result. Clearly this and the SEV factors are interrelated, and the team should be flexible in its evaluation of them. Each cause has its own verification method(s) and DET values.

Questions to be addressed are

- How would this failure mode detected given the cause listed? Would the operator or others be able to understand what the problem is and how to address it in time to avoid worst-case results?
- Are there multiple means for the failure mode to be detected? If so, all the methods of detection should be accounted in determining the DET value.
- Are the detection methods formalized as part of the process? Are they automatic or built-in? Manual? Or occasional?

The DET factor then is answering the question "How likely could the cause be detected in time or in a place where the effects could be mitigated?" If multiple means of detection were used, then all those methods would be included in evaluating the DET rank value.

Step 3: Risk Priority Number (RPN)

The RPN may be calculated by multiplying or adding the three values (SEV, OCC, and DET). If the RPN is calculated by multiplying the SEV, OCC, and DET values together, and the three factors can vary from 1 to 10, then the product of these three values will vary from 1 to 1000. If the RPN is calculated by addition of the three values, then the resultant is a number between 3 and 30. By using multiplication, it is a simple matter to separate the important from the unimportant while recognizing the importance of all the three factors as equal contributors to overall risk to the program. The RPN values can then be sorted by highest to lowest value, providing insight into the most important possible issues. These meet the criteria of highest impact, most likely to occur, and least likely to be detected.

8.5.4 Performing PFMECA, Improvement Actions

8.5.4.1 Prioritization of Potential Issues

Prioritization is not simply to select from the high-value RPNs. But rather it is to consider the higher valued items first. It is imperative to review the recommended actions through the lens of the business priorities: Are the recommendations actionable? Are they cost-effective? Are there other mitigation factors, such as human safety to be taken into account? Often some of the lower-value RPN items have simple low-cost solutions. These should be considered for follow-up, as well. These ideas, combined with the prioritization and documented rationale from the PFMECA, provide an excellent basis for making optimal decisions.

8.5.4.2 First Address Root Cause(s)

Clearly the first place to look for effective improvements is to address the source of the issue—the root cause. For instance, if the cause stems from the lack of tooling to assure proper alignment, then designing the equipment so that special alignment is not required would remove the risk altogether. If this is not possible or too costly, then the next step would be to look at alignment tools in the process to decrease the risk of a mistake. "Improved training" is often chosen as a catchall phrase in corrective actions. That is not to say that training is not important. It is. Certainly, if inadequate training is a factor, it must be addressed. However, often the addition of clarifying pictures or photographs, simplified designs, or steps and ensuring the correct tools and fixturing on hand are much more effective actions than general verbiage about signs or slogans, coaching, or training.

"Inspection" is also a useful tool for catching defects but is often overused as a corrective action. A saying among quality engineers is "You can't inspect quality in!" In other words, inspections have limited impact in finding and correcting defects. Corrective actions that prevent defects—that make it easier to do right and harder to do wrong—are much more effective than added inspection steps.

Certainly such items as improved training and inspections can always be considered as effective additions to selected improvement actions.

8.5.4.3 Present Recommendations to Management

Once the potential improvement recommendations are developed, the next step is to assure they are well defined in terms that management—program, production, and design management—can understand and use. As with any engineering action items, responsible personnel, scope of effort, and schedule must be defined to assure the recommendations are properly handled and reported to those making the decisions. These recommendations, then, must be delineated in terms of their estimated costs, so that the decision-makers can evaluate the "bang for the buck" in terms of risk.

8.5.4.4 Follow-Through

As follow-up to this effort, once improvements have been implemented or to serve as "what-if" studies, further PFMECA updates may be made, using the initial results as a starting place, and then observing the results of reconsidering the risk factors in light of the actions taken.

8.5.5 Performing PFMECA and Reporting Results

A completed PFMECA is not done until the results are documented in a report. The results reported should lead to design improvements through design changes that eliminate hazards and high severity failure modes. The hazards and high severity failure mode results must be sorted by RPN value to allow prioritization of the follow-up actions. Once these actions have taken place and their effectiveness evaluated, the PFMECA can then be reviewed for changes to the existing factors to measure RPN improvements [1].

The completed matrix represents a good review of the scope and level of detailed covered, but this is not enough. For the results to be properly reported, a more complete assessment of the ground rules, priorities, and decisions in the performance of the PFMECA serves as a record of the complete effort. The importance of reporting results is paramount:

- To communicate in writing the decisions, conclusions, and recommendations developed by the team.

- To serve as a record of the underlying rationale behind the possible concerns and improvements developed.
- The help track progress after improvement actions have occurred.
- As actions are taken and risks lowered, there will be new "#1" RPN elements. As follow-up actions are accomplishments, this analysis can serve as a road-map for the next priorities and considerations.
- Future PFMECAs can benefit from findings and insights reported from the initial PFMECAs.

8.6 Conclusion

FMEAs and FMECAs may be applied in a number of flexible ways at different points in the system, hardware, and software development lifecycle. FMEA and FMECA methodologies are widely recognized tools in the reliability and safety analysis toolbox especially applied in system failure risk assessments and should be considered a critical analysis practice for any system safety engineering program.

Acknowledgments

The author wishes to acknowledge the contributions from Robert Stoddard and Joseph Childs in providing much of the source material for this chapter, which was originally published in our book, *Design for Reliability*, Wiley 2012 [1].

References

[1] Raheja, D. and Gullo, L., *Design for Reliability*, John Wiley & Sons, Inc., Hoboken, NJ, 2012.

[2] IEEE 1413.1-2002, *IEEE Guide for Selecting and Using Reliability Predictions Based on IEEE 1413*, The Institute of Electrical and Electronics Engineers, Inc., New York.

[3] Philippe, K., Architectural Blueprints: The "4+1" View Model of Software Architecture, November 1995, http://www.cs.ubc.ca/~gregor/teaching/papers/4+1view-architecture.pdf (Accessed on August 2, 2017).

[4] IEEE-Std-1633-2008, *Recommended Practice on Software Reliability*, The Institute of Electrical and Electronics Engineers, Inc., New York, 2008.

Suggestions for Additional Reading

Carlson, C., *Effective FMEAs*, John Wiley & Sons, Inc., Hoboken, NJ, 2012.

Modarres, M., *What Every Engineer Should Know about Reliability and Risk Analysis*, Marcel Dekker, Inc., New York, 1993.

O'Connor, P.D.T., *Practical Reliability Engineering*, 3rd Edition, John Wiley & Sons, Ltd, Chichester, 1992.

U.S. Department of Defense, *Military Standard: Procedures for Performing a Failure Mode Effects and Criticality Analysis*, MIL-STD-1629A, Notice 3, August 4, 1998 (Cancelled), U.S. Department of Defense, Washington, DC.

U.S. Department of Defense, *Military Handbook, Electronic Reliability Design Handbook*, MIL-HDBK-338B, U.S. Department of Defense, Washington, DC, 1998.

9

Fault Tree Analysis for System Safety

Jack Dixon

9.1 Background

After Fault Tree Analysis (FTA) was invented at Bell Laboratories by H. A. Watson in 1961 [1], it has become a very popular and powerful analysis technique. FTA was initially applied, by Watson, to the study of the Minuteman Launch Control System. Dave Hassl, of Boeing, later expanded its use to the entire Minuteman Missile System. He presented an historic paper on FTA at the first System Safety Conference in Seattle in 1965 [2].

Since the early days of FTA, it has been applied to numerous complex systems in essentially every industry. The following list of some typical applications of FTA illustrates the widely varied use of the FTA technique:

- Nuclear power [3]
- Civil aircraft
- Air traffic control systems
- Rail systems
- Apollo program
- Space shuttle

Design for Safety, First Edition. Edited by Louis J. Gullo and Jack Dixon.
© 2018 John Wiley & Sons Ltd. Published 2018 by John Wiley & Sons Ltd.

- International space station
- Chemical plants
- Radioactive waste disposal
- Fuzes and safe-and-arm devices for weapon systems [4]
- Offshore aquaculture of seaweed [5]

The nuclear industry contributed much to the science and math of fault trees by creating techniques and computer programs to analyze fault trees. After courses in FTA were presented to many Nuclear Regulatory Commission (NRC) personnel during the late 1970s, the NRC published the first FTA handbook, NUREG-0492 [6].

Since these early days of FTA, many improvements have been made including algorithms to mathematically evaluate the fault tree and computer programs to plot and analyze trees.

While FTA is included here as a hazard analysis tool, it really is a root cause analysis tool since it is used to determine the cause of a hazard, or undesirable event, that has been identified by some other hazard analysis technique. It is an important and much used technique. FTA can be used to evaluate an identified hazard or to analyze an accident. It is also used in reliability analysis.

9.2 What Is a Fault Tree?

A fault tree is a representation in tree form of the combination of causes (failures, faults, errors, etc.) contributing to a particular undesirable event. It uses symbolic logic to create a graphical representation of the combination of failures, faults, and errors that can lead to the undesirable event being analyzed. The purpose of FTA is to identify the combinations of failures and errors that can result in the undesirable event. The fault tree allows the analyst to focus resources on the most likely and most important basic causes of the top event.

 Paradigm 4: Prevent Accidents from Single and Multiple Causes

FTA is a deductive (top-down) analysis technique that focuses a particular undesirable event and is used to determine the root causes contributing to the occurrence of the undesirable event. The process starts with identifying an undesirable event (top event) and working backward through the system to determine the combinations of component failures that will cause the top event.

Using Boolean algebra, the fault tree can be "solved" for all the combinations of basic events that can cause the top event. This results in a qualitative analysis of the fault tree. The fault tree can also be used as a quantitative analysis tool. In this case, failure rates or probability of occurrence values are assigned to the basic events, and the probability of the occurrence of the top level, undesirable event can be calculated.

9.2.1 Gates and Events

Logical gates are used, along with basic events, to create the fault tree. The standard symbols used in the construction of fault trees and the descriptions of these symbols are shown in Figure 9.1.

These gates, events, and other symbols are used to develop the fault tree.

9.2.2 Definitions

Some basic definitions are needed to be able to understand FTA.

Cut set: A cut set is a combination of hardware and software component failures that will cause system failure.

Minimal cut set: A minimal cut set is a smallest combination of hardware and software component failures that, if they all occur, will cause the top event to occur [6].

Failure: A failure is a basic abnormal occurrence in a hardware or software (HW/SW) component. Failure is the inability of a function to meet its requirement specification. For example, a relay fails closed when it should be open under normal operation according to its specification.

Fault: A fault is an undesired state of a function composed of hardware and software component that occurs at an undesirable time. For example, a relay closes when it is supposed to be open, not because the relay failed closed, but because it was "commanded" to close at the wrong time.

Primary fault: A primary fault is any fault of a HW/SW component that occurs in an environment for which the component is qualified. For example, a pressure tank, designed to withstand pressures up to and including a pressure p_o, ruptures at some pressure $p \leq p_o$ because of a defective weld [6].

Secondary fault: A secondary fault is any fault of a HW/SW component that occurs in an environment for which it has not been qualified. In other words, the component fails in a situation that exceeds the conditions for which it was designed. For example, a pressure tank, designed to withstand pressure up to and including a pressure p_o, ruptures under a pressure $p > p_o$ [6].

Command fault: A command fault involves the proper operation of an HW/SW component, but at the wrong time or in the wrong place; for example, an arming device in a warhead train closes too soon because of a premature or otherwise erroneous signal origination from some upstream device [6].

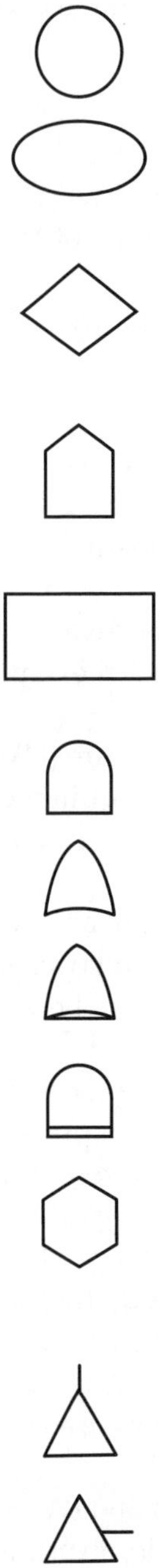

Primary event symbols

BASIC EVENT – A basic initiating fault requiring no further development

CONDITIONING EVENT – Specific conditions or restrictions that apply to any logic gate (used primarily with PRIORITY AND and INHIBIT gates)

UNDEVELOPED EVENT – An event that is not further developed either because it is of insufficient consequence or because information is unavailable

EXTERNAL EVENT – An event that is normally expected to occur

Intermediate event symbols

INTERMEDIATE EVENT – A fault event that occurs because of one or more antecedent causes acting through logic gates

Gate symbols

AND – Output fault occurs if all of the input faults occurs

OR – Output fault occurs if at least one of the input faults occurs

EXCLUSIVE OR – Output fault occurs if exactly one of the input faults occurs

EXCLUSIVE AND – Output fault occurs if all of the input faults occur in a specific sequence (the sequence is represented by a CONDITIONING EVENT drawn to the right of the gate)

INHIBIT – Output fault occurs if the (single) input fault occurs in the presence of an enabling condition (the enabling condition is represented by a CONDITIONING EVENT drawn to the right of the gate)

Transfer symbols

TRANSFER IN – Indicates that the tree is developed further at the occurrence of the corresponding TRANSFER OUT (e.g., on another page)

TRANSFER OUT – Indicates that this portion of the tree must be attached at the corresponding TRANSFER IN

Figure 9.1 Fault tree symbols

Exposure time: The exposure time is the time the system is operated, thus exposing the components of the system to failure. The longer the exposure time period, the higher the probability of failure.

Undesirable event: The undesirable event is the top-level event. It is the subject of the FTA. It is typically an event that has been identified by some other type of hazard analysis. For example, in a weapon system, the undesirable event may be "Inadvertent Detonation of Warhead."

9.3 Methodology

The safety analyst begins with the undesirable event that has been identified by earlier hazard analysis and identifies the immediate causes of the event. Each of those causes is evaluated further until the most basic causes have been identified. This fault tree model represents the relationship between the basic causes and the undesirable event. There may be a number of combinations of basic causes that can cause the top event. These combinations are known as cut sets. A minimal cut set is the smallest combination of basic causes that can cause the topic event to occur if that combination of basic causes occurs at the same time.

Fault tree diagrams consist of gates and events connected in tree from. AND and OR gates are the two most commonly used gates in FTA. To illustrate the use of these gates, Figure 9.2 shows two simple fault trees consisting of two input events that can lead to an output, or top, event. If the occurrence of either input event can cause the output event, then these input events are connected using an OR gate. If, however, both input events must occur to cause the output event to occur, then they are connected by an AND gate.

Construction and analysis of fault trees involves following a process:

- **Step 1. Define the system:** This step involves familiarization with the system being analyzed including its design and its operation.
- **Step 2. Define the top, undesirable event that is to be analyzed:** This event typically comes from a previously conducted hazard analysis, or in the case of an accident, it would be a short description of the accident (e.g., explosion in chemical plant X).

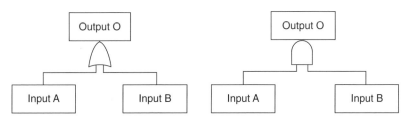

Figure 9.2 Simple fault trees

- **Step 3. Define the boundaries:** The analyst must define the boundaries of the system including subsystems, functions, environments, and so on.
- **Step 4. Construct the fault tree:** After selecting the undesired event, we can now begin to construct the fault tree by using AND and OR gates, or other gates that may apply. First start with the next level under the top event—the intermediate events and their combination that can lead to the top event. This process is repeated downward through as many levels as necessary until the root cause for each branch is identified or a point is reached where no further decomposition is unnecessary. Each branch of the fault tree will end with either a basic event or an undeveloped event.
- **Step 5. Solve the fault tree:** This step requires the application of Boolean algebra to solve for the cut sets that can cause the top event, resulting in a qualitative evaluation of the fault tree. If a quantitative evaluation is desired, or required, then failure rate data must be provided for each of the basic events, and the probability of the top event must be calculated.
- **Step 6. Evaluate the results:** Using the results from the previous step, the analyst must evaluate the importance of each of the cut sets and determine what mitigation needs to be applied to the design of the system to eliminate the potential for the occurrence of the top-level event. Alternatively, if a quantitative assessment has been made of the fault tree, then the analyst must determine if the results are within acceptable limits, or if design modifications must be made to improve the probability of occurrence.
- **Step 7. Report the results:** Document the FTA for delivery to the customer or to file for possible future reference.
- **Step 8. Update the fault tree:** The FTA will need to be updated if any errors are found in the tree or, more likely, if any design changes are made to the system either as a result of the FTA itself or changes made to the design for other reasons.
- **Step 9. Report the results:** Document the FTA for delivery to the customer or to file for possible future reference.

This process is shown in Figure 9.3.

Successful fault trees are drawn using a set of basic rules. These rules are adapted from NUREG-0492. [F]

- **Ground Rule 1.** Describe exactly each event by writing statements that are entered in the event boxes as faults; state precisely what the fault is and when it occurs (e.g., motor fails to start when power is applied).
- **Ground Rule 2**. If the answer to the question "Can this fault consist of a component failure?" is "yes," classify the event as a "state-of-component fault." If the answer is "no," classify the event as a "state-of-system fault."
- **Ground Rule 3**. No miracles rule. If the normal functioning of a component propagates a fault sequence, then it is assumed that the component functions normally.

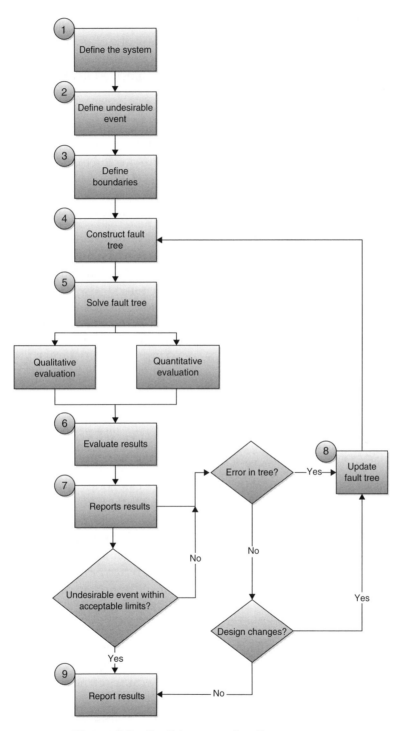

Figure 9.3 Fault tree construction process

- **Ground Rule 4**. Complete-the-gate rule. All inputs to a particular gate should be completely defined before further analysis of any one of them is undertaken.
- **Ground Rule 5**. No gate-to-gate rule. Gate inputs should be properly defined fault events, and gates should not be directly connected to other gates.

To elaborate on more the actual construction of the fault tree, the following is a summary of the thought process that goes into creating each of the lower levels of the fault tree. As the tree is created, the analyst must determine which gate is appropriate to use at each level. To do this, there are three approaches that are used. These approaches have been adapted from NUREG-0492 [6].

State-of-Component/State-of-System Approach

Expanding of Ground Rule 2 above, the analyst starts by asking the question "Can this fault consist of a component failure?" If the answer is "Yes," classify the event as a "state-of-component fault." If the answer is "No," classify the event as a "state-of-system fault." If the fault event is classified as "state-of-component fault," add an OR gate below the event and look for primary, secondary, and command modes. If the fault event is classified as "state-of-system fault," look for the minimum necessary and sufficient immediate cause or causes. A "state-of-system" fault event may require an AND gate, an OR gate, an INHIBIT gate, or possibly no gate at all. As a general rule, when energy originates from a point outside the component, the event may be classified as "state-of-system" fault (e.g., motor fails to start when power is applied to its terminals is a "state-of-component" fault. Motor inadvertently starts is a "state-of-system" fault).

Primary, Secondary, and Command Approach

In this approach the faults are classified into three categories: primary, secondary, and command. Each fault category is described on the fault tree with an event symbol. The event symbols used in the example of Figure 9.2 shows a basic event for the primary fault, an intermediate event for the command fault, and an undeveloped event for the secondary fault. An OR gate is used to connect these three fault categories to the top event that is an undesired output. If two or more of these are present as causes at the level being analyzed, then an OR gate is needed. See Figure 9.4.

Immediate, Necessary, and Sufficient Causes Approach

Beginning at the top level, the analyst must determine the immediate, necessary, and sufficient causes for the occurrence of the top event. These are typically NOT the basic causes of the event, but they are the immediate causes of the event. These immediate, necessary, and sufficient are now treated as the next-level events as the analyst proceeds down the tree. At each level, the immediate, necessary, and sufficient causes are determined along with the appropriate logical gate that defines the combination of the immediate, necessary, and sufficient causes that result in each level of event.

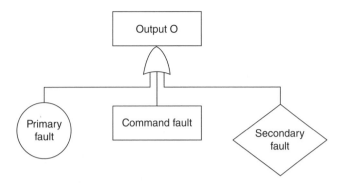

Figure 9.4 Primary, secondary, and command approach

This process is completed at each level of the tree until the analyst reaches "basic" failure that cannot be broken down any further (e.g., resistor fails open).

This fault tree construction process is replicated from the top level, through all the intermediate levels, until the analyst reaches the basic causes associated with the undesirable event.

9.4 Cut Sets

After the fault tree is constructed, it must be solved. A major product of FTA is the determination of cut sets. As defined earlier, a cut set is a combination of HW/SW component failures that will cause system failure. For very simple trees, as in Figure 9.5, cut sets can be determined by sight.

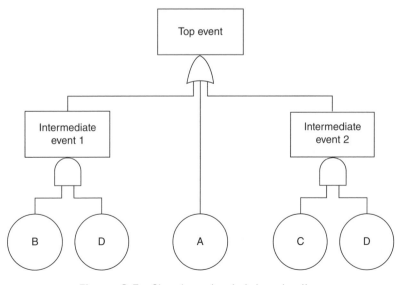

Figure 9.5 Simple cut set determination

The cut sets are

A

B,D

C,D

As fault trees become more complex, the reduction of the tree to cut sets becomes more complicated. Boolean algebra must be used. It is assumed that the reader is familiar with the rules of Boolean algebra. The following simple tree illustrates the calculation of cut sets and minimal cut sets in a slightly more complicated tree (Figure 9.6).

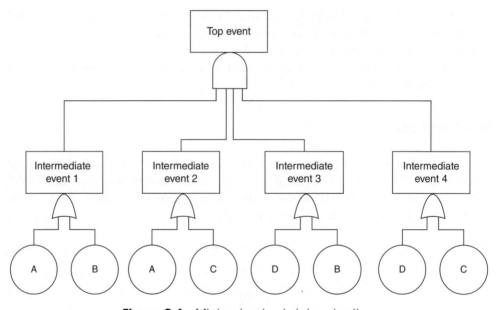

Figure 9.6 Minimal cut set determination

The cut sets are

$$(A + B)$$
$$(A + C)$$
$$(D + B)$$
$$(D + C)$$

But that's not the end of the story. We must determine the minimal cut sets. Recalling the definition from an earlier section, a minimal cut set is a smallest combination of HW/SW component failures that, if they all occur, will cause the top event to occur [6].

So we write a Boolean expression for the cut sets we found for the tree in Figure 9.6:

$$(A+B)\cdot(A+C)\cdot(D+B)\cdot(D+C)$$

Using the rules of Boolean algebra, we reduce this expression to it minimal cut sets as follows:

$$[A+(B\cdot C)]\cdot[D+(B\cdot C)]$$
$$(B\cdot C)+(A\cdot D)$$

Therefore the minimal cut sets are

$$(B\cdot C)$$
$$(A\cdot D)$$

As can be seen from this simple example, solving fault trees manually using Boolean algebra can get cumbersome very quickly. Because of this, an algorithm, called MOCUS, was developed and subsequently computerized to generate cut sets and minimal cut sets easily when the fault tree became large [7].

The steps of MOCUS algorithm are as follows:

Step 1: All gates and basic events are identified (labeled) by letters, numbers, or a combination.

Step 2: Put the name of the top-level gate in the first row, first column position in a matrix.

Step 3: Replace the gate name with its inputs. If the gate is an AND gate, place the inputs in the row of the matrix. If the gate is an OR gate, place the inputs in the column of the matrix.

Step 4: Iterate step 3 replacing the gates with their inputs working down through the fault tree. Continue this process of substitution until only basic events are left. The remaining basic events will be all of the cut sets to be analyzed.

Step 5: Using the rules of Boolean algebra, remove all non-minimal cut sets from the list. The final list will be the minimal cut sets.

Figure 9.7 shows an example of a simple tree to that we can apply the MOCUS algorithm (adapted from Ref. [7]).

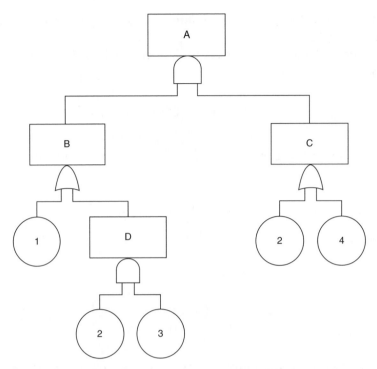

Figure 9.7 Simple fault tree for illustrating MOCUS algorithm

Using the steps outlined above for applying the MOCUS algorithm, we have the following.

The top-level event, A, corresponds to a gate A. Output from gate A is input to the top-level event, A. Put the name of the top-level event or gate A in the first row, first column position in a matrix.

A		

Replace the A gate with its inputs. Since A is an AND gate, place the inputs in the row of the matrix.

B	C	

Replace the B gate with its inputs. Since B is an OR gate, place the inputs in the column of the matrix. Note gate C is carried along in both entries since it was paired with gate B.

1	C	
D	C	

Replace the C gate with its inputs. Since C is an OR gate, place the inputs in the column of the matrix. Note that basic event 1 and intermediate event D are duplicated since each of the inputs to gate C is paired with both Events 1 and D.

1	2	
D	2	
1	4	
D	4	

Replace the D gate with its inputs. Since D is an AND gate, place the inputs in the row of the matrix.

1	2	
2	2	3
1	4	
2	4	3

Using the rules of Boolean algebra, remove all non-minimal cut sets from the list. In this case, the duplicate 2 can be removed from the second row, and the 2, 4, and 3 cut sets can be removed since only the inputs 2 and 3 are required to cause the top event, and they are already an identified cut set. The final list contains all the minimal cut sets.

1	2	
2		3
1	4	

MOCUS can be used manually to reduce a fault tree to its minimal sets, but will eventually get to be cumbersome once the fault tree becomes large and complex. At that point automated analysis will be necessary.

9.5 Quantitative Analysis of Fault Trees

There are several methods to calculate the probability of occurrence of the top undesirable event.

First, in all cases the failure rates of the basic events must be determined. This is typically provided by reliability calculations and/or data. The reliability of a HW/SW component is given as

$$R = e^{-\lambda t}$$

where

R = reliability
λ = HW/SW component failure rate
t = exposure time

The probability of failure is

$$P_f = 1 - R = 1 - e^{-\lambda t}$$

Once all the failure rates are determined, then the probability of the top event occurring can be calculated. One method to do this is to start at the bottom of the tree and calculate the probabilities for each gate moving up the tree to the top event. To do this, the following rules apply to the common AND and OR gates:

For an AND gate the probability of failure is

$$P_f = P_A \times P_B$$

For and OR gate the probability of failure is

$$P_f = P_A + P_B - (P_A \times P_B)$$

This method does not work, however, if there are multiple occurrences of events or branches within the tree. If these are present, then a cut set method must be used to obtain accurate results.

For the cut set method of probability calculation, once the minimal cut sets are obtained, if desired, the analyst can perform quantitative analysis of the fault tree by evaluating probabilities. The quantitative evaluations are most easily performed in a sequential manner by first determining the component failure probabilities, then determining the minimal cut set probabilities, and finally determining the probability of the top event occurring by summing up the probabilities of each of the minimal cut sets using the appropriate gate equations for determining the cut set probability.

As with determining cut sets and minimal cut sets, these calculations get very tedious and difficult for anything but the smallest fault trees. Automated means must be used.

9.6 Automated Fault Tree Analysis

Over the years numerous computer programs have been developed to assist the safety engineer in creating and analyzing fault trees. While the authors are not endorsing any particular products, a few examples of off-the-shelf FTA tools currently on the market are listed here for the reader's convenience.

CAFTA

CAFTA stands for computer-assisted fault tree analysis. It is an event tree and FTA tool for analyzing complex systems. CAFTA is a product developed by Electric Power Research Institute (EPRI), Palo Alto, California. Rolls-Royce Controls and Data Services Inc. is a licensed seller of CAFTA. Further information can be found at https://www.controlsdata.com/civil-aero/cafta

FaultTree+

FaultTree+ is a fault tree software package created by Isograph that has been incorporated into their Reliability Workbench 11 product that also does other types of safety and reliability analyses. Further information can be found at https://www.isograph.com/software/reliability-workbench/fault-tree-analysis/

ITEM Software

ITEM Software offers an ITEM Toolkit with a fault tree module with a Graphical User Interface (GUI) that allows construction of fault trees. The software can decompose system level failures into combinations of lower level events and Boolean gates to model their interactions. Further information can be found at http://www.itemsoft.com/fault_tree.html

Windchill FTA

PTC Windchill FTA (formerly Relex Fault Tree) provides a FTA tool with an intuitive graphical representation of fault trees and event trees plus analytical tools to evaluate the risk and reliability of complex processes and systems. Further information can be found at http://www.crimsonquality.com/products/fault-tree/

RELIASOFT BlockSim

RELIASOFT BlockSim is a system reliability and maintainability analysis software tool. It has a graphical interface that allows the modeling of the simplest or most complex systems and processes using Reliability Block Diagrams (RBDs) or FTA, or a combination of both. Further information can be found at http://www.reliasoft.com/BlockSim/index.html

9.7 Advantages and Disadvantages

FTA is a proven, popular, and powerful analytical technique that has been used in many different industries for many years. It has the flexibility to be conducted at various levels of design from conceptual to very detailed. It can provide valuable insight into safety problems using either a qualitative or quantitative form of analysis. FTA also provides a very visible presentation of the particular safety or accident analysis. It has been automated by readily available commercial software.

While FTA is a very powerful tool, its biggest disadvantage is that it can very easily become unwieldy, time-consuming, and expensive.

9.8 Example

An example may help the reader better understand the FTA process. A pressure tank example from NUREG-0492 has been around for many years, and this example is based on that the one in that Ref. [6]. Figure 9.8 shows the pressure tank–pump–motor device and its associated control system that is intended to maintain the tank in a filled and pressurized condition. The operational modes are shown in Figure 9.9.

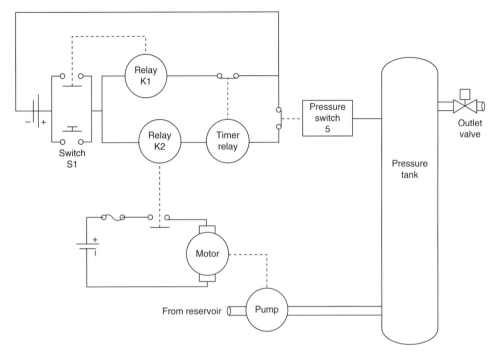

Figure 9.8 Pressure tank system

The function of the control system is to regulate the operation of the pump. The latter pumps fluid from an infinitely large reservoir into the tank. Assume that it takes 60 seconds to pressurize the tank. The pressure switch has contacts that are closed when the tank is empty. When the threshold pressure has been reached, the pressure switch contacts open, de-energizing the coil of relay K2 so that relay K2 contacts open, removing power from the pump, causing the pump motor to cease operation. The tank is fitted with an outlet valve that drains the entire tank in an essentially negligible time; the outlet valve, however, is not a pressure relief valve. When the tank is empty, the pressure switch contacts close, and the cycle is repeated.

Initially the system is considered to be in its dormant mode: switch S1 contacts open, relay K1 contacts open, and relay K2 contacts open; that is, the control system is de-energized. In this de-energized state the contacts of the timer relay are closed. The tank is also assumed to be empty and the pressure switch contacts are therefore closed.

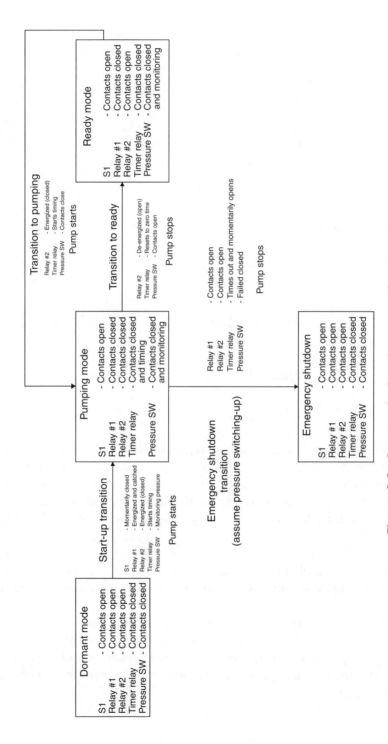

Figure 9.9 Pressure tank system operational modes

System operation is started by momentarily depressing switch S1. This applies power to the coil of relay K1, thus closing relay K1contacts. Relay K1 is now electrically self-latched. The closure of relay K1 contacts allows power to be applied to the coil of relay K2, whose contacts close to start the pump motor.

The timer relay has been provided to allow emergency shutdown in the event that the pressure switch falls closed. Initially the timer relay contacts are closed and the timer relay coil is de-energized. Power is applied to the timer coil as soon as relay K1 contacts are closed. This starts a clock in the timer. If the clock registers 60 seconds of continuous power application to the timer relay coil, the timer relay contacts open (and latch in that position), breaking the circuit to the K1 relay coil (previously latched closed) and thus producing system shutdown. In normal operation, when the pressure switch contacts open (and consequently relay K2 contacts open), the timer resets to 0 seconds.

The undesired event for this example is

Rupture of Pressure Tank after the Start of Pumping

To simplify things considerably we will ignore plumbing and wiring failures and also all secondary failures except the one of principal interest: "tank rupture after the start of pumping."

Using the rules and process describe above, the final fault tree is shown in Figure 9.10.

Since we are not expanding further the other secondary failures, we can simply omit from the diagram. Thus this tree can be further simplified:

The E's are fault events:

El: Pressure tank rupture (top event).

E2: Pressure tank rupture due to internal overpressure from pump operation for $t>60$ seconds that is equivalent to K2 relay contacts closed for $t>60$ seconds.

E3: EMF on K2 relay coil for $t>60$ seconds.

E4: EMF remains on pressure switch contacts when pressure switch contacts have been closed for $t>60$ seconds.

E5: EMF through KI relay contacts when pressure switch contacts have been closed for $t>60$ seconds, which is equivalent to timer relay contacts failing to open when pressure switch contacts have been closed for $t>60$ seconds.

The simplified tree is shown in Figure 9.11.

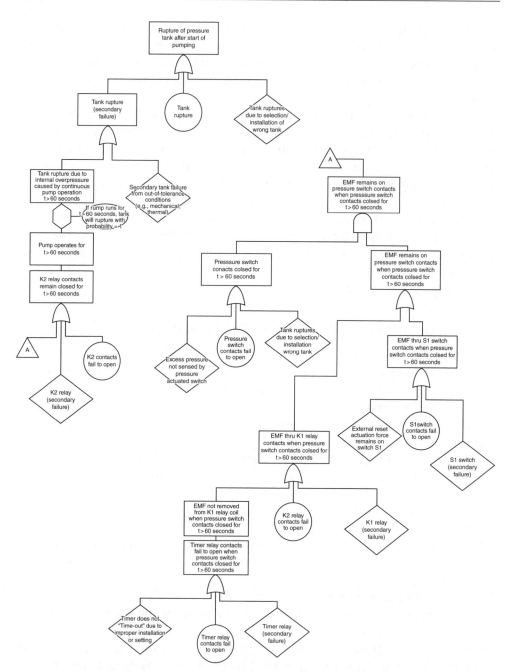

Figure 9.10 Pressure tank rupture fault tree example

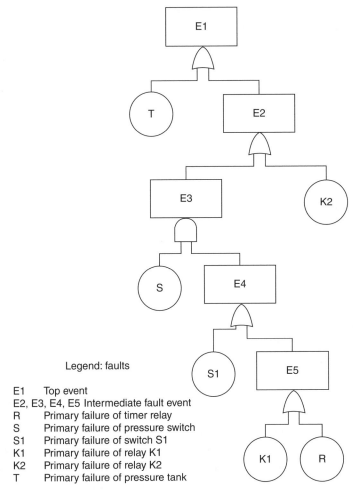

Figure 9.11 Simplified (reduced) fault tree for pressure tank example

Using the simplified fault tree of Figure 9.11, we can express the top event as a Boolean function of the primary input events by starting at the top of the tree and working down substituting as we go:

$$E1 = T + E2$$
$$= T + (K2 + E3)$$
$$= T + K2 + (S \cdot E4)$$
$$= T + K2 + S \cdot (S1 + ES)$$
$$= T + K2 + (S \cdot S1) + (S \cdot ES)$$
$$= T + K2 + (S \cdot S1) + S \cdot (K1 + R)$$
$$= T + K2 + (S \cdot S1) + (S \cdot K1) + (S \cdot R)$$

This expression of the top event in terms of the basic inputs to the tree is the Boolean algebraic equivalent of the tree itself. This reduces to five minimal cut sets:

$$K2$$
$$T$$
$$S \cdot S1$$
$$S \cdot K1$$
$$S \cdot R$$

Each of these cut sets defines the event or combination of events that will initiate the top event of the tree.

We can now make a qualitative assessment of the results. Qualitatively, the leading contributor to the top event is the single relay K2 because it represents a primary failure of an active component. Therefore, the safety of the system would be considerably enhanced by substituting a pair of relays in parallel for the single relay K2. Actually, however, the system contains a much more serious design problem—the controls are being monitored instead of the parameter of interest (pressure in this case). It should be just the other way around! Thus, the most obvious way to improve the system would be to install a pressure relief valve on the tank and remove the timer.

Next, we can make a quantitative assessment of the results. This requires estimates of failure probabilities for the components. Table 9.1 provides assumed values for the failure probabilities for the components in the system.

Because a minimal cut is an intersection of events, the probabilities associated with the five minimal cut sets are obtained by multiplying the appropriate component failure probabilities (assuming independence of failures):

Table 9.1 Failure probabilities for pressure tank example

Component	Symbol	Failure probability
Pressure tank	T	5×10^{-6}
Relay K2	K2	3×10^{-5}
Pressure switch	S	1×10^{-4}
Relay K1	K1	3×10^{-5}
Timer relay	R	1×10^{-4}
Switch S1	S1	3×10^{-5}

$$P(T) = 5 \times 10^{-6}$$
$$P(K2) = 3 \times 10^{-5}$$
$$P(S \cdot K1) = \left(1 \times 10^{-4}\right)\left(3 \times 10^{-5}\right) = 3 \times 10^{-9}$$
$$P(S \cdot R) = \left(1 \times 10^{-4}\right)\left(1 \times 10^{-4}\right) = 1 \times 10^{-8}$$
$$P(S \cdot S1) = \left(1 \times 10^{-4}\right)\left(3 \times 10^{-5}\right) = 3 \times 10^{-9}$$

The probability of the top event, E1, is what is to be estimated. Since the top event is expressed as the union of the minimal cut sets, the probability of the top event can be approximated as the sum of the individual minimal cut set probabilities, provided these probabilities are small. This is typically an accurate approximation for basic event probabilities below 0.1. The approximation is termed the "rare event approximation." Using this approach for this example, the minimal cut set probabilities are simply summed to provide the probability of the top event occurring as

$$\mathbf{P(E1) \approx 3.5 \times 10^{-5}}$$

9.9 Conclusion

FTA is a powerful technique that can be used either qualitatively or quantitatively to analyze a system. It can be used as a tool to analyze the probability of a particular design or as a tool for performing trade-offs of different designs. It can be applied to any type of system, but caution must be used since it can also be a very costly analysis to conduct.

References

[1] Watson, H.A. (1961) *Launch Control Safety Study*, Section VII Vol1, Bell Laboratories, Murray Hill, NJ.
[2] Hassl, D., Advanced Concepts in Fault Tree Analysis, Presented at the System Safety Symposium, June 8–9, 1965, Seattle, WA.
[3] Rasmussen, N.C., U.S. Nuclear Regulatory Commission; U.S. Atomic Energy Commission (1975) *Reactor Safety Study: An Assessment of Accident Risks in U. S. Commercial Nuclear Power Plants*, WASH-1400 (NUREG-75/014), U.S. Nuclear Regulatory Commission, Springfield, VA.
[4] Larsen, W. (1974) *Fault Tree Analysis*, Picatinny Arsenal, Dover, NJ.
[5] Sulaiman, O.O., Sakinah, N., Amagee, A., Bahrain, Z., Kader, AS.S.A., Adi, M., Othman, K., and Ahmad, M.F. Risk and Reliability Analysis Study of Offshore Aquaculture

Ocean Plantation System, Presented at 8th International Conference on Marine Technology, Terengganu, Malaysia, October 20–22, 2012.

[6] Vesely, W.E., U.S. Nuclear Regulatory Commission, Division of Systems and Reliability Research (1981) *Fault Tree Handbook*, NUREG-0492, U.S. Nuclear Regulatory Commission, Washington, DC.

[7] Fussell, J.B., Henry, E.B., and Marshall, N.H. (1974) *MOCUS: A Computer Program to Obtain Minimal Sets from Fault Trees*, Aerojet Nuclear Company, Idaho Falls, ID.

Suggestions for Additional Reading

Ericson, C. (2011) *Fault Tree Analysis Primer*, CreateSpace, Inc., Charleston, NC.

Raheja, D. and Allocco, M. (2006) *Assurance Technologies Principles and Practices*, John Wiley & Sons, Inc., Hoboken, NJ.

Raheja, D. and Gullo, L.J. (2012) *Design for Reliability*, John Wiley & Sons, Inc., Hoboken, NJ.

Roland, H.E. and Moriarty, B. (1990) *System Safety Engineering and Management*, John Wiley & Sons, Inc., New York.

10

Complementary Design Analysis Techniques

Jack Dixon

10.1 Background

While there are over a hundred different hazard analysis techniques that have been described in the literature and used for various projects, we have focused on the most common in this book.

In this chapter we cover several additional popular hazard analysis techniques including

- Event trees
- Sneak Circuit Analysis (SCA)
- Functional Hazard Analysis (FuHA)
- Barrier Analysis (BA)
- Bent Pin Analysis (BPA)

We also provide brief highlights of a few additional techniques that are less often used:

- Petri nets
- Markov Analysis (MA)

Design for Safety, First Edition. Edited by Louis J. Gullo and Jack Dixon.
© 2018 John Wiley & Sons Ltd. Published 2018 by John Wiley & Sons Ltd.

- Management Oversight Risk Tree (MORT)
- System-Theoretic Process Analysis(STPA)

The "Suggestions for Additional Reading" at the end of each chapter should be consulted for discussions of additional, less often used, analysis techniques.

10.2 Discussion of Less Used Techniques

Among the many hazard analysis options there are, many can be used to either complement or supplement the more common ones we discussed in Chapter 7. The techniques covered in this chapter are a selection of analysis techniques that provide a different and usually more specific focus than the more all-encompassing ones presented earlier. The techniques discussed here are typically used to investigate particular types of hazards in greater detail than the more general techniques.

 Paradigm 7: Always Analyze Structure and Architecture for Safety of Complex Systems

10.2.1 Event Tree Analysis

The Event Tree Analysis (ETA) is a graphical technique that can be used in support of both hazard analysis during the design process and analysis of accidents that have happened. It describes the sequence of events that could, or did, lead to an accident. It can be used as either a qualitative or quantitative analysis. ETA is similar to decision trees used in business.

10.2.1.1 Purpose

The purpose of the ETA is to determine the sequence of events following an undesirable initiating event and the probability of the various consequences that may result from the initiating event. The ETA not only is used to evaluate the consequences of the initiating event but also can be used to assess the likelihood that the initiating event will develop into a mishap. It can be used to evaluate the effectiveness of the protection systems and hazard controls that have been implemented and the safety procedures prescribed for the system.

10.2.1.2 Process

Assuming we are conducting hazard analysis versus accident investigation, the first thing to do when constructing an ETA is to choose an initiating event, which may be a system, component, equipment failure, human error, or some external event such as fire, explosion, and so on. This event will be undesirable and will usually come from other analyses such as PHA or HHA. Once the initiating event has been chosen, the analyst builds accident scenarios that may be possible. To do this, the intermediate events must be identified, and they are often referred to by other names—protection systems, safeguards, safety systems, barriers, mitigations, hazard control success or failure, and pivotal events. Consider intermediate events (usually associated with protection systems) and the success or failure of each intermediate event. Each branch can either fail or succeed. These are used to build a logical sequence of events that can happen after a particular incident, or initiating event, has occurred. Each branch of the event tree leads to some consequence that is identified by the analyst. Thus the event tree identifies a list of all consequences that can occur from one initiating event, as illustrated in Figure 10.1.

Note that, by tradition, failure is always shown on the lower branch and success is on the upper branch.

Once the event tree is constructed, the failure probabilities of each of the intermediate event must be determined. This information can come from numerous sources including Fault Tree Analysis (FTA), reliability prediction models or analysis, or any source of reliability failure rate data, such as internal field data, internal test data, supplier data, and handbook data (e.g., MIL-HDBK-217F, *Reliability Prediction of Electronic Equipment* [1]). Since each intermediate event is conditional

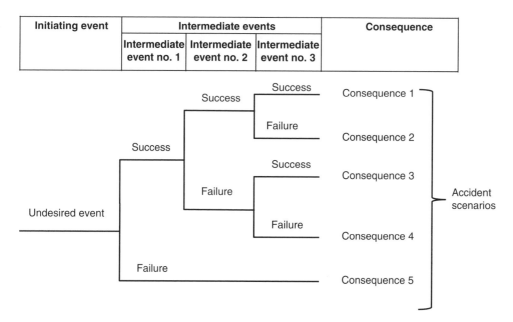

Figure 10.1 Generic event tree

on the occurrence of the previous intermediate event, the probabilities are conditional probabilities. They are displayed on each branch. Note that for each branch

$$\text{Probability of success} = 1 - \text{probability of failure}$$

The probability of each consequence is calculated by multiplying together all the probabilities along each path in the event tree, leading to a particular consequence. Figure 10.2 shows a generic event tree with probabilities assigned to each branch and the calculation of the probabilities for each consequence.

Once the event tree is completed, it can be used to evaluate the risk associated with each consequence. If the outcome of the analysis indicates that the risk associated with any particular path is unacceptable, design modifications, additional safeguards, or additional procedures can be added to mitigate the risk.

 Paradigm 4: Prevent Accidents from Single and Multiple Causes

10.2.1.3 Alternate Uses

The ETA technique can be used during accident investigation to help analyze and understand an accident.

Another variation is to use the event tree to represent a sequence of actions/ decisions that an operator is supposed to carry out in the event of an undesirable

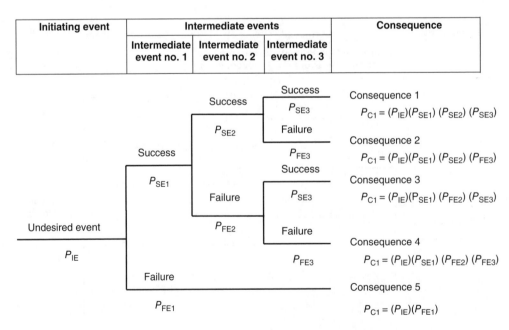

Figure 10.2 Generic event tree with probabilities

incident. For example, a leak of hazardous material from a vessel in a chemical plant may require an operator to accomplish certain tasks to prevent a larger catastrophe. These actions, in addition to any other protective devices, could be included in the event tree.

10.2.1.4 Advantages/Disadvantages

ETA is a graphical model and, therefore, provides a visual way of understanding the consequences and the paths to the consequences of an initiating event. The procedure can include the effects of hardware, software, and human responses. While ETA can be accomplished at any stage of design, it needs details concerning the design and the probabilities of failures of various components, so it is usually conducted later in the design process. Any changes identified may cause extensive redesign and substantial costs.

Timing issues may cause problems with event tree usage. Certain events are dependent on timing or sequences of activities and the results of other events. This timing issue can happen if a failure, which may be analyzed to occur during a certain event within a specified period of time, changes its time of failure occurrence based on stresses and physical properties and no longer is probable to occur during that certain event. This timing issue can be a significant disadvantage in performing the ETA, unless ETAs are performed for different timing scenarios.

Also, common cause failures may result in errors in the ETA.

The ETA can become very complex. For a large system, such as a power plant, many event trees will have to be developed in order to analyze all the possible undesirable events that might occur.

10.2.2 Sneak Circuit Analysis

The original Sneak Analysis (SA) was SCA. The procedure has been in use for almost 50 years, but remains somewhat of a mysterious art. The first major computer-aided version was developed for the NASA Apollo program in 1968 by the Boeing Company [2]. This analysis focused on electrical circuitry. The goal of SCA is to uncover hidden (sneak) paths or conditions in electrical circuitry that either cause undesired functions to occur, cause desired functions to occur at the wrong time, or inhibit desired functions from occurring as they are designed to do. These sneak paths happen even though no component has failed. SCA uses topological patterns and clues to assist the analyst in finding these sneak paths.

Limited public domain information on SCA is available and the techniques for conducting SCA are largely proprietary. Because of this, SCA is a technique that is not used widely in the practice of system safety. Due to the proprietary nature of the clues, it typically requires a special contractor to perform the analysis. It is also time consuming and, therefore, costly to perform. Because of the large effort required to do SCA and its associated costs, it is usually only applied to a very limited class of safety-critical items such as safe-and-arm devices and subsystems of missiles and rockets. We cover it here for completeness, but only from a top-level overview standpoint.

Although SCA was developed to focus on electrical circuitry, the technique has since been expanded to cover other areas such as pneumatic, hydraulic, and process flows and software using the same topological approach under the more general term, SA.

10.2.2.1 Purpose

The purpose of SCA is to find latent paths that either cause undesired functions to occur or inhibit desired functions from occurring. These sneak paths occur, while all components are functioning correctly.

10.2.2.2 Process

Conceptually, SCA is simple. However, the companies that do SCA have developed proprietary clues to facilitate the analysis, and they do not share them. They have also developed proprietary software to assist in the conduct of SCA. The process requires the conversion of schematic diagrams into topological diagrams, and then, using clues, the analyst must search for sneak paths.

The earliest work by Boeing on SCA broke sneak circuits into four categories based on different sneak conditions that may exist:

- Sneak Path - A sneak path may cause current or energy to flow along an unexpected route.
- Sneak Timing - Sneak timing may cause current or energy to flow or inhibit a function at an unexpected time.
- Sneak Indication - A sneak indication may cause an ambiguous or false display of system operating conditions.
- Sneak Label - A sneak label may cause incorrect stimuli to be initiated. [2]

The following are the steps required to conduct SCA:

Step No. 1: Gather data
Gather applicable design data including electrical schematics, interconnect diagrams, functional diagrams, and so on.

Step No. 2: Code data
Since most SCA is done using computer programs, the schematics are coded, using special proprietary rules, into the computer program.

Step No. 3: Create network trees

Convert the coded electrical schematics into network trees, which are simplified versions of the electrical circuitry. These network trees are oriented such that the power supplies are at the top and the ground is at the bottom.

Step No. 4: Identify topological patterns

There are five topological patterns that the analyst must identify in the network trees, as shown in Figure 10.3.

Step No. 5: Perform analysis

Perform analysis of the topological patterns by applying the proprietary clues to each node. These clues help the engineer identify any defects in the design based on the type of topological found in the network trees. Here are a few typical clues from the early Boeing work [2]:

- Power-to-Power Path - Different voltage sources/levels may be switched into parallel through the tree (unintentional path).
- Ground-to-Ground Path - The ground points of the tree may be at different absolute potentials. For example, even a ground bus may "float" off vehicle ground due to distributed impedances, thus changing the effective load voltages.
- Reverse Current Flow - Switching modes may allow current reversal through the cross branch of the H.
- Ambiguous Indicators - Attempts to monitor multiple modes or functions with a single indicator endangers the fidelity of the status display.

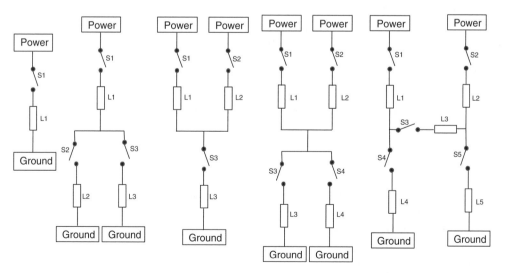

Figure 10.3 Sneak circuit analysis topological patterns

- Misleading Labels - Labels should denote all directly controlled functions for each switch, circuit breaker, and power source in the tree. Circuit breaker and switch labels in the areas of system interconnections are particularly prone to label discrepancies.

Step No. 6: Generate report
The computer program typically produces several reports based on the analysis:

- Drawing errors
- Design concerns
- Sneak paths
- Network trees
- Inadvertent operational modes
- Ambiguous indications

10.2.2.3 SCA Variation

Instead of conducting SCA after the fact when all the circuits have been designed, the designers could address the sneak paths during the early design process, thus eliminating the sneak paths early and without having to conduct expensive SCA or at least reducing the SCA effort needed in the end. This approach also allows corrective changes to be made early when they are less intrusive and expensive to implement.

Several sneak path design rules adopted from *Sneak Circuit Analysis for the Common Man* [3] are shown in the succeeding text. The reader is referred to the reference for further details.

Problem: Sneak paths involving multiple power sources and/or multiple ground returns.
Solution: Structure circuits so that all current for a given load flows from one power source to one ground return.
Problem: Sneak paths caused by interrupting current at the ground side of a load.
Solution: Do not place current interrupting elements (e.g., switches, relays, circuit breakers, fuses) in ground return paths.
Problem: Sneak paths caused by connectors (or other current interrupting devices) located on the ground side of a load.
Solution: When placement of a connector at the ground side of the load is required, keep the supply and ground return paths symmetrical.
Problem: Sneak paths caused by power and ground connectors.
Solution: Avoid the use of separate connectors for providing power and ground return lines to a circuit.
Problem: Sneak paths caused by selecting alternate paths in "wired-OR" circuits.
Solution: The "wired-OR" can be used only where the effect to be produced by the alternate conditions is exactly the same. When some conditions are intended to cause additional or modified effects, isolation must be provided.

Problem: Sneak timing due to momentary loss of power to volatile computer memory and other essential loads during switchover to an alternate power source.

Solution: For small memories, use break-before-make switching and sufficient capacitance to maintain the voltage during switchover. For large memories or to protect against a short on the main supply, use make-before-break switching and diode isolation.

Problem: Sneak label causing an action opposite to the one intended to occur when a switch is toggled.

Solution: Label switches according to the action performed in addition to the object being controlled.

10.2.2.4 Advantages/Disadvantages

SCA is very labor intensive and, therefore, costly to perform. Because very little information is in the public domain, the SCA technique has to largely be performed by specialized contractors that have developed their own proprietary clues and computer programs, thus making it difficult for the average safety engineer to conduct SCA independently. However, SCA remains a powerful technique that can find latent sneak paths that are otherwise hard to find through manual analysis.

While it would be best to do SCA early in order to identify problems early in the development process where they can be corrected inexpensively, because SCA is costly to perform, it is usually done only one time, after production-level drawings and information is available. When sneak paths are found at that point, they are expensive to fix.

10.2.3 Functional Hazard Analysis

A function is defined as "a specific or discrete action that is necessary to achieve a given objective (e.g., an operation that the system must perform to accomplish its purpose or a maintenance action that is necessary to restore the system to operational use)" [4]. Functional Hazard Analysis (FuHA) is a technique that evaluates each function of a system to determine hazards that may arise from malfunction, which could be a failure of the function or a function not operating at the proper time. Not all functional failures lead to hazardous conditions.

FuHA is usually conducted early in the development when system functions are known, but their detailed implementation has not yet been undertaken. It can be performed on the whole system, an individual subsystem, or on a system-of-systems.

10.2.3.1 Purpose

The purpose of FuHA is to analyze all system functions to identify system level hazards that might result from a malfunctioning of each function. The intention of the FuHA is to find the critical hazards. It is a starting point for a more detailed analysis. Its focus is on the consequences of the failure of the function and the severity of that failure, not on the details of the causes of the failures.

10.2.3.2 Process

First, the analyst must obtain a list of all functions performed by the system under analysis. This can usually be obtained from the concept of operations documentation and from the functional flow block diagrams that are created early in the development of the project.

Each function is then analyzed to determine the effect of each possible failure mode on the system. The following functional failure types should be considered for each function under analysis:

- Fails to operate: Function does not happen/perform when given the appropriate input.
- Operates early/late: Function performs earlier or later than it should have; if too late, function could be out of sequence.
- Operates out of sequence: Function occurs before or after the wrong function and without receiving the appropriate inputs.
- Unable to stop operation: Function continues even though the thread should move on to the next function.
- Degraded function or malfunction: Function does not finish or only partially completes; it generates improper output [5].

The analyst determines whether each functional failure produces a hazard. If so, that risk is assessed and recommended actions are developed. Further safety analysis is planned.

10.2.3.3 Worksheet

An FuHA is usually conducted using a matrix similar to many of the analyses. A typical format for the FuHA is shown in Figure 10.4.

Functional Hazard Analysis											
System: _____ Preparer: _____ Date: _____											
Identifier	Function	Hazard	Effects	Causal Factors	IRAC	Risk	Recommended Action	FRAC	Risk	Cmts	Status

Figure 10.4 Functional hazard analysis format

Column descriptions:

Header information: Self-explanatory. Any other generic information desired or required by the program could be included.

Identifier: This can simply be a number (e.g., 1, 2, 3, ... as hazards are listed). Or, it could be an identifier that identifies a particular subsystem or piece of hardware that will be in the system with a series of numbers as multiple hazards are identified with that particular item (e.g., motor 1, motor 2, etc.).

Function: This is a description of what function of the system is being evaluated. The list of functions should be a complete list of all functions in the system.

Hazard: This is the specific potential hazard that has been identified as a result of the functional failure. Note that there may be multiple hazards due to multiple ways in which the function can fail.

Effects: This is a short description of the possible resulting mishap. What is the effect(s) resulting from the potential hazard—death, injury, damage, environmental damage, and so on.

Causal factors: This should be a brief description of what caused the function to fail and how the resulting potential hazard could result in a mishap.

IRAC: This is a preliminary qualitative assessment of the risk.

Risk: This is a qualitative measure of the risk for the identified hazard, given that no mitigation techniques are applied.

Recommended action: Provide recommended preventive measures to eliminate or mitigate the identified hazards. Hazard mitigation methods should follow the preferred order of precedence provided earlier in a previous chapter of this book.

FRAC: This is a final qualitative assessment of the risk, given that mitigation techniques and safety requirements identified in "Recommended Action" are applied to the hazard.

Comments: This column should include any assumptions made, recommended control, requirements, applicable standards, actions needed, and so on.

Status: States the current status of the hazard (open, monitored, closed).

10.2.3.4 Advantages/Disadvantages

FuHA is relatively easy to perform and follows the approach of many other hazard analyses. If conducted early, it can provide useful data for controlling hazards before they are designed into the system or in suggesting further analysis. Since FuHA is focused on functions, it typically will not identify all hazards associated with the system, so other types of analysis should be conducted.

10.2.4 Barrier Analysis

The concept of BA is fairly simple. Identify an energy source, identify the target, define the flow of energy, and identify barriers that prevent the energy flow. See Figure 10.5. BA can be used for either as a system safety design analysis tool or as an accident investigation tool.

10.2.4.1 Purpose

The purpose of BA is to evaluate barriers to prevent hazardous energy flow to the target in order to prevent an accident. The target can be personnel or equipment.

10.2.4.2 Process

To conduct BA the analyst identifies the energy source and the energy flows that may be harmful to a target. Barriers are then identified or inserted to block the energy flow from reaching the target, thereby injuring personnel or damaging equipment. Of course, in any system there may be many energy sources, many barriers, and many targets.

Energy sources can be identified with the help of an energy checklist. Energy sources can include electrical, mechanical, radiological, sound, and so on. Barriers can be "hard" barriers such as walls, shielding, insulation, and so on. or they can be "soft" barriers such as procedures, training, warnings, and so on.

The process begins with the identification of all the energy sources in the system. Next, the energy flow paths within the system are identified. Then, the analyst must identify the targets and their vulnerability to the energy source. Next, the safety barriers are, or should be, between the energy source and the target. Also, at this point, the evaluation of the adequacy of the barriers must be assessed. This would include the effectiveness of the barriers as well as what happens in the event of barrier failures. Next the risk to the target with and without the barrier(s) is assessed. At this point, a determination of the adequacy

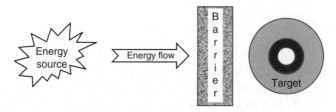

Figure 10.5 Barrier separates energy source from target

of the barriers and the acceptability of the risk is made. If the barriers prove to be inadequate for the purpose of protecting the target or the risk is unacceptable, then recommendations must be made to improve the safety by improving the barriers, adding more barriers, removing or reducing the energy source, or providing more analysis.

10.2.4.3 Worksheet

A worksheet is usually used to document BA. A typical format for the BA is shown in Figure 10.6.

Barrier Analysis										
System: _____										
Preparer: _____										
Date: _____										
Energy Source	Hazard	Target	IRAC	Risk	Barrier	Recommended Action	FRAC	Risk	Cmts	Status

Figure 10.6 Barrier analysis format

Column descriptions:
Header information: Self-explanatory. Any other generic information desired or required by the program could be included.
Energy source: This is the type of energy source present (e.g., flammable gas) and the amount, location, and any other useful descriptive information.
Hazard: This is the specific potential hazard that has been identified (e.g., fire, explosion).
Target: This describes the target or targets that can be affected by the energy source and result in a mishap (e.g., personnel, equipment, environment, etc.).
IRAC: This is a preliminary qualitative assessment of the risk.
Risk: This is a qualitative measure of the risk for the identified hazard, given that no mitigation techniques are applied.
Barrier: This is a description of all the hard and/or soft barriers that are in place or need to be added.

Recommended action: Provide recommended preventive measures to eliminate or mitigate the identified hazards. Hazard mitigation methods should follow the preferred order of precedence provided earlier in a previous chapter of this book.

FRAC: This is a final qualitative assessment of the risk, given that mitigation techniques and safety requirements identified in "Recommended Action" are applied to the hazard.

Comments: This column should include any assumptions made, recommended control, requirements, applicable standards, actions needed, and so on.

Status: States the current status of the hazard (open, monitored, closed).

10.2.4.4 Advantages/Disadvantages

The BA technique is a relatively simple method to grasp and use, and it is quick, effective, and inexpensive.

The BA is limited by the analyst's ability to identify all the energy sources present. And it does not identify all hazards that might be present in the system, but only those associated with identified energy sources.

10.2.5 Bent Pin Analysis

BPA, sometimes referred to as Cable Failure Matrix Analysis (CFMA), is a specialized analysis technique that is used to analyze all possible combinations of failures that can occur within a cable connector due to pins being bent. Any connector presents the potential for bent pins to occur, typically during disconnecting/connection activities involved with assembly or maintenance operations. Unwanted connections of pins due to bending can cause malfunctions of the system, inadvertent operations, or hazardous results.

10.2.5.1 Purpose

The purpose of BPA as a technique is to systematically determine the susceptibility of a system to bent pins in a connector. While the technique is often used as a reliability tool to identify all types of failures that can be attributed to bent pins within the connector, it has real value to system safety in the analysis of safety-critical circuits within the connector.

10.2.5.2 Process

The first step in conducting a BA is to develop a matrix of all possible bent pin combinations. The pin length and pin spacing should be considered when determining the credible pin-to-pin connections that are possible. Note that there are some combinations of pins that are not feasible because the particular bent pin cannot reach another pin or the case. These combinations are therefore excluded from the analysis. See Figure 10.7 for an example matrix.

Next, the analyst uses the information about each pin (i.e., voltage present, signal, purpose, etc.) to determine the results of each feasible combination of pin-to-pin or pin-to-case (ground) connection. Each result is evaluated to determine if the combination presents a hazard. If so, the risk associated with that hazard is assessed.

BPA usually only considers pin-to-pin rather than multiple pins bending and contacting a third pin. Multiple pin bending is relatively rare, but may need to be considered in designs involving very high consequences.

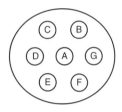

Pin	A	B	C	D	E	F	G	Case
A		X	X	X	X	X	X	
B			X				X	X
C				X				X
D					X			X
E						X		X
F							X	X
G								X

Figure 10.7   Connector configuration and matrix of possible pin-to-pin combinations

10.2.5.3 Worksheet

A typical worksheet format for the BPA is shown in Figure 10.8.

	Bent Pin Analysis											
	System: _____ Preparer: _____ Date: _____											
Identifier	Bent Pin Combination	Pin Data	Circuit Condition	Effects	Hazard	IRAC	Risk	Recommended Action	FRAC	Risk	Cmts	Status

Figure 10.8   Bent pin analysis format

Column descriptions:

Header information: Self-explanatory. Any other generic information desired or required by the program could be included.

Identifier: This can simply be a number (e.g., 1, 2, 3, ... as hazards are listed). Or, it could be an identifier that identifies a particular subsystem or piece of hardware that will be in the system with a series of numbers as multiple hazards are identified with that particular item (e.g., motor 1, motor 2, etc.).

Bent pin combination: This is an identification of the combination of bent pins being analyzed.

Pin data: This is information regarding what is present of the pin (e.g., data, voltage, etc.).

Circuit condition: This column indicates whether the result of the pin-to-pin (or case) connection results in a short that has an effect either upstream or downstream or an open that has an effect downstream.

Effects: This is a short description of the possible resulting effect(s) resulting from the bent pin contact.

Hazard: This is the specific potential hazard that results from the bent pin contact.

IRAC: This is a preliminary qualitative assessment of the risk.

Risk: This is a qualitative measure of the risk for the identified hazard, given that no mitigation techniques are applied.

Recommended action: Provide recommended preventive measures to eliminate or mitigate the identified hazards. Hazard mitigation methods should follow the preferred order of precedence provided earlier in this chapter.

FRAC: This is a final qualitative assessment of the risk, given that mitigation techniques and safety requirements identified in "Recommended Action" are applied to the hazard.

Comments: This column should include any assumptions made, recommended control, requirements, applicable standards, actions needed, and so on.

Status: States the current status of the hazard (open, monitored, closed).

10.2.5.4 Advantages/Disadvantages

The BPA technique is not particularly complicated or hard to learn, but the process of reviewing each pin on each connector within a system can be laborious and tedious. Although a time-consuming process, it is a necessary one as part of a thorough risk assessment.

10.3 Other Analysis Techniques

There are many techniques used in system safety analysis. We have covered the most popular ones in this and the previous three chapters. Here is a brief summary of some other techniques the reader might come across. These techniques are less often used and are typically used only in very special situations.

10.3.1 Petri Nets

Petri Net Analysis (PNA) is used to model systems, subsystems, and so on at an abstract level. It is a mathematical technique that uses graphical symbols to describe the system being analyzed. The technique uses state transitions within the model as changes to the system occur. It is best used for software-intensive systems and can also include hardware and humans in the modeling. It is used to analyze properties like reachability. That is, hazardous states are included in the model, and the analysis can determine whether that hazardous state can be reached. It can also be used to help understand recoverability and fault tolerance and to analyze software timing concerns. Because Petri nets can be defined in mathematical terms, the analysis can be automated.

PNA is best applied to a system very early during concept development. It is always best to identify software safety issues early in the design process.

PNA is one of the more difficult techniques to learn and use. A graduate-level understanding of mathematics and computer science is needed to be able to effectively apply the technique. The technique also requires a very detailed knowledge of the process that is to be modeled. Because of the tedious nature of this particular analysis, applying it to large complex systems becomes very cumbersome and costly.

10.3.2 Markov Analysis

MA is a statistical technique that uses state transition diagrams to describe the operational states and failed states of the system, subsystem, or component. The model is used to calculate the probability of reaching the various states of the system. It is used to study timing, failure, and repair in a system. The processes modeled are random processes where the future state is dependent only on the present state. Once the graphical version of the state diagram is drawn, it can be translated into a series of differential equations from which probabilities can be determined.

The MA analysis starts with drawing the state diagram. The state is represented by a circle. A connecting line with an arrowhead is used to represent the transition between two states. Each line can have either a failure rate (λ) or a repair rate (μ) associated with it. To illustrate, Figure 10.9 shows a Markov model for a one-component system with repair.

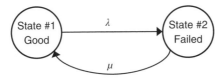

Figure 10.9 Markov state transition model for one-component system with repair

As the system grows, the number of states and state transitions grows quickly. For example, for a two-component system with repair, there would be states for each combination of component failures and transitions between all of them, and more equations would be required to solve the problem. It is easy to see how the model would get cumbersome very quickly. The models developed to support MA become large and unwieldy for complex systems.

MA does not identify hazards. It is used to model state transitions that can provide insight into the operation of the system. Therefore the use of MA in system safety is somewhat limited. It is good for early evaluation of concepts or for critical subsystems that need detailed analysis of timing, failures, and repairs.

10.3.3 Management Oversight Risk Tree (MORT)

The MORT concept of was developed in the early 1970s by the US Atomic Energy Commission (now Nuclear Regulatory Commission) [6]. While it is used predominately for accident investigation, it can also be used as a hazard analysis technique. MORT is a root cause analysis procedure that is used in determining causes and contributing factors of accident or mishaps. MORT typically focuses on energy-related hazards. Therefore BA is an essential preparation for MORT analysis.

The analyst starts with a predefined logic tree. This tree "…is simply a diagram which arranges safety program elements in an orderly and logical manner. It presents a schematic representation of a dynamic, idealized (universal) safety system model using Fault Tree Analysis methodology" [7]. The top of the predefined tree is shown in Figure 10.10.

This predefined tree contains "…over 1,500 'basic events' (i.e., causative problems or preventive measures related to an ideal safety system). These, in turn, underlie nearly 100 different generic problems identified in successively broader areas of management and accident prevention" [7].

A complete, modern version of the MORT chart can be found in *MORT— Management Oversight and Risk Tree* by the Noordwijk Risk Initiative Foundation [8].

Using the predesigned tree as a guide, the analyst draws an accident or system-specific new tree using the FTA rules with color-coding specific to MORT. The color codes are green when there is no problem, red when there is a problem, and blue when more data is needed to complete the analysis. Items that are not applicable are crossed out. The process is repeated until all items have complete data and are completely analyzed. Risk is assessed where hazards have been identified.

The primary advantage of MORT lies in its use for accident/incident investigation because the predefined tree provides a systematic way of evaluating

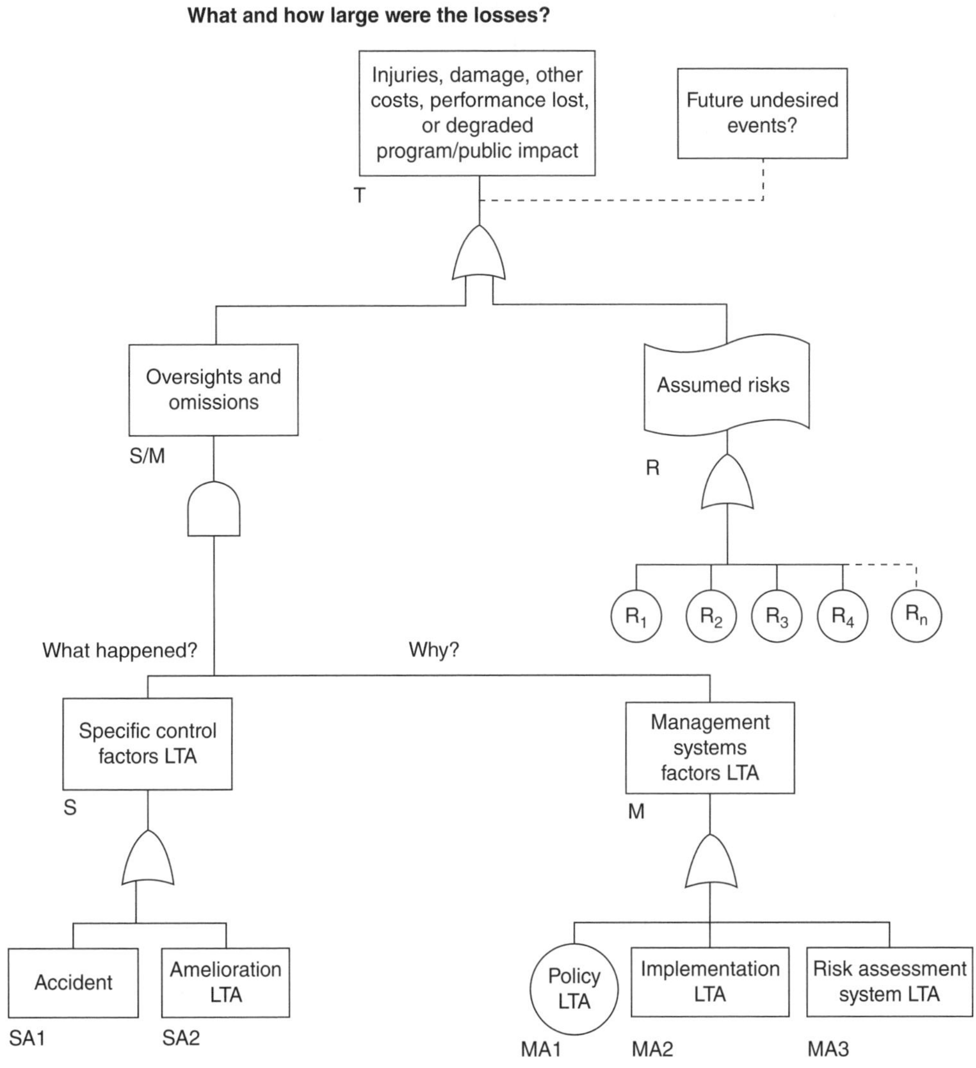

What and how large were the losses?

Figure 10.10 MORT top events [7]

all the factors that might contribute to an accident, control as well as management failings, or oversights. A disadvantage to using MORT for either accident investigation or hazard analysis is the fact that it is cumbersome, tedious, time-consuming, and, therefore, costly.

10.3.4 System-Theoretic Process Analysis

The STPA technique is a relatively new method developed by Nancy Leveson, MIT professor and author of two popular books on system safety: *Safeware: System Safety and Computers* and *Engineering a Safer World* [9].

STPA was developed in order to address more complex, software-intensive, socio-technical systems not typically covered by other techniques. New causal factors can be included using STPA including software flaws, component interaction accidents, cognitively complex human decision-making errors, and social, organizational, and management factors contributing to accidents [9].

STPA uses a functional control diagram and the requirements, the system hazards, and the safety constraints and safety requirement for the component. It has two main steps [9]:

Step 1. Identify the potential for inadequate control of the system that could lead to a hazardous state.

a. A control action required for safety is not provided or not followed.
b. An unsafe control action is provided.
c. A potentially safe control action is provided too early or too late, that is, at the wrong time or in the wrong sequence.
d. A control action required for safety is stopped too soon or applied too long.

Step 2. Determine how each potentially hazardous control action identified in Step 1 could occur.

a. For each unsafe control action, examine the parts of the control loop to see if they could cause it. Design controls and mitigation measures if they do not already exist or evaluate existing measures if the analysis is being performed on an existing design. For multiple controllers of the same component or safety constraint, identify conflicts and potential coordination problems.
b. Consider how the designed controls could degrade over time and build in protection, including the following:
   i. Management of change procedures to ensure safety constraints are enforced in planned changes.

ii. Performance audits that were the assumptions underlying the hazard analysis are the preconditions for the operational audits and controls so that unplanned changes that violate the safety constraints can be detected.

iii. Accident and incident analysis to trace anomalies to the hazards and to the system design.

The information provided in the first step of STPA can be used to identify the necessary constraints on component behavior to prevent the identified system hazards, that is, the safety requirement. In the second step of STPA, the information required by the component to properly implement the constraint is identified as well as additional safety constraints and information necessary to eliminate or control the hazards in the design or to design the system properly in the first place. This second step identifies the scenarios leading to the hazardous control actions that violate the component safety constraints [9].

Once the causal analysis is completed, each of the causes that cannot be shown to be physically impossible must be checked to determine whether they are adequately handled in the design (if the design exists) or design features added to control them if the design is being developed with support from the analysis [9].

An additional step in the STPA is to consider how the designed controls could degrade over time and to build in protection against it. To accomplish this protection, degradation mechanisms must be identified and mitigated, performance audits could be used, and/or management of change procedures should be developed. If design changes are made, the STPA must be updated. If accidents or incidents occur with the system, the STPA should be updated to determine why the controls were not effective. If humans are part of the system, they can be treated in the same way as automated components in Step 1 of STPA. The causal analysis and detailed scenario generation for human controllers, however, is much more complex. If humans are included, they require an additional process model [9].

System dynamics modeling is used to show the relationship among the contextual factors and unsafe control actions and the reasons the safety control structure migrated toward ineffectiveness over time. Most modeling techniques provide only direct relationships (arrows), which are inadequate to understand the indirect relationships between causal factors. System dynamics provides a way to show such indirect and nonlinear relationships. For this effort, system dynamics models [are] created to model the contextual influences on the behavior of each component. Then the models [are] combined to assist in understanding the behavior of the system as a whole and the interactions among the components [9].

The primary advantage of this technique is that it can deal with more complex, software-intensive, socio-technical systems not typically covered by other techniques. It is, however, new and still being developed, and, if being used to analyze organizational, management, or social systems, it requires knowledge of system dynamics [9].

References

[1] United States. Department of Defense (1991) *Reliability Prediction of Electronic Equipment*, MIL-HDBK-217F, Department of Defense, Washington, DC.

[2] Rankin, J. P. and White, C. F. (1970), *Sneak Circuit Analysis Handbook*, Boeing D2-118341-1, Boeing Company, Houston, TX.

[3] Miller, J. (1989) *Sneak Circuit Analysis for the Common Man*, Rome Air Development Center, U.S. Air Force Systems Command, Griffiss Air Force Base, NY.

[4] Blanchard, B. S. and Fabrycky, W. J. (1998) *Systems Engineering and Analysis*, Prentice Hall, Upper Saddle River, NJ.

[5] Adam Scharl, Kevin Stottlar, and Rani Kady (2014) NAVSEA NSWCDD-MP-14-00380, Functional Hazard Analysis (FHA) Methodology Tutorial, Presented at International System Safety Training Symposium, St. Louis, MO, August 4–8, 2014., http://issc2014.system-safety.org/83_Functional_Hazard_Analysis_Common%20Process.pdf. (Accessed August 24, 2017).

[6] Johnson, W. G. (1973) *The Management Oversight and Risk Tree: MORT*, U.S. Atomic Energy Commission, Washington, DC.

[7] United States. Department of Energy, (1992) *MORT User's Manual*, U.S. Department of Energy, Washington, DC.

[8] The Noordwijk Risk Initiative Foundation, MORT: Management Oversight and Risk Tree, 2002, http://nri.eu.com/NRI2.pdf (Accessed August 9, 2017).

[9] Leveson, N. G. (2011) *Engineering a Safer World*, The MIT Press, Cambridge, MA.

Suggestions for Additional Reading

Ericson, C. A. (2005) *Hazard Analysis Techniques for System Safety*, John Wiley & Sons, Inc., Hoboken, NJ.

Leveson, N. G. (1995) *Safeware: System Safety and Computers*, Addison-Wesley Publishing Company, New York, NY.

Raheja, D. and Allocco, M. (2006) *Assurance Technologies Principles and Practices*, John Wiley & Sons, Inc., Hoboken, NJ.

Raheja, D. and Gullo, L. J. (2012) *Design for Reliability*, John Wiley & Sons, Inc., Hoboken, NJ.

Stephans, R. A. (2004) *System Safety for the 21st Century*, John Wiley & Sons, Inc., Hoboken, NJ.

11

Process Safety Management and Analysis

Jack Dixon

11.1 Background

The basis of Process Safety Management (PSM) is a continuous effort to prevent catastrophic accidents involving hazardous processes that involve dangerous chemicals and energies. It applies management principles and analytic techniques to reduce risks to processes during the manufacture, use, handling, storage, and transportation of chemicals. A primary focus of PSM is on hazards related to the materials and energetic processes present in chemical production facilities. PSM could also be a primary focus in facilities that handle flammable materials, high voltage devices, high current load devices, and energetic materials, such as rocket motor propellants.

PSM is addressed in specific standards for the general and construction industries. The US Occupational Safety and Health Administration (OSHA) was created in 1970 to ensure safe and healthy working conditions for workers by establishing safety and health-related standards and enforcing them. It has issued the PSM of Highly Hazardous Chemicals standard (29 CFR 1910.119) [1], which contains requirements for the management of hazards associated with processes using highly hazardous

Design for Safety, First Edition. Edited by Louis J. Gullo and Jack Dixon.
© 2018 John Wiley & Sons Ltd. Published 2018 by John Wiley & Sons Ltd.

chemicals. OSHA also provides training, education, and assistance to the industry. After a series of major catastrophic incidents involving chemical plants, it began its foray into chemical process safety in the early 1980s. The largest precipitating event occurred in 1984 with the release of Methyl Isocyanate (MIC) at Union Carbide's Bhopal, India, facility, which resulted in over 3800 deaths and thousands of injuries [2]. This incident got the attention of the entire world and of OSHA. After this incident and another similar accident in 1985, involving the same chemical at a Union Carbide plant in West Virginia that injured 135 people, OSHA began inspecting all US facilities that produced MIC. These inspections led OSHA to realize that current standards were insufficient and better management systems, work practices, and protective systems were needed.

Accidents continued to happen in other chemical plants. The following is a list of similar chemical accidents during the three-year time period, 1988–1991, in different US locations, with the number of fatalities for each accident:

- 1988—7 deaths in Louisiana
- 1989—23 deaths in Texas
- 1990—17 deaths in Texas
- 1991—5 deaths in Louisiana
- 1991—8 deaths in Louisiana
- 1991—9 deaths in South Carolina

In 1990, OSHA proposed the PSM standard to respond to these accidents. Later that year the congress passed the amendments to the Clean Air Act that required OSHA to promulgate a rule on chemical process safety. The final rule, 29 CFR 1910.119 "Process Safety Management of Highly Hazardous Chemicals," was published in February 1992 and provides requirements for preventing or minimizing the consequences of catastrophic releases of toxic, reactive, flammable, or explosive chemicals. These releases may result in toxic, fire, or explosion hazards [1].

The PSM standard applies to processes with more than a specified quantity of a highly hazardous chemical, a flammable liquid, or a flammable gas. While PSM is a legal requirement for certain hazardous chemicals in certain quantities, the process can be applied to any processing situation.

11.2 Elements of Process Safety Management

The major objective of process safety management of highly hazardous chemicals is to prevent unwanted releases of hazardous chemicals especially into locations which could expose employees and others to serious hazards. An effective process safety management program requires a systematic approach to evaluating the whole process. Using this approach the process design, process technology, operational and maintenance activities and procedures, non-routine activities and procedures, emergency preparedness plans and procedures, training programs, and other elements which impact the process are all considered in the evaluation. ...Process safety management is the pro-active identification, evaluation and mitigation or prevention of chemical releases that could occur as a result of failures in process, procedures or equipment. [3]

There are 14 elements established by OSHA that are critical to the success of PSM:

1. **Employee participation.** Employers must develop a written plan for the implementation of employee participation. The employers must consult with employees on the conduct of Process Hazard Analyses (PrHAs) and the other elements of PSM and must provide them access to the PrHAs and to all other information developed for the PSM standard.

2. **Process safety information.** Complete and accurate written information concerning process chemicals, process technology, and process equipment is essential to an effective PSM program and to a PrHA. This information should include information on the chemicals, the process intermediates, fire and explosion characteristics, reactivity hazards, safety and health hazards, corrosion and erosion effects, Material Safety Data Sheets (MSDS), inventory levels, block flow diagrams, process flow diagrams, piping and instrument diagrams, applicable codes and standards, and any deviations from the codes and standards. See Figure 11.1 for an example of process flow diagram [4].

3. **PrHA**
   A Process Hazard Analysis (PrHA), sometimes called a process hazard evaluation, is one of the most important elements of the process safety management program. A PrHA is an organized and systematic effort to identify and analyze the significance of potential hazards associated with the processing or handling of highly hazardous chemicals. A PrHA provides information which will assist employers and employees in making decisions for improving safety and reducing the consequences of unwanted or unplanned releases of hazardous chemicals. A PrHA is directed toward analyzing potential causes and consequences of fires, explosions, releases of toxic or flammable chemicals and major spills of hazardous chemicals. The PrHA focuses on equipment, instrumentation, utilities, human actions (routine and non-routine), and external factors that might impact the process. These considerations assist in determining the hazards and potential failure points or failure modes in a process. [3]

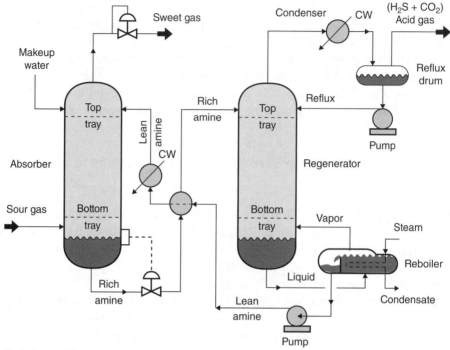

Typical operating ranges

Absorber : 35–50°C and 5–205 atm of absolute pressure
Regenerator : 115–126°C and 1.4–1.7 atm of absolute pressure
at tower bottom

Figure 11.1 Example process flowchart [4]

(*Note: The term process hazard analysis is often abbreviated as PHA, but since the acronym PHA is more commonly used for preliminary hazard analysis, which we discuss in Chapter 7, we will use PrHA as the acronym for process hazard analysis.*)

4. **Operating procedures.** Operating procedures are often referred to as the Standard Operating Practices (SOPs). These should describe all tasks that must be performed, the data that is to be recorded, what operating conditions are to be maintained, what samples are to be collected, and the safety and health precautions that must be taken to maintain worker safety and public safety. If changes are made to a process, the operating procedures must be updated to document the changes. SOPs should state what is done. If the tasks change, then SOPs could be informally revised immediately with pencil and ink changes (also known as red-lined changes) and the formal SOP document revised at a later date following the formal draft update, reviews, and approvals.

5. **Training.** Training must include all employees, maintenance personnel, and contractor employees that are involved with highly hazardous chemicals. The training must impart to the trainee the full understanding of the safety and health hazards of the chemicals and processes involved. This understanding is critical for the protection of trainees, their fellow workers, and the public in nearby communities.
6. **Contractors.** Contract employees must also perform their work safely and must be screened, trained, and/or certified so that they can accomplish the job without compromising the safety and health of employees or the public.
7. **Prestart-up safety review.** A prestart-up safety review must be performed for new facilities and for modified facilities.
8. **Mechanical integrity.**

   Equipment used to process, store, or handle highly hazardous chemicals needs to be designed, constructed, installed and maintained to minimize the risk of releases of such chemicals. This requires that a mechanical integrity program be in place to assure the continued integrity of process equipment. Elements of a mechanical integrity program include the identification and categorization of equipment and instrumentation, inspections and tests, testing and inspection frequencies, development of maintenance procedures, training of maintenance personnel, the establishment of criteria for acceptable test results, documentation of test and inspection results, and documentation of manufacturer recommendations as to meantime to failure for equipment and instrumentation. [3]

9. **Hot work permit.** The employer must issue a hot work permit for any hot work operations conducted on or near a covered process. (Note: Hot work is any work that involves burning, welding, and using fire- or spark-producing tools or that produces a source of ignition when flammable material is present.)
10. **Management of change.** Change includes all modifications to equipment, procedures, raw materials, and processing conditions other than "replacement in kind." These changes must be properly managed by identifying and reviewing them prior to implementation of the change.
11. **Incident investigation.** Incident investigation is the process of identifying the underlying causes of incidents and implementing steps to prevent similar events from occurring. The intent of an incident investigation is to learn from past experiences and avoid repeating past mistakes. The incidents that employers should investigate are the types of events that result in or could reasonably have resulted in a catastrophic release. Some of the events are sometimes referred to as "near misses," meaning that a serious consequence did not occur, but the possibility exists that it could have occurred and is possible to occur in the future [3].

12. **Emergency planning and response.** The employer must identify what actions employees are to take when there is an unwanted release of highly hazardous chemicals. These plans should include consideration of evacuations, evacuation routes, alarms, emergency response organizations and procedures, how to deal with minor emergencies or incidental releases, training needs, mutual aid agreements, and coordination and communication with outside emergency responders.

13. **Compliance audits.** Employers need to select a trained individual or assemble a trained team of people to audit the PSM system and program. The audit should include an evaluation of the design and effectiveness of the PSM system and a field inspection of the safety and health conditions and practices to verify that the employer's systems are effectively implemented [3]. Appropriate corrective actions must be taken to resolve findings of the compliance audits.

14. **Trade secrets.** All information necessary to demonstrate compliance with PSM requirements must be made available upon request. While the protection of trade secrets is important, employers must make all relevant information available to those persons responsible for:
    a. Compiling the process safety information
    b. Assisting in the development of process hazard analyses
    c. Developing the operating procedures
    d. Conducting incident investigations, emergency planning, and response
    e. Performing compliance audits without regard to possible trade secret status of such information

Confidentiality agreements, such as Nondisclosure Agreements (NDAs) or Proprietary Information Agreements (PIAs), that warn not to disclose certain types of information may be required.

11.3 Process Hazard Analyses

While all of the fourteen elements of PSM are important, we will emphasize and focus on PrHA in this section, since it ties in directly with the system safety engineering processes discussed throughout this book.

The OSHA PSM standard states,

The employer shall perform an initial process hazard analysis on processes covered by this standard. The process hazard analysis shall be appropriate to the complexity of the process and shall identify, evaluate, and control the hazards involved in the process. Employers shall determine and document the priority order for conducting process hazard analyses based on a rationale which includes such considerations as extent of the process hazards, number of potentially affected employees, age of the process, and operating history of the process. [1]

OSHA further requires that the PrHA be performed by a team with expertise in engineering and process operations.

The PSM standard requires that every PrHA address each of the following:

- Process hazards
- Identification of previous incidents
- Engineering and administrative controls
- Consequences of failure of engineering and administrative controls
- Facility siting
- Human factors
- Qualitative evaluation of possible safety and health effects of failure of controls on employees
- Actions to be taken

OSHA suggests that one or more of the following methodologies be used to determine and evaluate the hazards of the process being analyzed:

- What-if analysis
- Checklist
- What-if/checklist
- Hazard and Operability Study (HAZOP)
- Failure Modes and Effects Analysis (FMEA)
- Fault Tree Analysis (FTA)
- An appropriate equivalent methodology

Paradigm 1: Always Aim for Zero Accidents

As with all good systems engineering practices, we aim for zero accidents in all phases of the system development process. This goal is applicable for not only the design of the system but also the design of the development processes, the manufacturing processes, and all other processes required for the installation, operation, maintenance, and sustainment of the system. A very valuable method for systems engineering that is applied by safety engineers for PrHA is the what-if analysis to ensure zero accidents during the course of process implementation and execution. What-if analysis requires that certain types of questions are asked and answers are determined for each question, which may be grouped into categories. One example of a category of questions is the what-if scenario, which is a speculation about how a set of circumstances might occur and how to react to those circumstances. For example, a safety engineer might ask a system designer how the system will respond to a given situation involving a set of inputs, operational conditions, and a sequence of events that require intervention from a human. If the goal is to find the situations that can cause accidents, the various types of scenarios will be explored in the what-if scenario analysis considering all possible sets of inputs, operational conditions, and sequences of events to determine if the scenario could lead to an accident where humans are harmed.

11.3.1 What-If Analysis

What-if analysis involves a group of experts brainstorming what could happen if various things go wrong or deviate from the design of the process. It can be applied to any process and at any stage in the life of the process. Examples of what-if questions include

What if chemical tank containing X ruptures? (X represents a type of hazardous chemical.)
What if the seals of pump Y leak? (Y represents a unique designator for a particular pump.)
What if valve Z fails closed? (Z represents a unique designator for a particular valve.)

The brainstorming process should be structured by the subject, process flow, or some other means; otherwise the "what-if" questions might be random and incomplete. Another obvious limitation to the "what-if" analysis techniques is that its results are dependent upon the experience of the participants.

A typical format for a "what-if" analysis is shown in Figure 11.2.

What-If Analysis				
Process: _____				
Participants: _____				
Date: _____				
What-if	**Consequence**	**Safety Controls**	**Scenario**	**Comments**

Figure 11.2 What-if analysis worksheet

Column descriptions:

Header information: Self-explanatory. Any other generic information desired or required to describe the process or system could be included.

What-if: This is a statement of "what happens if something is wrong?" (e.g., "what-if"…wrong chemical is delivered to processing tank 1? What if valve X is closed when it should be open?).

Safety controls: This is a description of what should be in place to mitigate the problem identified (e.g., proper material handling procedure, temperature monitor, etc.).

Scenario: The scenario describes the sequence of events that could occur as a result of the "what-if."

Comments: This column should include any assumptions made, recommended control, requirements, applicable standards, actions needed, and so on.

11.3.2 Checklist

Checklists are probably one of the easiest and most versatile methods that can be applied to a process in any stage of its like. They are commonly used to assess compliance with codes and standards. In this case, checklists can be used as an aid for identifying hazards in the process and for this purpose should be applicable to and tailored for the specific industry, process, or facility. Checklists for hazard identification can start with existing checklists from various sources and/or then be tailored to apply to the particular process at hand.

Checklists have limitations. If they are borrowed from a handbook, their applicability may be questionable or incomplete. The quality and completeness of their creation is limited by the expertise and experience of the people creating them. And the checklist may help to indicate that particular hazards exist, but they shed no light on the accident scenario that may result from the hazard. See Chapter 6, Appendix A, and Appendix B of this book for a more detailed coverage of checklists.

11.3.3 What-If/Checklist Analysis

The what-if/checklist analysis combines the brainstorming features of the what-if analysis with the more systematic checklist approach. The team can use the what-if analysis to brainstorm the types of accident scenarios that might occur and then use checklists to fill in any gaps. The what-if analysis worksheet along with completed checklists can be used to document the results of the what-if/checklist analysis.

11.3.4 Hazard and Operability Study

HAZOP is an analytical technique for identifying and analyzing hazards and operational concerns. The technique was first developed for the chemical industry, but it can be, and has been, applied to many diverse types of systems. HAZOP is used to identify deviations from the intended design.

HAZOP is conducted by a team. Members of the team composed of experts from various disciplines such as engineering, operations, safety, HAZOP process, and so on. By using a team, the analysis process is enhanced since the various team members bring different perspectives to the analysis so that creativity and new ideas can flow. In HAZOP analysis, the team focuses on specific points of a process called "study nodes," process sections, or operating steps. These are points in the system that have distinct boundaries (e.g., a pipe between two tanks) where deviations from the design might occur. The team uses "guide words" to assist in analyzing the causes and consequences of deviations from the design. These guide words are applied at the study nodes and are combined with specific process parameters to help identify any potential deviations. A couple of examples of using guide words combined with parameters to identify deviations area shown in Table 11.1.

Table 11.1 Examples of guide word–parameter–deviations in HAZOP

Guide word	Parameter	Deviation
No	Flow	No flow in pipe X
Less	Temperature	Low temperature in reactor vessel
More	Pressure	Pressure in vessel A causes rupture

HAZOP must be performed in a systematic manner to reduce the possibility of omissions. At each study node the team must consider all deviations that could happen with each process parameter. All possible deviations for each node must be analyzed before the team moves on to the next node.

Documentation for a HAZOP analysis typically uses a tabular format similar to what we have seen in other analyses techniques. A suggested format is shown in Figure 11.3.

HAZOP Analysis

System: _____

Participants: _____

Date: _____

Identifier	Item	Parameter	Guide Words	Causes	Consequences	Recommended Action	Comments

Figure 11.3 HAZOP analysis worksheet

Column descriptions:

Header information: Self-explanatory. Any other generic information desired or required to describe the process or system could be included.

Identifier: This can simply be a number (e.g., 1, 2, 3, ... as hazards are listed). Or, it could be an identifier that identifies a particular subsystem or piece of hardware, study node, and so on that will be in the system with a series of numbers as multiple hazards are identified with that particular item (e.g., motor 1, motor 2, etc.).

Parameter: This is a physical or chemical property associated with the process (e.g., temperature, pressure, etc.).

Guide words: These are the simple words used by the team to provide clues and stimulate brainstorming to identify process hazards (e.g., no, more, less, etc.).

Causes: This should be a brief description of why a deviation may occur (e.g., loss of containment, loss of power, etc.).

Consequences: This is a short description of the possible results of a deviation.

Recommended action: Provide recommended preventive measures to eliminate or mitigate the identified hazards. Hazard mitigation methods should follow the preferred order of precedence provided earlier in the book.

Comments: This column should include any assumptions made, recommended control, requirements, applicable standards, actions needed, and so on.

HAZOP can result in very thorough analysis, but it can be extremely time consuming on a large system.

11.3.5 Failure Modes and Effects Analysis

FMEA is used to determine the effects of various failure modes of components, subsystems, or functions may have on a system or process. The reader is referred to Chapter 8 (*Failure Modes, Effects, and Criticality Analysis for System Safety*) for a complete discussion of FMEA. Especially relevant to the evaluation of hazardous processes is the section on process FMECA.

11.3.6 Fault Tree Analysis

FTA is a deductive method that uses Boolean logic symbols (i.e., AND gates, OR gates) to identify the combination of basic failures that cause an undesirable event (top event). The basic failures can be component failures, equipment failures, or human errors. FTA begins with the top event and identifies the causes and their logical relationships between the causes and the top event. The resulting fault tree is a graphical representation of the relationships between basic events and the top undesirable event. A separate fault tree must be developed for each top event. The reader is referred to Chapter 9 (*Fault Tree Analysis for System Safety*) for a complete discussion of FTA.

11.3.7 Equivalent Methodologies

Other methods of performing hazard analysis can be used in place of or as a supplement to the analyses prescribed by the OSHA PSM standard. Many of these techniques are described in detail elsewhere in this book. The following are other methods that might be considered when analyzing hazardous chemical processes:

- From Chapter 7 (*System Safety Hazard Analysis*):
  - Preliminary Hazard Analysis (PHA)
  - Subsystem Hazard Analysis (SSHA)
  - System Hazard Analysis (SHA)
  - Operating & Support Hazard Analysis (O&SHA)
- From Chapter 10 (*Complementary Design Analysis Techniques*):
  - Event Tree Analysis (ETA)
  - Functional Hazard Analysis (FuHA)
  - Barrier Analysis (BA)

11.4 Other Related Regulations

While the intent of this chapter is not to cover every Environmental, Safety, and Heath (ESH) regulation, there are several regulations that relate to the general discussion of process safety and provide insight into their history and interrelationships. They will be briefly described in this section with recommended references provided for further reading.

11.4.1 US Legislation

Prior to the 1970s, there were a number of laws passed that attempted to regulate air and water pollution. The early water-related legislation attempted, but not too successfully, to address the concerns presented by accidents. This early legislation consisted of the following:

- The *Rivers and Harbors Act of 1899* was directed toward water pollution prevention.
- The *Oil Pollution Act of 1924* prohibited the discharge of oil into navigable waters.
- The *Federal Water Pollution Control Act of 1948* established a national policy for the prevention, control, and abatement of water pollution.
- The *Oil Pollution Act of 1961* regulated the discharge of oil into the water.

- The *Clean Water Restoration Act of 1966* extended coverage of discharge pollution from only coastal waters to rivers, too.
- The *Solid Waste Disposal Act of 1965* began to address the problems associated with increasing amounts of solid wastes.

More significant progress began to be made on health, safety, and emergency response during the 1970s with the passage of the *Clean Air Act* and the *Clean Water Act*. Concerns regarding hazardous and toxic chemicals led to the passage of the *Resource Conservation and Recovery Act (RCRA) of 1976* and the *Comprehensive Environmental Response, Compensation, and Liability Act (CERCLA) of 1980*. RCRA addressed the conservation and recovery of materials through recycling. It also requires contingency planning to minimize hazards to health and environment from fire, explosions, and unplanned releases of hazardous waste. CERCLA addressed abandoned hazardous waste sites, known as Superfund sites.

Also in the early 1970s, both OSHA and the Environmental Protection Agency (EPA) were created. OSHA issued the PSM standard, which has been the major focus of this chapter. In addition, the EPA, under provisions of the Clean Air Act, developed the Risk Management Program (RMP) regulations that are designed to reduce the risk of accidental releases of toxic, flammable, and explosive chemicals.

The EPA has issued 40 CFR 68, *Chemical Accident Prevention Provisions* [5]. This regulation contains requirements similar to the OSHA PSM standard. Any facility that stores or uses a certain hazardous chemicals of certain amounts must comply not only with OSHA's PSM regulations but also with very similar US EPA RMP regulations. The EPA has published a model RMP plan for an ammonia refrigeration facility [6] that provides excellent guidance on how to comply with either OSHA's PSM regulations or EPA's RMP regulations. While this regulation is similar to PSM, it also emphasizes the prevention of accidental releases with requirements of worst-case release scenario analysis, alternative release scenario analysis, and off-site consequence analysis. [7, 8]

Later, the Pollution Prevention Act of 1990 was issued. Its focus was to provide a pollution prevention hierarchy to reduce the generation of waste, pollutants, and chemicals that can create health, safety, and accident management problems. This hierarchy was declared

...to be the national policy of the United States that pollution should be prevented or reduced at the source whenever feasible; pollution that cannot be prevented should be recycled in an environmentally safe manner, whenever feasible; pollution that cannot be prevented or recycled should be treated in an environmentally safe manner whenever feasible; and disposal or other release into the environment should be employed only as a last resort and should be conducted in an environmentally safe manner. [9]

11.4.2 European Directives

The Seveso Directives are the main European Union (EU) legislation dealing specifically with the control of on-shore major accident hazards involving dangerous substances. They are named after the Seveso disaster in Seveso, Italy, in 1976, which resulted in the highest known exposure to dioxin. The Seveso III Directive became effective on June 1, 2015, replacing the older Seveso II Directive. These regulations are known as Control of Major Accident Hazards (COMAH) regulations in the United Kingdom and are enforced by the Health and Safety Executive (HSE) in the United Kingdom. The intent of the regulation is the prevention of major accidents involving dangerous substances and the mitigation of their effects on people and the environment.

11.5 Inherently Safer Design

Process safety strives to minimize risk through a variety of methods that attempt to reduce the hazardous effects of a process and the likelihood of an accident. The previous sections in this chapter detail this approach. However, there is another related approach that deserves some attention, which is Inherently Safer Design (ISD). The main proponent for ISD has been Trevor Kletz. His paper, *What you don't have, can't leak*, published in 1978, codified the principle of ISD [10]. Inherent safety is associated with the intrinsic properties of the process (e.g., the use of reduced amounts of chemicals, the substitution of safer chemicals, etc.). The essence of ISD lies in the avoidance and removal of hazards rather than in trying to control them by adding on barriers, protective systems, or procedures. While the ISD approach as developed by Kletz has not been widely adopted in the chemical process industry, the approach essentially follows the "order of precedence" presented in Chapter 7. This problem is summarized by Heikkilä [11]:

> The concept of inherently safer plant has been with us now for many years. But in spite of its clear potential benefits related to Safety, Health and the Environment (SHE), as well as the costs, there has been few applications in chemical plant design. But as Kletz [12] has written there are hurdles to be overcome. Inherently safer design requires a basic change in approach. Instead of assuming e.g. that we can keep large quantities of hazardous materials under control we have to try and remove them. Changes in belief and the corresponding actions do not come easily.

The traditional attitude in plant design is to rely much on the added-on safety systems. Reactions from industry can be expressed by two questions: "How do I know if my process is designed according to inherently safe principles?" and "Can the influence of a process change on the inherent safety of a plant be measured?" The plants are designed in a tight time schedule by using standards and so called sound engineering practice. Lutz [13] has realized that inherent safety alternatives has become a requirement in companies that understand that inherently safer plants have lower lifetime costs and therefore are more profitable. Chemical process industry in general overlooks the simplicity of designing to eliminate the hazard at the earliest opportunity. The result is controls being engineered near the end-point of the design and capitalization process. With this approach add-on systems become the only opportunities for process safety and pollution controls. Systems added late in design require continual staffing and maintenance throughout the life of the plant greatly adding to the lifetime costs as well as repetitive training and documentation upkeep.

 Paradigm 1: Always Aim for Zero Accidents

The typical approach to safety in the chemical industry, as well as other industries in general, is to reduce risk by lowering the probability of incidents and by mitigating the consequences of incidents. While this approach is important and is usually effective, it focuses on controlling hazards rather than on eliminating them. Inherent safety methods that try to eliminate hazardous chemicals or reduce the hazards are an essential approach, and they are particularly important in today's world where terrorists may attempt to cause the release of toxic or flammable chemicals by defeating the traditional safety protections.

The ISD philosophy is to permanently eliminate or reduce the hazards, thus avoiding, or reducing, the consequences of an accident. This approach applies to the entire life cycle including design, construction, operation, storage, transportation, and disposal. Any chemical process will be inherently safer if the hazards associated with the chemical and plant operations are eliminated, or reduced, and the elimination, or reduction, will be permanent; therefore the process will always be safer. It is always better to avoid hazards than try to control them.

The approaches to the design of inherently safer processes and plants can be grouped into four major strategies [14]:

1. Minimize – Reduce quantities of hazardous substances.
2. Substitute – Replace a material with a less hazardous substance.

3. Moderate – Use less hazardous conditions, a less hazardous form of a material, or facilities that minimize the impact of a release of hazardous material or energy.
4. Simplify – Design facilities which eliminate unnecessary complexity and make operating errors less likely and which are forgiving of errors that are made.

Minimize

The "minimize" strategy means that the amount of hazardous chemical used or the energy used in the process should be minimized. This could be accomplished via some new technology, by the reduction of the amount of a hazardous chemical used, and by the reduction of on-site chemical inventory through optimization of purchasing, shipping, and/or transportation scheduling.

Substitute

The "substitution" strategy relies on the replacement of hazardous materials or processes with alternate ones that are not hazardous or are less hazardous. It is important that this strategy be implemented in the design phase, since substitution at later stages is more difficult and costly. The main methods of substitution involve changing the chemistry of the reactions to less energetic or hazardous ones or by changing the chemicals, especially solvents, to less volatile choices.

Moderate

The "moderate" strategy requires the use of materials under less hazardous conditions. Moderation can be accomplished by lowering temperatures of the process, lowering energy inputs, and reducing pressures, as examples. Dilution is another means of implementing the "moderate" strategy. The "dilution" strategy reduces the hazards by lowering the boiling points of hazardous materials. This can be accomplished by reducing the pressure under which the chemical is stored or by reducing the atmospheric concentration if an accidental release does occur.

Simplify

The "simplify" strategy can be achieved by designing a simpler process by eliminating as much complexity as possible, thus reducing the chances of errors in design, in implementation, and in operation. Not only will the less complex design most likely be safer, but it will also probably be cheaper.

Paradigm 5: If the Solution Costs Too Much Money, Develop a Cheaper Solution

 Paradigm 7: Always Analyze Structure and Architecture for Safety of Complex Systems

ISD seems to be catching on.

Inherent safety has been recognised as a desirable principle by a number of national authorities, including the US Nuclear Regulatory Commission and the UK Health and Safety Executive (HSE). In assessing COMAH (Control of Major Accident Hazards Regulations) sites the HSE states "Major accident hazards should be avoided or reduced at source through the application of principles of inherent safety." The European Commission in its Guidance Document on the Seveso II Directive states "Hazards should be possibly avoided or reduced at source through the application of inherently safe practices." In California, Contra Costa County requires chemical plants and petroleum refineries to implement inherent safety reviews and make changes based on these reviews. [15]

11.6 Summary

This chapter focused mostly on PSM and the analyses associated with it. Particular attention is given to the PSM of Highly Hazardous Chemicals standard (29 CFR 1910.119). Although PSM is a legal requirement for certain regulated hazardous chemicals that are toxic, flammable, or explosive and that are above specified threshold quantities and/or flammable gasses or liquids in quantities above 10,000 lbs, the process and analysis approach can be applied to any processing situation. Likewise, the principles of ISD can be applied not only to the chemical industry but also to any process, system, or facility.

References

[1] U.S. Occupational Safety and Health Administration (OSHA) (2012) Process Safety Management of Highly Hazardous Chemicals, 29 CFR 1910.119, OSHA, Washington, DC.

[2] Long, L. (2009) History of Process Safety and loss prevention in the American Institute of Chemical Engineers, *Process Safety Progress*, 28(2), 105–113.

[3] U.S. Occupational Safety and Health Administration (OSHA) (2012) *Compliance Guidelines and Recommendations for Process Safety Management (Nonmandatory)*, Appendix C to 29 CFR 1910.119, OSHA, Washington, DC.

[4] Wikipedia Definition, https://en.wikipedia.org/wiki/Process_flow_diagram (Accessed on August 9, 2017).

[5] U.S. Environmental Protection Agency (2011) Chemical Accident Prevention Provisions, 40 CFR 68, U.S. Environmental Protection Agency, Washington, DC.

[6] U.S. Environmental Protection Agency (1996) *Model Risk Management Program and Plan for Ammonia Refrigeration*, Science Applications International Corporation, Reston, VA.

[7] U.S. Environmental Protection Agency (2009) *Risk Management Program Guidance for Offsite Consequence Analysis*, U.S. Environmental Protection Agency, Washington, DC.

[8] U.S. Environmental Protection Agency (1999) References for Consequence Analysis Methods, U.S. Environmental Protection Agency, Washington, DC.

[9] U.S. Environmental Protection Agency (1990) *Pollution Prevention Act of 1990*, U.S. G.P.O, Washington, DC.

[10] Kletz, T.A. (1978) What You Don't Have, Can't Leak, Chemistry and Industry, May 6, 1978.

[11] Heikkilä, A. (1999) Inherent Safety in Process Plant Design an Index-Based Approach, Dissertation for the degree of Doctor of Technology, Valtion teknillinen tutkimuskeskus (VTT), Technical Research Centre of Finland.

[12] Kletz, T.A. (1996) Inherently Safer Design: The Growth of an Idea. *Process Safety Progress*, 15(1), 5–8.

[13] Lutz, W.K. (1997) Advancing Inherent Safety into Methodology. *Process Safety, Progress*, 16(2), 86–88.

[14] Center for Chemical Process Safety (2009) *Inherently Safer Chemical Processes: A Life Cycle Approach*, American Institute of Chemical Engineers, New York, NY.

[15] Wikipedia Definition, https://en.wikipedia.org/wiki/Inherent_safety (Accessed on August 9, 2017).

Suggestions for Additional Reading

Center for Chemical Process Safety (2008) *Guidelines for Hazard Evaluation Procedures*, John Wiley & Sons, Inc., Hoboken, NJ.

Center for Chemical Process Safety (2012) *Guidelines for Engineering Design for Process Safety*, John Wiley & Sons, Inc., Hoboken, NJ.

Kletz, T. and Amyotte, P. (2010) *Process Plants: A Handbook for Inherently Safer Design*, CRC Press, London.

Raheja, D. and Gullo, L.J. (2012) *Design for Reliability*, John Wiley & Sons, Inc., Hoboken, NJ.

Redmill, F., Chudleigh, M., and Catmur, J. (1999) *System Safety: HAZAOP and Software HAZOP*, John Wiley & Sons, Ltd, West Sussex.

12

System Safety Testing

Louis J. Gullo

12.1 Purpose of System Safety Testing

Empirical observations, practical experiences, and physical knowledge of a system's safe performances, unsafe conditions, undesirable operations, risky problematic functional effects, and strange anomalous behaviors may emanate from system safety testing. This empirical knowledge is actual experience data that cannot be found in any other way. There is no substitute for system safety testing when collecting first-hand authentic evidence during initial and subsequent system test events run under nominal and worst-case conditions using assorted mission scenarios and use cases. A person who understands how a system performs in test and actual mission scenarios will be able to design better for safety for that particular system and to improve the design for safety if an accident, mishap, or safety-critical failure occurs later in the system life cycle.

The primary reason we perform system safety testing is to verify and validate that the system performs safely, that safety functions perform as intended, and that safety-critical design requirements were implemented correctly and safely. Electrical safe circuits (e.g., electrostatic discharge strap with a path to ground through a $1\,M\Omega$ resistor), safety devices (electrical, mechanical, electromagnetic, electromechanical, electrooptical, etc.), and functional circuits integrated with other circuits designed for safe operations are intended to work as intended in preventing mishaps and in lowering or eliminating the risk of system hazards.

Design for Safety, First Edition. Edited by Louis J. Gullo and Jack Dixon.
© 2018 John Wiley & Sons Ltd. Published 2018 by John Wiley & Sons Ltd.

The safety testing data and results provide qualitative and quantitative measures for risk assessments and risk reduction or risk elimination in hazard analysis.

While performing or witnessing demonstrations and testing of a system's operating characteristics, a person can verify and validate the system performance requirements, system safety architecture, and system safety design features and topology. When system safety testing discovers a catastrophic hazard that could cause personal harm or death when deployed in a customer use environment, the test usually pays for itself many times over. This means that the cost in performing the system safety test is justified, proving a huge Return on Investment (ROI). For instance, if the cost of planning and performing a safety test is $200,000 including the cost of a design change to correct a safety-critical fault identified during the test, and the estimated cost of damages and liabilities due to the severity of the failure effects that may occur in a customer use application is $1M, this means the ROI is 500%. This estimated cost of $1M is the potential cost that is mitigated or eliminated by performing the test and immediately changing the design to correct and prevent the safety-critical fault. The ROI from the testing could be perceived as being priceless when you attempt to put a value on human life, realizing the potential population size that is at risk, and on the multiple liabilities associated with the risk of harm to humans or loss of human life over a long period of time. In most rational minds, the ROI is a "no-brainer."

12.1.1 Types of System Safety Tests

There are different types of system safety testing. For the purposes of this book, demonstrations are considered a type of test, even though some engineering documentation may state that demonstrations and tests are considered different in certain situations and types of programs. System safety tests can be characterized as formal and informal test events. Demonstrations may be put on for show as formal events. These formal demonstrations prove system performance and safety to high-ranking officials, Very Important Persons (VIPs), and dignitaries. Demonstrations like this are one type of formal test event, which is conducted following a successful completion of many sequential formal and informal development, integration, and Operational Tests (OTs). Informal tests are performed before formal testing. Informal tests provide confidence that formal tests will be successfully conducted. There are many iterations of informal tests that occur before one formal test occurs. In informal tests, we learn how the system performs when functioning as designed, integrated with all subsystems and components and how well the system was designed when the initial system power-up test occurs. Informal tests verify how the system performs when applied under nominal and worst-case conditions (including required environmental stress conditions). We learn by experiencing faulty conditions (including mission-critical failures and safety-critical failures). We learn if the system has probabilistic nondeterministic faults by performing and repeating same test scenarios multiple times. It is perfectly OK to experience hazards and failures during informal tests, which are performed to uncover the hazards and failures before a formal test event, so if hazards and failures must occur, they should occur during informal tests. Formal tests occur to verify the system design requirements are satisfied, to meet or exceed system specifications, and to ensure the system performs as the customer intended. There should be neither hazards discovered nor failures occurring during formal test events.

We use various types of system safety testing to surface hazards early to prevent accidents later. We use a suite of tests (e.g., test regime and test utilization) to make sure the causes of test failures and hazards are eliminated and never occur again. Types of Development Testing (DT) performed by the system developer or product developer includes Design Verification Test (DVT) or Design Qualification Test (DQT), which are performed to verify the specification design requirements of the system or product before the design transitions into manufacturing and production. DVT and DQT are used to create an Acceptance Test Procedure (ATP) that is used in production testing. The ATP verifies that the product was assembled properly and usually relies on costly external Test Equipment (TE), which is designed specifically for the product to be tested in manufacturing test operations and is designed usually using Commercial Off-the-Shelf (COTS) hardware and software from suppliers that specialize in TE development for system manufacturers and Original Equipment Manufacturers (OEMs). The TE that leverages supplier COTS products integrated with OEM-designed TE is very specialized for unique product applications. This type of TE is referred as Special Test Equipment (STE). One problem with STE is that it is very costly. To eliminate the need for costly STE, TE developers realize the benefits of embedding diagnostic test functions into the system or product. TE developers that take advantage of embedded test strategy design their test circuitry and functions using different methods for self-test or Built-In-Test (BIT). Another term used in the industry is Built-in-Self-Test (BIST).

Referencing Paradigm 1 from Chapter 1, we use system safety testing to aim for zero accidents in customer applications and missions.

Paradigm 1: Always Aim for Zero Accidents

This paradigm applies when a system is deployed, launched, released to the public, and put into service. It applies during the system development process, production operations, and system testing and when conducting formal DT and OTs where the customer may or may not be involved (formal and informal test events). The goal is to find causes of accidents or safety-critical failures early in the system and subsystem development processes using the various verification methods, such as electrical circuit design analyses, Models and Simulations (M&S), tests, and inspections. If hazards exist in the design, following subsystem development processes using the various verification methods, then they should be found as early as possible during system development integration using the various system verification methods including informal system test. We want to prevent accidents in customer-use applications by preventing test escapes following informal system test. We should not discover hazards in formal system test. A hazard found in formal test is a test escape from the system development verification processes. Test escapes are hazards that should have been caught in an informal test or during system DT or Integration and Test (I&T). Test escapes were either not detected during the system and subsystem development verification processes or they were detected, but not fixed, and "escaped" to the next level of system test, such as OT, formal system test, or an end-user application. Independent Verification and Validation (IV&V) is a type of formal OT usually performed by representatives from the government for the military customer community. System safety testing is part of IV&V testing and is mandated by certain types of safety review boards that ultimately decide when a system is ready for deployment and use in the field or fleet.

Formal tests are tests performed for customer-specific purposes to verify and validate requirements are met. These may be performed by the OEM, by the OEM customers, by the OEM suppliers, or by third-party test houses contracted by the OEM or the customer. There are many organizations that focus on the safety and security of systems for different technologies and marketplaces, such as Information Technology (IT), transportation, energy, defense, and medical systems.

Sometimes, end-user customers are not directly involved in formal testing, so independent agencies and organizations are involved to watch on behalf of the general public as end-user customers. These types of formal testing may be referred to as regulatory compliance testing or non-regulatory compliance testing. For example, National Information Assurance Partnership (NIAP) is a joint initiative between the National Security Agency (NSA) and National Institute of Standards and Technology (NIST) responsible for security testing needs of both IT consumers and producers and promotion of the development of technically sound security requirements for IT products and systems, as well as appropriate measures for evaluating those products and systems. NIST is a non-regulatory federal agency under the Department of Commerce. NIST is the US National Measurement Institute. More on regulatory compliance testing will be discussed later in this chapter.

12.2 Test Strategy and Test Architecture

It is important to understand the test strategy and test architecture. Test strategy is an integrated roadmap for executing all test-related activities (e.g., development, production, and operational informal and formal testing) to achieve performance, quality, safety, reliability, cost, and schedule milestones on a program. Test architecture is the detailed design implementation of the tests that are to be performed on a particular system or product during the life of a program. It determines how the program's system or product test strategy is to be implemented using a combination of external support equipment, STE, and embedded system test capabilities and BIT.

Embedded test is the ability of the system or product to perform internal measurements, provide test results, and develop parametric information on the state of the system or product. It can be used to determine pass/fail criteria of a system or product and diagnostics for troubleshooting or maintenance and provide large parametric data sets in support of Statistical Process Control (SPC) analyses. SPC data processing may be useful during system or product development and production test as well as in customer-use environments. SPC capabilities added to the external STE and embedded test strategy and test architecture enable cost-effective Preventive Maintenance (PM) and Condition-Based Maintenance (CBM) using Prognostics and Health Monitoring (PHM) to minimize total Life Cycle Costs (LCCs) over the product life cycle.

The cost of test should be a major consideration during test strategy development. Test strategy is important to reduce cost to become more competitive in the marketplace. It is important to test early in development and production using the "ten-to-one rule." The cost to test, detect a defect, and correct the defect increases 10 times between development and production and between production and postdelivery in the customer application.

BIT is a subset of embedded test. Its categories include power-up BIT, continuous BIT, and maintenance BIT. Test design requirements and specifications should be written based on the test strategy and the test architecture including what categories of BIT are to be employed for different purposes. Detailed designs of STE and embedded electrical test circuitry, BIT test software and firmware, and test interface circuitry should be created from the test strategy and the test architecture.

BIT signals, whether they are designed into a system for PHM or not, may be classified into three types [1]:

a. A-BIT or S-BIT—Automatic BIT or Start-Up BIT—initiated on start-up of the product or system, and usually involves a full functional test to ensure all critical requirements are met. A-BIT is not as thorough in terms of test coverage as an I-BIT, but more thorough than a C-BIT. There is usually no difference in A-BIT designs comparing a system with PHM or a system without PHM capabilities.
b. C-BIT—Continuous BIT which runs in the background during normal product or system operation, and usually is a cursory set of tests requiring minimum

processing power to ensure the majority of critical functions are operating. C-BIT is the least amount of test coverage of the three types of BIT. C-BIT for a PHM system will have more design analysis features than a system without PHM. The design analysis features for C-BIT in a PHM system will include statistical data analysis and prognostic reasoners to detect when a failure is imminent.

c. I-BIT—Initiated BIT is manually started by a maintainer or operator during normal system functions by exercising specially programmed diagnostics and prognostics to determine causes of errors, faults or failures, and assist maintainers in the fault detection and isolation of failures. I-BIT is the most thorough form of BIT with the highest percentage of test coverage or fault coverage. When running I-BIT, normal system operation may be suspended, or may be slowed down to allocate processing power to the enhanced BIT programs. In I-BIT, maintainers should have access to C-BIT and A-BIT logs where errors and fault data are stored and easy to retrieve for further analysis of the product or system. There is usually no difference in I-BIT for a system with PHM or without PHM. [1]

The utilization of A-BIT, C-BIT, and I-BIT may be accomplished with the implementation of two types of sensors for system health monitoring. The two types of sensors for automatic stress monitoring are mechanical and electrical [1].

Mechanical sensors
Thermal sensors (such as thermocouples), vibration sensors (such as accelerometers), and stress and strain gauges are a few examples of mechanical sensors [1].

Electrical sensors
Voltage sensors, current sensors, charge sensors, magnetic sensors, impedance sensors (including resistance sensors to determine opens circuits, short circuits, low or high resistance, and reactance sensors to determine low or high capacitive or inductive parameters), power and power density sensors, frequency sensors, noise sensors, and timing sensors are a few examples of electrical sensors [1].

Many times, trade studies are conducted to determine the best test strategy and test architecture. Affordability for the OEM during the development and production cycle, and for the customer over the system or product life cycle, is considered when making program decisions of the test strategy and test architecture. The following is a short notional list of criteria for affordability and test design trade studies in determining the best test strategy and test architecture:

- Test allocation and test utilization to levels of hardware, assembly, and software and cost to implement each option
- BIT and external support STE complexity and cost considerations
- Selection of different categories of BIT for different purposes

- Reuse of reference architecture or common architecture compared to new custom architecture
- Degree of test coverage, fault coverage, fault detection, and fault isolation capabilities
- Types of test coverage per function considering combinations of BIT and external support STE
- Fault tolerant test design capabilities
- System or product maintenance concept (e.g., organizational maintenance to discard on failure vs. repair down to piece part at depot level)
- Test execution for safety design features
- Environmental screening methods during development and production
- Types of sensors for automatic stress monitoring
- Multiple iterations of development tests (repeatability of results)
- Retest criteria after failure identification
- Retest criteria after repair and corrective action
- System check-out during periodic PM
- System certification and recertification

These criteria need to be optimized for the best test strategy and test architecture to avoid expensive systems that are not affordable for customers and result in poor ROI over the entire system or product life cycle. Using data from LCC models, reliability models, test coverage models, and Design Margin Analysis (DMA), you can trade different test strategy and test architecture options to find the best test utilization, system performance, and lowest-cost path forward. For example, consider a trade study whether to perform a BIT test at the box level or at the Circuit Card Assembly (CCA) level. Embedded test architecture enables direct component input/output (I/O) stimulation, faster test data accessibility, and selective data extraction for external support TE to support the entire product life cycle enabling a cost-effective test compared to using STE. The cost of BIT at the CCA level, where each CCA in a box has a 90% fault coverage requirement, will be much higher than that at the box level with a 90% fault coverage requirement. If the maintenance concept is to throw away a CCA upon box-level test fault detection, it would not be cost-effective to conduct extensive BIT for the CCA. In this case, box-level BIT should be performed to demonstrate the 90% requirement for a reduced test design cost. This does not mean there would be no BIT implemented at the CCA level. The BIT would be cursory-level diagnostics with detection of only certain critical faults, such as power distribution errors. The BIT at CCA level would not perform detailed fault isolation diagnostics to an electrical device piece part or component level, but it could provide a summary health check to the box level to determine which CCA within the box is experiencing an error and should be removed and replaced with another CCA from CCA spares stock.

Paradigm 8: Develop a Comprehensive Safety Training Program to Include Handling of Systems by Operators and Maintainers

As part of the test strategy, a system safety training program is developed for certifying system operators and system maintainers before they use the system in the field applications. Safety training requires knowledge of the components of a system and an understanding not only of the components and subsystems but also of the entire system, to become aware of the potential of any and all hazards and failure risks in using and maintaining a system. The safety certification of the person operating or maintaining the system should not be limited to using and maintaining only certain components or subsystems, but rather the certification should include all features, functionality, and repair actions (preventive action and corrective actions) for that system. The operator and maintenance personnel must realize the risks associated with the total system that may include hidden hazards. Scenarios must be developed to provide instructors and students with a realistic understanding of the whole system and how to protect and prevent system hazards.

For example, as stated earlier in Chapter 1, a system with a single power source may negatively impact mission performance and affect increased workload for system operators and maintainers. For safety-critical work, adding a second power source is a positive step to mitigate risk of a hazard and workload to restore systems following a power outage. However, if all sources of power are lost (prime, secondary, and emergency), due to a common-mode fault, the total system will not work until a correction is made. This common-mode event requires all available operators and maintainers to step up efforts to get systems back to an operational state. These personnel should be trained in the operation and maintenance of the total interconnected system, addressing potential worst-case scenarios including total system power loss and the hazards associated with such a loss. System maintenance training should leverage established system safety test methods used during system development and integration test as a risk mitigation method if widespread loss of a complex system occurs. It is imperative that thorough and regularly scheduled maintenance and system test training of the system operators and maintainers be incorporated into the system operation and maintenance procedures to ensure all critical system personnel gain confidence in their abilities to correctly operate and maintain the system at all times given all likely scenarios.

12.3 Develop System Safety Test Plans

It is critical that system safety testing is performed. This testing only happens through efficient and effective planning. Many times, system safety testing is not planned and performed as a separate test on its own. Safety test plans leverage other types of tests to accomplish safety objectives. System safety testing is accomplished as part of other testing to verify and validate system performance requirements, while a safety engineer witnesses the test performance and collects data that is applicable to the verification and validation of system safety requirements. The various methods for system safety testing should be documented in a System Safety Test Plan (SSTP), which should be integrated within the subsystem and system development test plans and should include a schedule of safety test events, as part of a larger system test plan, for those tests uniquely performed for system safety requirements verification and for those tests accomplished as part of the various subsystem and system development test events. All unique system safety tests and those subsystem and system development test events leveraged by system safety must be synchronized with the system program schedule.

MIL-STD-882E [2] is a military standard document that does not require SSTPs but has a task associated with Test and Evaluation (T&E) participation (Task 303). Task 303 is to participate in the T&E process to evaluate the system, verify and validate risk mitigation measures, and manage risks for test events. The contractor shall participate in T&E planning, support the preparation of test event safety releases, conduct post-test event actions, and maintain a repository of reports. The objective is to eliminate the hazards or reduce the associated risks when the hazards cannot be eliminated for both the system and the test events.

In accordance with MIL-STD-882E [2], Task 303, T&E planning shall include the following:

- Participation in the preparation and updating of the T&E Strategy (TES) and the T&E Master Plan (TEMP) to include hazard considerations and identification of when hazard analyses, risk assessments, and risk acceptances shall be completed in order to support T&E schedules.
- Participation in the development of test plans and procedures to include hazard considerations that support the following:
  - Identification of mitigation measures to be verified and validated during a given test event with recommended evaluation criteria
  - Identification of known system hazards present in a given test event, recommended test-unique mitigations, and test event risks
  - Preparation of the safety release
  - Analysis of hazards associated with TE and procedures
  - Government completion of applicable environmental analysis and documentation pursuant to DoD service-specific National Environmental Policy Act (NEPA) and Executive Order (EO) 12114 requirements in T&E planning schedules
  - Documentation of procedures for advising operators, maintainers, and testers involved in the test event of known hazards, their associated risks, test-unique mitigation measures, and risk acceptance status

MIL-STD-882E [2] Task 303 ensures that the following post-test event actions are conducted to:

- Analyze test results to assess effectiveness of mitigation measures as tested.
- Analyze test results to identify and assess new system hazards and to potentially update risk assessments for known hazards.
- Analyze incident, discrepancy, and mishap reports generated during test events for information on hazards and mitigation measures and ensure mitigation measures are incorporated in future test plans as appropriate.
- Document new or updated system related hazard information in the Hazard Tracking System (HTS) as appropriate.

MIL-STD-882E [2] Task 303 further states that a repository of T&E results must be maintained. Government representatives will be allowed access to the safety test data repository, and the full safety data repository will be provided to the government at the end of the contract. The repository shall include the following:

- Hazards identified during test events
- Verification and validation of mitigation measures
- Incident, discrepancy, and mishap reports generated during test events with information on corrective actions

MIL-STD-882E [2], System Safety Standard Task 401, states the following safety verification requirements for defense-contracting organizations:

- Define and perform tests and demonstrations, or use other verification methods on safety-significant hardware, software, and procedures to verify compliance with safety requirements.
- Define and perform analyses, tests, and demonstrations, develop models, and verify the compliance of the system with safety requirements on safety-significant hardware, software, and procedures (e.g., safety verification of iterative software builds, prototype systems, subsystems, and components).
- Induced or simulated failures shall be considered to demonstrate the acceptable safety performance of the equipment and software.
- When analysis or inspection cannot determine the adequacy of risk mitigation measures, tests shall be specified and conducted to evaluate the overall effectiveness of the mitigation measures.
- Specific safety tests shall be integrated into appropriate system T&E plans, including verification and validation plans.
- Where safety tests are not feasible, the contractor shall recommend verification of compliance using engineering analyses, analogies, laboratory tests, functional mock-ups, or M&S.

- Review plans, procedures, and the results of tests and inspections to verify compliance with safety requirements.
- Safety verification results shall be documented and a report submitted that includes the following:
  - Test procedures conducted to verify or demonstrate compliance with the safety requirements on safety-significant hardware, software, and procedures
  - Results from engineering analyses, analogies, laboratory tests, functional mock-ups, or M&S used
  - T&E reports that contain the results of the safety evaluations, with a summary of the results provided

In many industries, just witnessing testing for purposes other than system safety is not adequate to uncover the hazard risks. Special testing for safety must be performed above and beyond the testing performed to verify and validate system operation and performance characteristics. A category of these types of special safety tests are compliance testing. Various types of compliance tests exist. These compliance tests are required by certain regulatory agencies in order for a product to be used in a particular application where the regulatory agency governs the acceptable use of products by customers it is serving to protect.

12.4 Regulatory Compliance Testing

In terms of general context, compliance means an adherence and intent to conform to rules, specifications, requirements, policies, standards, and laws [3]. A policy is an intentional set of written principles and statements of intent that guide decisions and result in expected outcomes and conclusions. Examples of principles are fundamental laws, doctrines, normative rules, codes of conduct, and states of nature underlying observable evidence witnessed by many people. A policy is implemented through the use of written plans, procedures, and protocols. Policies are enacted and adopted by a governance body within an organization, group, or entity. They assist senior management of organizations in conducting subjective decision-making where they must consider the relative merits of a number of factors and criteria before reaching decisions and establishing courses of action. Regulatory compliance describes the objectives an organization, group, or entity aspires to achieve as they ensure they are cognizant of and take steps toward satisfying any relevant laws, rules, policies, and regulations.

Due to the increasing number of regulations and need for operational transparency, organizations are increasingly adopting the use of consolidated and harmonized sets of compliance controls. This approach is used to ensure that all necessary governance requirements can be met without the unnecessary duplication of effort and activity from resources. [3]

When system safety testing is not adequate, the customer is going to find safety hazards in their systems and products, which should have been found by system and product developers and manufacturers. Mechanisms and procedures have been established to allow customers of consumer products to report their safety findings and potential hazards to a government organization tasked with collecting the data and tracking the results. Consumer Product Safety Commission (CPSC) [4] is charged with protecting the public from unreasonable risks of injury or death associated with the use of the thousands of types of consumer products under the agency's jurisdiction. Deaths, injuries, and property damage from consumer product incidents cost the nation more than $1 trillion annually. CPSC is committed to protecting consumers and families from products that pose a fire, electrical, chemical, or mechanical hazard. CPSCs work to ensure the safety of consumer products—such as toys, cribs, power tools, cigarette lighters, and household chemicals—contributed to a decline in the rate of deaths and injuries associated with consumer products over the past 40 years [4, 5].

Federal law requires manufacturers and importers to test many consumer products for compliance with consumer product safety requirements. Based on passing test results, the manufacturer or importer must certify the consumer product as compliant with the applicable consumer product safety requirements in a written or electronic certificate. Certificates are required to accompany the applicable product or shipment of products covered by the certificate, and a copy must be provided to retailers, to distributors, and, upon request, to the government [3].

Various organizations build and operate test facilities to conduct regulatory compliance testing and product safety testing. Two of these compliance-testing organizations that are widely known are Underwriters Laboratory (UL) [5] and Edison Test Labs (ETL) [6], a division of Intertek. Both are independent testing and consulting companies that provide professional engineering services. They have established teams of quality assurance and product safety experts with comprehensive knowledge of the regulatory requirements impacting consumer product performance and safety. These test facility organizations, such as UL, help system and product development companies develop testing programs and conduct testing on their behalf for a wide range of international regulations. These test facility organizations provide global regulatory testing capabilities that help system and product development companies ensure compliance for every country in which their customers and products exist.

UL provides the following four examples of national and international regulations that it can support [7]:

- Registration, Evaluation, Authorization and Restriction of Chemicals (REACH). This European Union (EU) regulation addresses the production and use of chemical substances and their potential impacts on both human health and the environment.

- Consumer Product Safety Improvement Act (CPSIA). UL is registered with the CPSC as an accredited laboratory for CPSIA third-party testing [7].
- Restriction of Hazardous Substances (RoHS) Directive. This directive restricts (with exceptions) the use of six hazardous materials in the manufacture of various types of electrical equipment. It is closely linked with the Waste Electrical and Electronic Equipment Directive (WEEE), which sets collection, recycling, and recovery targets for electrical equipment and is part of a legislative initiative to solve the global toxic electronic waste problem.
- China's mandatory textiles and apparels standard (GB 18401) [7].

Many of these test facility organizations, such as UL and ETL, are accredited. Laboratory accreditation provides formal recognition to competent laboratories, providing a means for customers to identify and select reliable testing services.

As an example of the types of laboratory accreditations, UL [8] is accredited by these multinational organizations [7]:

- ANSI-ASQ National Accreditation Board (ACLASS)
- The United Kingdom Accreditation Service (UKAS)
- The China National Accreditation Service for Conformity Assessment (CNAS)
- The California Medical Association (CMA)
- The Hong Kong Laboratory Accreditation Scheme (HOKLAS)
- Japan Ministry of Health, Labour and Welfare

As a result of their accreditation, these test facility companies use special product label markings (e.g., UL, CSA, ETL, and CE) or other forms of compliance markings on products that they test and certify. As an example, publicly available information from UL states

UL is a global independent safety science company offering expertise across three strategic businesses: Commercial & Industrial, Consumer and UL Ventures. Our breadth, established objectivity and proven history mean we are a symbol of trust, enabling us to help provide peace of mind to all. The UL Mark is the single most accepted Certification Mark in the United States, appearing on 22 billion products annually.

UL is actively involved in the development and revision of international product safety standards. It is considered an industry expert able to identify hazards early in the production cycle to reduce product development cost and increase speed to market. It also ensures that their customers meet current regulatory demands and are aware of emerging regulations [7, 8].

Some companies such as DLS [7] are certified UL agencies that are allowed to conduct product safety testing and mark the products that are proven to be UL complaint with a UL marking. "D.L.S. Conformity Assessment is fully approved to perform UL and cUL testing under the UL Third-Party Test Data Program (TPTDP). This program allows full Underwriters Laboratories approvals and listings using the D.L.S. value-added testing program, including UL factory evaluations and follow-up services" [9].

Besides UL markings, there is also Canadian Standards Association (CSA) markings. cUL is the Canadian UL mark of approval. Many products require a CSA/UL marking to be sold in the US and Canada. ETL Listed Mark [8] is another common compliance standard marking for products.

The ETL Mark is proof of product compliance to North American safety standards. Authorities Having Jurisdiction (AHJs) and code officials across the US and Canada accept the ETL Listed Mark as proof of product compliance to published industry standards. Retail buyers accept it on products they're sourcing. And every day, more and more consumers recognize it on products they purchase as a symbol of safety. [6]

CE Marking using the letters "CE" is the abbreviation of French phrase "Conformité Européene," which literally means "European Conformity." The term initially used was "EC Mark," and it was officially replaced by "CE Marking" in the Directive 93/68/EEC in 1993 [9]. "CE Marking" is now used in all EU official documents. "CE marking" is a process that applies to a wide variety of products and one in which manufacturers located in the EU or importers of goods into the EU must complete. If a product falls within the scope of at least one of the CE marking directives and is not specifically excluded, it must be CE marked. Once the applicable directives are identified, the product must be designed and tested to meet the directives. Once a company has demonstrated that they have fulfilled the essential requirements of the directive, a technical file is assembled, and a declaration of conformity is produced [10].

Other types of regulatory markings:

- Medical Device Directives
- Occupational Safety and Health Administration (OSHA)
- Qi compliance testing
- EN/IEC testing for the EU
- CB Global Testing Program
- CCC mark for China
- US Federal Aviation Administration (FAA) in the United States for flight safety system testing
- US Food and Drug Administration (FDA) for medical, drugs, and food safety

- US Federal Communications Commission (FCC) in the United States for Radio Frequency (RF) transmitters
- Joint Aviation Authorities (JAA), which is a cooperative of most European (EU and non-EU) civil aviation regulatory authorities
- European Aviation Safety Agency (EASA), which was created in 2002 by the European Commission (EC) and took over the functions of the JAA of the EU countries
- International Civil Aviation Organization (ICAO), a specialized agency of the United Nations (UN)
- Civil Aviation Authority (CAA), a generic term used in many countries, notably the UK and China (In China, CAA means Civil Aviation Administration.)
- Radio and Telecommunications Terminal Equipment (R&TTE) Directive Testing established by the EC within the EU
- Automotive airbag inspection and testing
- Fire Protection and Life Safety System Testing (or Fire and Life Safety System Compliance Testing)

Besides regulatory compliance testing for consumer and industrial grade systems and products, there are military defense standards and the North Atlantic Treaty Organization (NATO) standards for testing military- and space-grade systems and products. It is not the intent of this book to mention all the applicable military standards that govern compliance to safety performance and test requirements. MIL-STD-882E, which was previously discussed in this chapter as related to SSTPs, is an important document for requirements related to system and product safety for the military and space applications. There are many other military test and safety-related documents that apply to the major missile and weapon system developers and manufactures in the US Department of Defense (DOD).

For weapon system and missile developers and manufacturers, and their parent nations, to sell their defense systems to allied countries, there is an incentive for such organizations to be aware of the nonclassified international standards and protocols for storage, transportation, safety, and general T&E methods that apply to their systems in other countries. In the international marketplace for missile and weapon systems, with the vast number of specifications and standard documents that exist, the most prevalent of these standards and protocols are the NATO Standardization Agreements (STANAGs). These define processes, procedures, terms, and conditions for common military or technical procedures or equipment between the member countries of the NATO Alliance.

The main purpose of a STANAG is to provide common operational and administrative procedures and logistics, so one member nation's military may share their stored systems and stockpiled hardware (e.g., missiles, munitions, and weapons) with another NATO member nation's military. STANAGs are the basis for technical interoperability between a wide variety of test, evaluation, assessment, and analysis methodologies essential for NATO and Allied nations in their use of such information developed by another NATO member. Allies of NATO members will also disseminate NATO STANAGs within their own governments and military acquisition organizations.

Depending on the specific STANAG, either a majority of all 28 NATO nations or a subset specifically defined by the NATO Standardization Committee of the 28 nations is required for ratification. If the Standardization Committee defines a limited subset of the 28 NATO nations, then a percentage of that subset is required for ratification of a particular STANAG. If a nation ratifies a STANAG, that nation agrees to implement the requirements of the STANAG within their military involving procurements, logistics, transportation, and T&E.

One particular STANAG document worthy of mention here is STANAG 4404 [11], which defines the safety-critical computing system function protection requirements and guidelines, software analysis and test requirements, and interface design requirements, to name a few topics. STANAG 4404 [11] is the NATO document for safety design and lists requirements and guidelines for munition-related safety. This standard also includes requirements for the following areas of design for safety concern:

- System design requirements and guidelines
- Power-up system initialization requirements
- Computing system hardware requirements and guidelines
- Self-checking design requirements and guidelines
- Safety-critical computing system function protection requirements and guidelines
- Interface design requirements
- User interface
- Software design and coding requirements and guidelines
- Software maintenance requirements and guidelines
- Software analysis and testing
- Critical timing and interrupt functions

STANAG 4404 [11] system design requirements and guidelines that apply to the general system design include the following:

- Designed safe states
- Stand-alone computer
- Ease of maintenance
- Safe state return
- Restoration of interlocks

- I/O registers
- External hardware failures
- Safety kernel failure
- Circumvent unsafe conditions (e.g., fallback and recovery, simulators, positive feedback mechanisms, peak load conditions, and system error logs)

STANAG 4404 [11] provides design compliance checklists to assist the safety practitioner in determining the state of the safety-critical design for any system being assessed and uncover gaps in the design for safety requirements. Once the design safety assessment is completed, plans should be put into place to verify the satisfactory implementation of the design requirements and determine how to fill the gaps that were noted on the STANAG 4404 checklist.

There is value from a safety-critical system perspective in the participation by a nation's military organizations and their missile and weapon system manufacturers in the development of NATO STANAGs. The value is especially considerable as related to future STANAGs for Prognostics and Health Management (PHM) policies, design and quality standards, test procedures, certification of PHM procedures, PHM device communication standards and protocols, PHM production-level Hazards of Electromagnetic Radiation to Ordnance (HERO) safety tests and certifications, and noncommunication procedures to ensure noninterference with platform electronic communications and ability to not compromise a platform or other location's electromagnetic radiation emission silent conditions. The two major NATO action committees likely to be charged by the Standardization Committee with developing the appropriate STANAGs to codify and standardize the PHM concepts, procedures, standards, and tests, as mentioned earlier, are AC/326 Ammunition Safety Group and AC/327 Life Cycle Management Group (includes HERO) [12].

The value of PHM and its ancillary benefits with the resultant ROI depends on the cost to implement PHM, the type of PHM, the system safety and reliability, the time period in the life cycle to assess reliability and safety, the cost risk of liabilities that are eliminated with PHM capabilities, and the maintenance cost over the life cycle that is avoided if PHM is implemented. Referencing Paradigm 6 from Chapter 1, PHM is a continuously evolving system safety test capability used throughout the system life cycle to know the health state of the system at any period of time, to minimize sudden disasters by predicting imminent danger, and to aim for zero accidents in customer applications and missions.

12.5 The Value of PHM for System Safety Testing

Paradigm 6: Design for Prognostics and Health Monitoring (PHM) to Minimize the Number of Surprise Disastrous Events or Preventable Mishaps

PHM technology enhances the efficiency and effectiveness of system safety capabilities. PHM is an enabler for system reliability and safety. It must be deployed in safety-critical systems wherever possible. In today's complex systems, most of system failures are caused by fundamental limitations in the design strength to withstand exposure to stress environments. Mechanical degradation mechanisms surface and propagate with time leading to physical wear-out. Design strength must withstand worst-case customer use applications and environmental stress conditions. But what happens when the design strength does not withstand the accumulation of fatigue during repeated applications in stress conditions? There is a need for innovative solutions for discovering hidden problems and detecting latent defects, which initially appear as probabilistic nondeterministic anomalies and faults. Through the use of embedded sensors for automatic stress and health monitoring, Physics of Failure (PoF) modeling with machine learning, and embedded processors with algorithms for prognostic reasoners and Remaining Useful Life (RUL) calculations, PHM solutions conducted online in real-time to determine the right time to perform all forms of maintenance, such as corrective maintenance, CBM, predictive maintenance, PM Checks and Services (PMCS), and prescriptive maintenance, is a possibility.

12.5.1 Return on Investment (ROI) from PHM

It is impossible to justify the ROI associated with spending large sums of money to achieve near 100% test coverage by analyzing millions of combinations of events or faults that might occur in today's complex systems. These faults might be due to incomplete test coverage or fault coverage because of limited time and funds to provide 100% fault detection capability through BIT and/or external support TE. It is a given fact that 100% test coverage is not cost-effective and not justified by reasonable ROI for large complex systems composed of hundreds or thousands of electrical integrated circuit devices each with millions or billions of nanoscale components. There will always be a certain percentage of defective components, which are latent faults that are unknown unknowns. These represent unknown failure mechanisms with unknown probabilities of occurrence that have yet to be detected. A known unknown fault is a fault that was discovered, maybe once or a few times, but its root cause failure mechanism, the nature of its occurrence, and probability of occurrence are not fully understood. Any complex system will have an unknown number of failures or hazards (either unknown unknowns or known unknowns) that will remain hidden, latent, and unknown until they occur and are detected (as an initial occurrence or repetitively occur enough to determine a pattern). The unknown faults might become known faults with the right sensors and instrumentation collecting the right data at the right time, capturing the critical design parametric measurements, and characterizing and correlating the failure mechanism with the stress conditions that accelerated the mechanism. Many times, anomalies that occur are detected as unknown faults, most likely during an actual customer mission scenario when a function that is critical to a mission is disabled and it was never seen before. These are called unpredicted failures.

To counter such unpredicted failure or hazard risks, early prognostic warning indications with health monitoring and PoF models are needed, through embedded sensors to prevent a major mishap. The design process to incorporate these early warnings consists of postulating all the possible mishaps and designing intelligence to detect unusual behavior of the system. The intelligence may consist of measuring important features and making a decision on their impact. For example, a sensor input occasionally occurs after 30 ms instead of 20 ms as the sensor design timing requirement states. Is this an early warning indication of an imminent sensor failure? If the sensor fails, could that lead to a disaster? If so to both questions, then the sensor should be replaced before the failure manifests itself to a critical state.

The potential values of PHM to systems users, producers, logisticians, maintainers, and designers are simply the cost savings in knowing what, how much, and for how long environmental stressors (e.g., temperature, humidity, vibration, and pressure) will cause degradation to the safety, reliability, or performance properties of the individual subsystems and components of the system. It is important to have this knowledge of the overall system population and spare parts inventory population that have experienced these stressors so that CBM, use, upgrade, and ultimate retirement or end-of-life or end-of-service decisions for the population can be made. This information has an influence on future system design improvements as well. As field-use data is collected and knowledge of the stressors are gained, this will improve the system or product developers' abilities to modify their designs to reduce or eliminate the failure or hazard causes and effect associated with those stressors [12].

All components of a system are subjected to the environmental stressors. For example, let's consider the thermal and vibration stressors that affect the reliability of CCA solder joints. It is well known through existing PoF models that solder joints are weakened over time by temperature cycling and mechanical shock and vibration forces. Knowledge of the actual environments that a system experiences has a direct relationship to the programmatic cost drivers and needs for scheduled or periodic recertification actions, whether or not maintenance is performed on the individual systems, which are costly to the customers, or whether the systems should be decommissioned and taken out of service. These cost-driven considerations include necessary quantitative increases in the inventory of the system requiring periodic recertification to allow the System-of-Systems (SoS) platforms to go without individual systems as they are returned to a specialized facility for the recertification and maintenance. Other cost considerations are the logistics costs to exchange and transport the individual system to and from the customer SoS platform locations and the recertification locations and finally the cost of the recertification themselves [12].

Focusing on a subsystem example of the value of PHM for system safety testing, let's look at tactical rocket motors for missile systems. Today's highly advanced tactical rocket motors require constant design enhancements, reliability improvements, and technology advancements to demonstrate continuous improvements in missile system capabilities. These design enhancements, reliability improvements, and technology advancements are able to transition to engineering development and production of new solid rocket motors within tactical missile system designs when there is a compelling business case analysis of the cost benefits for the customer over the entire system life cycle. There is mature PHM technology that exists for other systems, such as full-scale rocket motors in strategic missile systems, commercial avionics, telecommunication networks, and assorted military and nonmilitary system applications. The majority of these PHM technologies lead to development of novel approaches to embedded sensors for the purpose of assessing stress conditions and predicting future failure events using PoF models that determine RUL and degradation wear-out failure mechanisms. The same PHM sensors that are used to model the RUL of the motor can also be used to monitor the current stresses on the motor and ascertain if a failure is imminent. Besides these value-added benefits, PHM provides other capabilities, such as asset tracking, logistics management, in-process monitoring, and Insensitive Munitions (IM) [12].

The bottom line of PHM is to improve the value-added benefits of new systems, or make enhancements to existing systems with embedded sensors that will improve the accuracy of safety assessments, physical data that pinpoint suspicious individual systems from an entire lot of systems, hazard and failure predictions, and service life extensions plans. The goal is to identify and remove a suspect missile system or component of the system, such as a rocket motor, from inventory and not have to pull an entire missile lot out of service, thus increasing PHM system ROI, system reliability, and system safety. This value may include prevention of loss in customer confidence if mission-critical failures occur or liabilities due to safe-critical failures that would be mitigated with PHM [12].

12.5.2 Insensitive Munitions

An ongoing challenge to missile system safety is IM requirements to prevent safety-critical failures by demonstrating "no response to IM stimuli" or "no effect caused by IM stimuli." CBM with PHM approaches serve dual uses in today's weapon systems. The CBM and PHM approaches ensure successful implementation of designs that demonstrate "no response" to IM stimuli as well as provide reduced manning and decreased cost for sustainment of weapon systems over their entire life. Knowledge of the actual stress conditions of each system during a potential unsafe event can be useful to determine if a hazard has been detected and whether an asset should be destroyed, or a system is able to be repaired in a safe manner and prolong its life. PHM capabilities with embedded sensors could be used to interrogate an asset just prior to and during a potential IM incident. These same sensors can then be used to trigger appropriate responses ranging from a simple alert via the wireless network or to prompt an active mitigation device such as an electronic passive device for initiating a thermal venting system. Just as the thermal IM stimuli sensors for PHM devices are likely to provide a cost-effective solution, other reasonable approaches for IM prevention may include mechanical shock and vibration sensors to react to induced events with varying timescales intrinsic to each category of IM stimuli [12].

The ability to measure environmental or platform induced stressors experienced by the metals, ceramics, polymers, and composite materials in a missile system and to predict changes in mass properties of these materials are important in relation not only to traditional age-limiting safety parameters in energetic materials, where catastrophic effects occur from IM stimulus, but also to understanding and predicting material changes that could potentially adversely influence the chemical and mechanical IM properties of the energetic devices as a whole subsystem. As IM by definition involves an SoS approach (primary and secondary/tertiary energetic materials, separation structures or components, and containment vessel), knowledge of changes to all aspects of the system is important in identifying, understanding, and predicting the effects of stressors to the overall system [12].

12.5.3 Introduction to PHM

PHM is an approach to protect the integrity of equipment and avoid unanticipated operational problems leading to mission performance deficiencies, degradation, and adverse effects on mission safety. Researchers have developed a variety of approaches, methods, and tools that are useful for these purposes, but applications to real-world situations may be hindered by the lack of visibility into these tools, the lack of uniformity in the application of these tools, and the lack of consistency in their demonstrated results. There is a need for documented and favorable guidance that will encourage practitioners to invest the resources necessary to put these techniques into practice. While specific application domains often require customized treatment for PHM application development, some core principles apply to all. IEEE P1856 is a draft standard document that describes those core principles and exemplifies their application within the electronics domain. IEEE P1856 is a new IEEE standard that is planned for publication in 2017 [13].

The goal of IEEE P1856 is to provide information for the implementation of PHM for electronic systems. This standard can be used by manufacturers and end users for planning the appropriate PHM methodology to implement and the associated life cycle operations for the system of interest. This standard aims to provide practitioners with information that will help them make business cases for PHM implementation and select proper strategies and performance metrics to evaluate PHM results. The overall aim is to provide a broad overview of PHM while at the same time providing significant details to assist the practitioner in making appropriate decisions [13].

Within the system health management community, there are several different interpretations of the term *prognostics*, such as predictive analysis, reliability prediction, damage accumulation prediction, or condition-based prediction. However, this standard covers all aspects of PHM of electronic systems, including definitions, approaches, algorithms, sensors and sensor selection, data collection, storage and analysis, anomaly detection, diagnosis, decision and response effectiveness, metrics, LCC of implementation, ROI, and documentation. This standard describes a normative framework for classifying PHM capability and for planning the development of PHM for an electronic system or product. The use of this standard is not required throughout the industry. This standard provides information to aid practitioners in the selection of PHM strategies and approaches to meet their needs [13].

The purpose of IEEE P1856 is to classify and define the concepts involved in PHM of electronic systems and to provide a standard framework that assists practitioners in the development of business cases and the selection of approaches, methodologies, algorithms, condition monitoring equipment, procedures, and strategies for implementing PHM of electronic systems [13].

A PHM system consists of the following capabilities [13]:

- Acquisition of object system data by means of health status monitoring sensors
- Data management
- Data processing algorithms and/or processes for diagnostics, health state estimation, and PHM

PHM system performance is measured for its contribution to achieving or improving object system goals. PHM system performance is measured in terms of metrics chosen from the following categories [13]:

- Accuracy
A measure of deviation of a prognostic or diagnostic output of current object system state from measured, observed, or inferred ground truth. Accuracy can be measured by comparing prognostic and diagnostic outputs from measured ground truth using several metrics such as, but not limited to, detection accuracy, isolation accuracy, and prognostic system accuracy. These terms are defined in the previous normative definitions section.
- Timeliness
A measure of how quickly the PHM system's functions produce their outputs in relation to the failure effects that they are mitigating. Overall timeliness includes the total time of state estimation and control for the PHM control loop. It can be subdivided into several functions, though in all cases the total PHM "control loop" timeliness from prediction or detection through response must be summed up to compare to the time required for the failure(s) being mitigated. If desired, comparisons for the prognostics function and the health management function can be made separately by grouping the relevant subfunctions. The timeliness aspects of the PHM subfunctions that contribute to the total PHM loop include detection time, diagnostic time, prognostic time, decision time, and response time. These terms are defined in the previous normative definitions section.
- Confidence
A measure of trust (or, conversely, a measure of uncertainty) in a PHM system's output. For detection and diagnostic functions, it is computed through, but not limited to, robustness, sensitivity, and uncertainty metrics. For prognostics it is specified by including the estimation of uncertainty in predictions and stability of predictions over time, in addition to sensitivity and robustness measures.
- Effectiveness
A measure of the PHM system's ability to preserve or attain relevant system goals. In general, effectiveness measures relate to the goals that the PHM has been put in place to preserve or attain. Typical effectiveness goals include system reliability, safety, performance, cost, or schedule. The sub-metrics of accuracy, timeliness, and confidence must all be combined for specific PHM mechanisms and for the PHM as a whole for the system to generate the PHM effectiveness measured against the relevant system goals, as allocated to PHM.
- Any additional metrics as defined in the PHM or object system specifications [13].

12.6 Leveraging Reliability Test Approaches for Safety Testing

It is important for SSTPs to leverage all available testing that is performed as part of the Systems Engineering Test Plans (SETPs). Many times SETPs include Reliability Test Plans (RTPs). Valuable data may be collected from these tests that benefit reliability and system safety engineering alike. Reliability tests are categorized as time-terminated tests or failure-terminated tests. Time-terminated reliability tests may result in a pass decision with no failures after completion of test execution over a required test duration. Failure-terminated tests are preferred over time-terminated tests since failure-terminated tests result in high confidence in the reliability test results and higher reliability data (e.g., failures over time or Time-to-Failure (TTF) data points) accuracy. In a failure-terminated reliability test, all Units Under Test (UUTs) are tested to failure and the TTF data is collected. The UUTs are repairable units containing multiple circuit card assemblies, each made up of multiple parts and components. All failures that occur in a reliability test will be analyzed to determine root cause failure mechanisms as well as the effects on the system if the failure is not corrected. These failure effects may lead to safety-critical effects or mission-critical effects. Assessments are performed to determine the severity and probability of occurrence of these effects to justify the cost to perform a design change to ensure the failure cause is eliminated and any associated hazards will not occur.

As stated in the paper published in the Millbank Quarterly, September 2013, titled, "Hospitals Still Far from Being Highly Reliable," Elizabeth Zhani [14] states that The Joint Commission [15] calls attention to traditional reliability methods to detect serious safety-related failures. The value of this paper was to create an awareness in the health-care and medical field of reliability testing. Too many health-care providers experience serious safety failures as routine and inevitable parts of daily work. To prevent hazards caused by these failures, affecting millions of patients each year, the article specifies a framework to assist hospitals in making progress toward high reliability and safety over long periods of time, comparable to commercial air travel, amusement park rides, and the nuclear power industry.

One of the 14 components of the high reliability framework that was described in the report was the need for hospital leadership to be committed to the ultimate goal of high reliability or zero patient harm. The hospital leadership section of the framework identifies specific roles for the board of trustees, the chief executive officer and all senior management (including nursing leaders), the engagement of physicians, the hospital quality strategy, its use of data on measures of quality, and the use of IT to support quality and safety improvement. "Although no hospital has been able to achieve high reliability, there are some very practical changes that can be made to improve safety and quality" says Dr. Chassin. "The time is now to start taking the steps needed to get from where we are today to where we want to be" [14].

Founded in 1951, The Joint Commission seeks to continuously improve health care for the public, in collaboration with other stakeholders, by evaluating health-care organizations and inspiring them to excel in providing safe and effective care of the highest quality and value. It evaluates and accredits more than 20,000 health-care organizations and programs in the United States, including more than 10,300 hospitals and home care organizations, and more than 6,500 other health-care organizations that provide nursing and rehabilitation center care, behavioral health care, laboratory and ambulatory care services. It currently certifies more than 2000 disease-specific care programs, focused on the care of patients with chronic illnesses such as stroke, joint replacement, stroke rehabilitation, heart failure, and many others. It also provides health-care staffing services certification for more than 750 staffing offices. An independent, not-for-profit organization, The Joint Commission is the nation's oldest and largest standards-setting and accrediting body in health care [14, 15].

12.7 Safety Test Data Collection

Safety test data collection must be planned early in the safety test planning process. The effectiveness of the test is determined by the data that is collected. It must be very clear to everyone involved in planning and conducting a test what data are to be collected before any testing starts. Data collection process begins as early in the system or product development process as possible. Data collection may occur with accumulation of design change records from engineering change requests made to a change control board. Data collection to calculate reliability usually occurs at the start of a reliability test event where TTF data for each unit in test is annotated. Data collection may also occur at the user application, under field or fleet conditions where users or customers collect the TTF data.

The method for data collection begins with test planning and selection of test samples. The test plans include the test requirements, the number of UUTs to select, a schedule for test start and ending, the Failure Reporting, Analysis, and Corrective Action System (FRACAS), the environmental test conditions, and a description of the OT for detection of failures using BIT or external TE. BIT is embedded test capability in the design of the system or product. The BIT or external TE should continuously monitor the UUTs so that the failure detection is almost immediately recorded close to the time a failure occurs. The selection of samples may be performed on a random basis, or be selected as the first articles from a production line. The number of samples is determined in the test plan. The more samples selected, the better the accuracy of the test due to the increased number of TTF data points that will be collected, but the more expensive the testing will be. The UUT serial numbers or unique identification numbers will be documented prior to the start of any test. All TTF data will correspond to these serial numbers. In a failure-terminated test, the test continues until all UUTs have a TTF recorded.

The following figure (Figure 12.1) provides an example of how the TTF data is tracked and collected during the course of a reliability test for each UUT. Figure 12.1 is a product or system UUT timeline histogram for test data collection.

Figure 12.2 is a method for binning the TTF data to conduct a best fit analysis. "X" represents the unit that failed for each of the 12 UUTs in the test. All 12 UUTs failed at least once. The test was designed to be a "failure-terminated" test so there will be at least 12 data points, one data point for each unit that fails. There were a total of 18 failures during the test. This means that several units failed more than once. Any unit that failed early in the test was repaired and returned to the test. The test continued until all UUTs failed at least once. Some units failed near the end of the test and did not return to the test following repair. The numbers below the Xs represents the assignments of data bins in time units of hours. For instance, the "X" over column 375 means the UUT failed between 251 and 375 hours, and the 2 Xs under column 1125 means 2 UUTs failed within the time durations 1001–1125 hours.

After the collection of the TTF data for each serial number, the data is entered into a tool to calculate the reliability metrics and plot graphs of the data points.

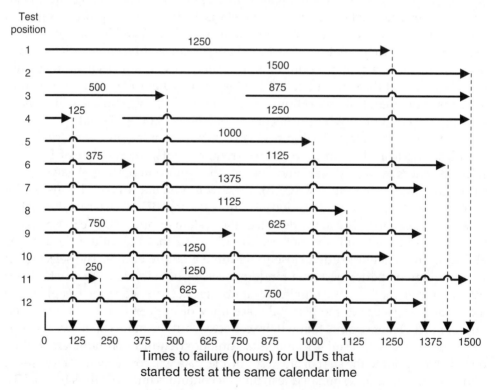

Figure 12.1   Test failure data by test position

									X		
									X		
			X	X				X	X		
X	X	X	X	X	X	X	X	X	X	X	X
125	250	375	500	625	750	875	1000	1125	1250	1375	1500

Figure 12.2 Test failure data by TTF

Historical distributions are fit to the data using the best fit approach. This curve fitting of distributions to the actual TTF data is useful in forecasting the probabilities of failure in the future. These historical distributions may be either a Probability Density Function (PDF) and a Cumulative Distribution Function (CDF). The better a PDF or CDF fits to the empirical TTF data, the better the accuracy of the reliability prediction or estimation. Confidence intervals and correlation factors are determined to understand how well the distribution fits the data. An example of a CDF graph is shown in Figure 12.3, and an example of a PDF curve is shown in Figure 12.4. More details on PDF are described in Chapter 14.

It is possible that the TTF data does not fit any one distribution very well. In this case, it usually means a Non-Homogenous Poisson Process (NHPP) where the failure intensity is not constant and design changes have occurred as failures are detected and isolated to a particular root cause.

There are alternatives to collection of TTF data that achieves the same purpose. Instead of TTFs, Cycles to Failure (CTFs) may be collected if the product being assessed is an actuator or a switch. Another alternative measurement is Miles to Failure (MTF) if the system being assessed is a moving vehicle such as an airplane or automobile.

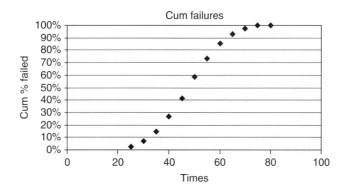

Figure 12.3 Example of a CDF

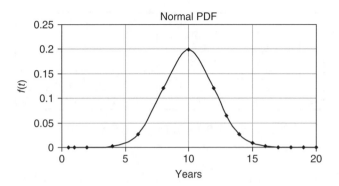

Figure 12.4 Example of a PDF

12.8 Test Results and What to Do with the Results

One of the most important aspects of testing is verification of the elimination or control of safety and health hazards. Test results need to be thoroughly reviewed by the safety personnel.

12.8.1 What to Do with the Test Results?

After testing is completed, there are things the safety engineer should be doing with the test results. First, the safety engineer needs to read them. Many times the testing is conducted, and the safety engineer is never provided with the results. If they are not forthcoming, the safety engineer should seek them out. Then, the HTS should be updated to reflect the closeout of hazards if mitigation is verified and updated to add new hazards if any are found during testing. And finally, the test results should be incorporated into the appropriate safety documentation such as Safety Assessment Reports (SARs), safety releases, or safety certification documents.

12.8.2 What Happens If the Test Fails?

Testing must include consideration of equipment and man-related failure:

- Are the failures related to mechanical, electrical, or chemical malfunctions?
- Are the failures the result of man/item incompatibility, inadequacy of procedural guidance, or inappropriate or inadequate training, selection, or orientation of personnel?

Test Incident Reports (TIRs) should be written and provided by the test organization (government or contractor) to provide the results of any incident occurring during testing. These should be reviewed for safety impact, and, if any new hazards are found, they should be added to the HTS, and work should begin to eliminate or mitigate the new hazard.

12.9 Design for Testability

Design for Testability (DFT) is a design technique that incorporates testability features to a hardware product design. The testability features make it easier to develop and apply development, manufacturing, and customer testing for the designed hardware and software. The purpose of development test is to verify the product design satisfies specification requirements, and all design defects are eliminated. The purpose of manufacturing test is to validate that the product hardware contains no manufacturing assembly and workmanship defects that could adversely affect the product's operation when used by customers.

Testability is a design characteristic that allows the status (operable, inoperable, or degraded) of an item to be determined and the isolation of faults within the item to be performed in a timely manner [16]. Testability analysis is the process of assessing the testability of a design in order to provide a quantitative measure of the design's fault detection and fault isolation capabilities and to optimize testability for a given test strategy.

Typical requirements might be Probability of Fault Detection (Pfd) at 95% for all failure modes of all Line Replaceable Units (LRUs) or Field Repairable Units (FRUs), and Probability of Fault Isolation (Pfi) at 95% for an ambiguity group of 3 for all failure modes of all Line Replaceable Units (LRUs) or Field Repairable Units (FRUs). An ambiguity group of 3 means that three LRUs are identified as potential causes of the failure. For prognostics, a requirement may be written which requires all LRUs or FRUs to report any high stress conditions 100 hrs prior to detection of 95% of any failure modes. The same BIT design capability used to detect and isolate faults may be able to detect and isolate imminent faults for prognostics capability by continuously monitoring the condition of the critical signals in situ to include design parametric stress measurements and data from sensors. [1]

12.10 Test Modeling

Test modeling is a design practice performed during the development of a product to identify fault coverage and functional coverage of the circuit. DSI eXpress [17] is a tool that can be used to perform test modeling. This tool can be used for the following:

- Perform test development for each function.
- Conduct BIT test allocation by analysis of design capabilities.
- Model interconnect and functions at the functional block level.
- Assess coverage by test type.
- Identify functional test gaps.

Test modeling should be performed as early as possible in the product development process. During the design process, a block diagram of the CCA should be modeled to verify test coverage, fault detection, fault isolation, and sound DFT approaches.

12.11 Summary

An SSTP is critical to ensure that all test events relevant to uncovering risks of safety hazards are accomplished and the data collected and analyzed. When a system program does not conduct testing for safety reasons to verify the system is safe, regulatory and compliance agencies may get involved. Sometimes, their involvement is cited in the system requirements and must be planned in an SSTP, or, at other times, their involvement is not planned and a result of customers or outside agencies getting involved to make sure systems are safe or unsafe risks discovered are corrected and prevented from ever happening again.

References

[1] Raheja, D. and Gullo, L., *Design for Reliability*, John Wiley & Sons, Inc., Hoboken, NJ, 2012.
[2] MIL-STD-882E, System Safety Standard Practice, U.S. Department of Defense, Washington, DC, 2012.
[3] Regulatory Compliance, https://en.wikipedia.org/wiki/Regulatory_compliance (Accessed on August 10, 2017).
[4] U.S. Consumer Product Safety Commission, MixBin Electronics Recalls 270k iPhone Cases, http://www.cpsc.gov/en/ (Accessed on August 10, 2017).
[5] U.S. Consumer Product Safety Commission, CPSC Testing-Certification, http://www.cpsc.gov/en/Business--Manufacturing/Testing-Certification/ (Accessed on August 10, 2017).
[6] Intertek, Edison Test Labs, http://www.intertek.com/marks/etl/ (Accessed on August 10, 2017).

[7] UL, Regulatory Testing, http://services.ul.com/service/regulatory-testing/ (Accessed on August 10, 2017).

[8] Underwriters Laboratories, UL 2016 Annual Report, http://www.ul.com/ (Accessed on August 10, 2017).

[9] DLS Electronic Systems, Inc., EMC, Wireless, Environmental, Product Safety Testing and Consulting, https://www.dlsemc.com/safety/ul/ul.htm (Accessed on August 10, 2017).

[10] CE, CE Definition, https://en.wikipedia.org/wiki/CE_marking (Accessed on August 10, 2017).

[11] STANAG 4404, NATO Standard on Safety Requirements and Guidelines for Munition Related Safety Critical Computing Systems, North Atlantic Treaty Organization, Brussels, 1997.

[12] Gullo, L., Villarreal, J., and Swanson, R., Value Added Benefits of Embedded Sensors in Tactical Rocket Motors: The Key to Successful PHM Implementation, Joint Army-Navy-NASA-Air Force (JANNAF) Propulsion Meeting, June 2015, Nashville, TN.

[13] IEEE P1856, Standard Framework for Prognosis and Health Management of Electronic Systems, The Institute of Electrical and Electronics Engineers, Inc., New York.

[14] Elizabeth, Z., Hospitals Still Far from Being Highly Reliable, Millbank Quarterly, September 17, 2013.

[15] The Joint Commission, https://www.jointcommission.org/ (Accessed on August 10, 2017).

[16] MIL-HDBK-2165, Testability Program for Electronic Systems and Equipments, U.S. Department of Defense, Washington, DC, 2014.

[17] DSI, eXpress: System Modeling for Diagnostic Design and Analysis, http://www.dsiintl.com/weblogic/Products.aspx (Accessed on August 10, 2017).

13

Integrating Safety with Other Functional Disciplines

Louis J. Gullo

13.1 Introduction

This chapter focuses on integrating safety engineers with other types of engineers to drive safety engineering principles and techniques during system or product development. The chapter emphasizes the need for system safety engineers to establish strong interfaces with other functional disciplines and discusses how safety engineers can build strong relationships within their program, product, systems engineering, and development teams. The functional disciplines are those people involved in different facets of systems engineering and program management. The interfaces with the various functional disciplines may be classified into three types. These types of interfaces are (1) internal interfaces with fellow employees of a development organization, (2) interfaces with customer organizations, and (3) interfaces with supplier organizations. Most of the contents of this chapter apply to all three types of functional discipline interfaces.

System safety engineers must be integrated with systems engineers, hardware and software development engineers, and other personnel in multidisciplinary engineering fields and non-engineering professionals within their internal development organizations. All these engineers and business professionals must operate

Design for Safety, First Edition. Edited by Louis J. Gullo and Jack Dixon.
© 2018 John Wiley & Sons Ltd. Published 2018 by John Wiley & Sons Ltd.

together if the organization is to be successful. It is imperative that engineering personnel perform their respective functions effectively and efficiently in a diverse development team environment writing good engineering development plans and schedules, generating requirements, conducting analyses and tests, documenting their analyses data and test findings, and creating analyses and test reports in a tight and cohesive partnership, forming an appropriate balance between safety, cost, schedule, and performance that results in a system that operates as intended by the customer and is void of system safety hazards.

13.1.1 Key Interfaces for Systems Safety Engineering

System safety engineering interfaces with multiple specialty engineering disciplines and non-engineering professionals. Systems engineering is the primary interface for a system safety engineer. In many organizations, the lead systems engineer on the program is an Integrated Product Team Lead (IPTL). System safety engineers will report to the system's IPTL, as do the other specialty engineers, which include reliability engineers, maintainability engineers, and supportability engineers, to name a few. The system safety engineer will interface with these engineers in their Integrated Product Team (IPT) on a regular basis, usually weekly. There are many other technical and human-centered disciplines that a system safety engineer interfaces with. Besides systems engineering, key interfaces for system safety engineering include, but are not limited to, the following disciplines:

- Chief engineers for a particular program or product line
- Electrical engineering (including circuit designers for analog, digital, RF, power, and component devices)
- Mechanical engineering
- Software engineering
- System Security Engineering (SSE) (including Information Assurance (IA), Operations Security (OPSEC), Communications Security (COMSEC), and anti-tamper engineering)
- Software assurance
- Cybersecurity engineering
- Reliability, Maintainability and Availability (RMA) engineering
- Product engineering
- Test engineering
- Interface design engineering
- Control engineering
- Human factors engineering (including Human System Integration (HSI))
- Regulatory and compliance engineering
- Logistics, sustainment, and supportability engineering

- Manufacturing and production engineering
- Operations research
- Performance engineering
- Program and project management (including planning, scheduling, and proposal management)
- Quality engineering (including design assurance, design for six sigma engineering, mission assurance, systems assurance, supplier quality, and product assurance)
- Business development (including customer engineering, application engineering and marketing)
- Contract and subcontract professionals
- Legal counsels and law professionals
- Configuration management
- Risk management
- Environmental Health and Safety Engineering (EHSE)
- Industrial engineering
- Materials engineering
- Supply chain engineering
- Field service engineering

System Safety Engineers (SSEs) interface with these disciplines within the system development phase during execution of various systems engineering processes. SSEs are typically most involved with the disciplines of reliability, maintainability, quality, logistics, human factors, software engineering, and test engineering. These disciplines should be involved in the hazard analysis, hazard control, hazard tracking, and risk resolution activities. A diagram showing more specific information on the various interfaces and data that are passed back and forth between safety and other functional disciplines is shown in Figure 13.1. This figure not only illustrates some of the common relevant interfaces that should be established within a system development team but also shows the data flow types and direction of data flows between SSEs and other disciplines.

13.1.2 Cross-Functional Team

As stated in Chapter 5, a cross-functional team is usually formed to identify missing functions in a specification, using a System Requirements Analysis (SRA), which is always conducted by a team with at least one member from each of the following engineering disciplines should be present: R&D, design, quality, reliability, safety, manufacturing, and supportability or field service. To find system design requirement flaws early, a cross-functional team has to view the system documentation and

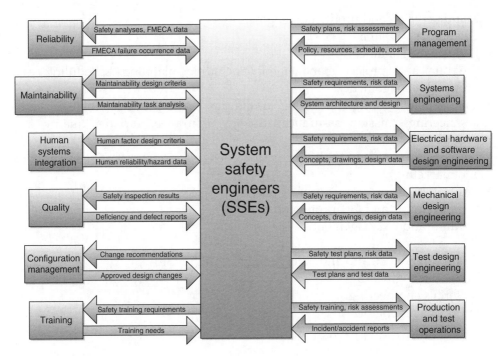

Figure 13.1 SSE interfaces to other functional disciplines

models from different angles. A cross-functional system team must be able to view the system requirements from many different perspectives to make sure all the different functional concerns are covered.

Integrating with other engineers involves showing off your value and proving your worth every day. An integrated team works together to uncover all unknowns that are relevant to system operational performance risks. It must be cognizant to develop anomaly detectors to find unknown conditions in the system that could affect system operational performance and determine the risks associated with each condition. Anomaly detectors should be developed to detect two types of anomalies. Type 1 is the unknown unknown anomalies and type 2 is the known unknown. Unknown unknown anomalies are those conditions that we have no idea what they are and that they even exist. The known unknowns are anomalies that we know about, but don't know how they are caused. Type 2 anomalies are conditions we have seen and have felt their effects, but have not determined how to prevent them. More data is needed before a fix can be found. This data collection requires a diverse background of engineers to ask the right questions and develop a means to capture the problem and develop possible solutions to correct it.

13.1.3 Constant Communication

There is no better way to interface with a variety of engineering and non-engineering disciplines then through constant communications. This is why agile development methods are becoming so prevalent in the system and software development world. Agile development requires daily communication within the teams, known as scrum. Scrums are an opportunity for each team member to share their accomplishments and near-term plans with their team members on a daily basis. Besides reporting status, the scrum encourages the sharing of existing or potential roadblocks so that they are known in advance and adjustments can be made quickly to prevent or minimize the cost, schedule, and performance effects of the roadblocks. Many team members will embrace open and candid communication in the scrums. Some may choose to wait to communicate more privately in one-on-one meetings with their scrum master or peers. Many members of the scrum will choose to communicate providing war stories to demonstrate through details that they understand what is being done by their team and others can learn from their shared experiences, and all will support the team with new ideas, thinking outside the box and being innovative in their approaches to work and to address challenges that may require novel solutions.

13.1.4 Digital World

We live in a digital world. Living a life immersed in technology has become a normal way of life for many people. We use digital mobile devices that are interconnected on digital communication networks. Our digital world follows a logical system for data handling and information flow. This digital logic system uses digital data processing following Boolean algebra, with defined inputs and combinational logic equations to calculate outputs in a truth table. Truth tables are practical logic aids in analyzing digital circuits, which allows a complete understanding of logic circuit performance. Electrical devices such as digital integrated circuits use Boolean algebra to process logical expressions and conditions to determine resultants. Digital inputs and outputs may be thought of as positive and nonpositive values. All digital data is coded into machine language as ones (1s) and zeroes (0s). These two values are prolific across our digital world no matter if we are looking at the state of the art for today's electronics technology or the way we conduct ourselves on social media and with interpersonal relationships.

Consider the correlation between how the technology works versus how we process data as humans. Decisions must result in yes or no outcomes, like going forward with changes or remaining with status quo. The questions to ask yourself can be boiled down into simply deciding if you take an action or not take an action. In other words, after you collect enough data, you decide to do something or not do something, no matter the type of action being contemplated. Concerning a personal relationship with a significant other, you decide together if you want your relationship to be legal as a married couple or continue as you are together, unmarried and living together. There is no partially married. You either are married or you are not married. Let us take this decision process a bit further considering many more types of digital data processing that we conduct all the time. There is right and wrong, black and white, light and dark, on and off, open and close, in and out, supported and unsupported, working together and working separately, for and against, and agreement and disagreement, such as being "on board" with decisions of the team or not being "on board." When you are "on board" with decisions of the team, regardless of whose idea it was, the team stands behind the decision and each team member speaks for the decision as if it were their own decision.

Did you ever hear the phrase "Close only happens with horseshoes and hand grenades"? There is no "close" for everything else in the digital world. In the legal world, there either is evidence of something being illegal or there is no evidence of something being illegal. This determination helps to build court cases and decides if a case should be brought to a court of law.

In the engineering world, this is also true. The engineering community must collect data and use this data to make decisions, such as making a design change based on evidence of a hazard or failure mode. The engineering team is constantly asked to make tough decisions. Sometimes a team decision is reached that does not have full team consensus. This is the result of an ineffective team. A team must strive to reach consensus. This consensus happens with negotiations. Everyone on the team should feel they are getting some of what they want in a decision. No one team member should feel they got everything they wanted in a decision. It must be a unified approach to solving problems, which means no one person is driving the solution. It must be a group decision and at the final conclusion for the decision, everyone agrees to the decision and will fully support it and drive to make it reality.

There is no middle of the road. There is no abstinence. You must make a commitment to your team and be "all in," so the team will be successful.

13.1.5 Friend or Foe

Think friend or foe when deciding on your relationship with your coworker. If your coworker is your friend, you get along with that person, you respect that person's opinion, and you will work together as part of an integrated team. If your coworker is your foe, you may not agree with one another on all things, but you respect each other and understand each other's position on a topic. You embrace diversity of thought freely. You know you are driving for a common cause and common goal. If you do not consider your coworker your friend, then the coworker is your enemy. Either you must work to turn that relationship around so your coworker becomes your friend or an attitude adjustment must occur for both of you if you are to be effective contributors within a team environment.

Determine who your friends are and who your enemies are. Keep your friends close and your enemies closer. Do not let your enemies continually take advantage of you. Approach your enemies with others that support you. There is strength in numbers.

Friend-or-foe decision starts when you first meet someone. Ever heard of "Judging a book by its cover"? You should recognize this since we all do it, whether we admit it or not. You meet a new engineer who joins your team and you look at how that person is dressed and how the person speaks. You decide from the moment you meet someone new whether that person is your friend or your foe. We decide at that initial moment whether you feel you can work with that person or you feel threatened by that person.

The friend-or-foe decision continues for many encounters between the two people that just met for several more encounters until the realization is made by each person whether the other is their friend or their foe. When you hear a person who has some issues with you on the job, say, things against you, they might say, "It was only business, don't take it personally." There can be no more of a wrong thing to say. It is all personal. If you are on a team, it is always personal. It is your personal decision to be on a team or not. Just because your boss assigns you to a team does not mean you cannot leave that team if you have interpersonal issues. Don't stay on the team if you feel devalued. These team members are your foe and you need to do something about this. No one should feel they need to suffer the stresses of being on a team without feeling being a valued and respected team member. If you do feel stressed and non-valued and respected, then leave the team, leave the location, leave the company, change careers, do whatever you do to gain your self-esteem, and avoid stresses of work, because it all adds up to your happiness in life and a long healthy life span. And that means business is most certainly personal. You spend more of your life in a work environment than in any other type of environment in your life. The majority of your personal life is in business environments, so you need to make that environment comfortable and friendly, and as stress free as much as possible.

If the team does not openly accept your contributions or backstabs you after accepting your thoughts, by secretly belittling you within a closed circle of acquaintances, this is a noneffective teamwork and can diminish the capability of the team and ultimately destroy a team. All team members need to feel included in a team from the initial introductions, and the team members need to do all they can to bring the new team member up to speed quickly to accelerate their learning curves and make the new team member a valued contributor. The faster the team is integrated, and all team members brought on quickly to make them feel that they are a valued member of the team, the faster the team becomes effective and efficient.

When new team members approach each other, they should have already built an understanding of the environment they work within, the roles of all team members, and the ways the team communicates. When new team members approach each other, they should be able to initially identify with each other as a friend and a partner. But what if something previously occurred between them in their past where they did not get along, and there are negative feelings between them? What if the feelings are strong, but these feelings can be suppressed to enable them to work together? As an analogy, consider the need for defense systems that identify oncoming military aircraft as friend or foe. This type of system was developed to prevent mistaken identity and allies shooting at each other. If a friendly aircraft interrogates an approaching aircraft with the sensors to determine the presence of a friendly aircraft beacon, the aircraft pilot knows not to shoot at the other aircraft. This type of system is called Integrated Friend or Foe (IFF). IFF is a safety-critical system that requires a multidisciplinary engineering approach to get it right. A failure of the IFF when used in a military aerospace context results in catastrophic effects and potential loss of life when two friendly aircraft commit fratricide and try to shoot each other down.

Another similar type of aircraft system that is used in commercial aircraft instead of military aircraft is the Traffic Collision and Avoidance System (TCAS) using a safety-critical system used for system safety functionality. TCAS is system that helps aircraft in flight avoid collisions and avoid conflicts with other aircraft. Wouldn't it be wonderful if such systems existed between people for rapidly ramping up mutually cooperation and friendliness to work together, facilitate communications, and avoid conflicts?

Next, let's consider what one company says about its code of conduct and core values. Raytheon has established a code of conduct that drives the company forward with high ethical business standards.

13.2 Raytheon's Code of Conduct

Raytheon's code of conduct (the "Code") and its company-wide ethics and business conduct program articulate Raytheon's expectations of its leaders, employees, suppliers, agents, consultants, and representatives to undertake actions that are consistent with Raytheon's guiding business values, relevant laws and regulations, and company policies and procedures. Raytheon has earned a reputation as a technology and innovation leader specializing in defense, civil government, and cybersecurity markets throughout the world. This reputation rests on the company's unwavering commitment to the highest standards of ethical business conduct. Employees, customers, suppliers, shareholders, and others all benefit from Raytheon's reputation as a company of integrity. Notice Raytheon's values as shown in Figure 13.2. Four of these values stress the importance of working together. These are trust, respect, collaboration, and accountability.

Employees are expected to treat fellow employees, customers, and business partners with respect and dignity. We value an inclusive workforce because it promotes diversity of thinking and helps us collaborate to achieve innovative solutions that meet the needs of our customers. We are committed to fair employment practices. Our employment-related decisions are made without regard to race, color, creed, religion, national origin, sex, sexual orientation, gender identity and expression, age, disability or veteran status. Raytheon leaders at all levels have a special obligation to encourage an open work environment and ethical culture where employees are treated respectfully and may raise issues or concerns without fear of retaliation. [1]

Employees need to be treated with dignity and respect by their managers. They need to treat one another with dignity and respect, and they should work together as a team where everyone's function and contributions to the team are respected and valued. Everyone needs to perform as a team player to earn the dignity and respect of the team.

Our Values

Trust
- » We take pride in our ethical culture, are honest, and do the right thing.

Respect
- » We are inclusive, embrace diverse perspectives and value the role we all play in our shared vision.

Collaboration
- » We fuel more powerful ideas, deeper relationships and greater opportunities to achieve shared objectives together.

Innovation
- » We challenge the status quo and act with speed and agility to drive global growth.

Accountability
- » We honor our commitments, anticipate the needs of our customers, serve our communities and support each other.

Figure 13.2 Raytheon's values

Now the big question, "What can you do if you are not treated with dignity and respect, and you are not valued as a contributor to the team?" As a starter, you should learn the seven habits of highly effective people and constantly work to practice these habits every day. Develop these habits so they become your paradigm, a paradigm to practice daily as you fulfill your role as a system safety engineer, or any other type of engineer, doing what matters to the team. What matters to the system safety engineer is successful employment of Design for Safety (DfS) methodologies that are accepted by the team.

13.3 Effective Use of the Paradigms for Design for Safety

Consider the 10 paradigms discussed in Chapter 1. These paradigms are imperative for safety engineers in order to design for safety. Examples of these DfS paradigms that are applicable to integrating with other functional disciplines are the following:

- Always aim for zero accidents.
- Be courageous and "Just say no."
- Taking no action is usually not an acceptable option.
- If you stop using wrong practices, you are likely to discover the right practices.

This chapter shows you how to be effective when using these paradigms.

The Seven Habits of Highly Effective People embody many of the fundamental principles of human effectiveness. They represent the internalization of correct principles upon which enduring happiness and success are based. But before we can really understand these Seven Habits, we need to understand our own 'paradigms' and how to make a 'paradigm shift'. Both the Character Ethic and the Personality Ethic are examples of social paradigms.

Character ethic represents basic principles of effective living, such as bravery, humility, temperance, fidelity, modesty, and integrity, to name a few principles. It is your core of being. Personality ethic refers to ones attitudes and behaviors, skills, and traits that are visible to the public, which facilitate human integration and interface. It is your means to achieve results quickly and easily without much work. Personality ethic is the symbol without the substance, while character ethic is all about the substance. The word "paradigm" originates from the Greek language. It was originally meant to be a scientific term, but now it is commonly used to mean a model, theory, assumption, or frame of reference.

In the more general sense, it's the way we 'see' the world—not in terms of our visual sense of sight, but in terms of perceiving, understanding, and interpreting. For our purposes, a simple way to understand paradigms is to see them as maps. We all know that 'the map is not the territory.' A map is simply an explanation of certain aspects of the territory. That's exactly what a paradigm is. It is a theory, an explanation, or model of something else. [2]

Paradigms are powerful because they create the lens through which we see the world. The power of paradigm shift is the essential power of quantum change, whether that shift is an instantaneous or a slow and deliberate process. [2]

Habits are powerful factors in our lives. Because they are consistent, often unconscious patterns, they constantly, daily, express our character and produce our effectiveness...or ineffectiveness. For our purposes, we will define a habit as the intersection of knowledge, skill, and desire. Knowledge is the theoretical paradigm, the 'what to do' and the 'why'. Skill is the 'how to do'. And desire is the motivation, the 'want to do'. In order to make something a habit in our lives, we have to have all three.

The seven habits of highly effective people, which are the fundamental principles of human effectiveness, are listed as follows [2]:

- Habit 1: Be proactive
- Habit 2: Begin with the end in mind
- Habit 3: Put first things first
- Habit 4: Think win/win
- Habit 5: Seek first to understand, then to be understood
- Habit 6: Synergize
- Habit 7: Sharpen the saw

Habits 1–3 provide for private victory. Habits 4–6 provide for public victory. Habit 7 is renewal and refreshing the habits. Moving from habit 1 to habit 6 is the development of a person from dependence to interdependence.

Think Win/Win (Habit 4 from the Seven Habits of Highly Effective People)

Win/Win is a frame of mind and heart that constantly seeks mutual benefit in all human interactions. Win/Win means that agreements or solutions are mutually beneficial, mutually satisfying. With a Win/Win solution, all parties feel good about the decision and feel committed to the action plan. Win/Win sees life as a cooperative, not a competitive arena. Win/Win is based on a paradigm that there is plenty for everyone, that one person's success is not achieved at the expense or exclusion of the success of others. Win/Win is the belief in the Third Alternative. It's not your way or my way; it's a better way, a higher way. [2]

Practice the seven habits, but especially Habit 5, which is "Seek first to understand, then to be understood," and Habit 6, which is "Synergize." These two habits are essential to truly be effective as a member of an engineering team where everyone depends on each other to do their job and make decisions together. Practice Habit 7, "Sharpening the saw," on a periodic basis, such an annually, to refresh yourself on the six other habits if you are not constantly applying them, and they have not yet become your paradigm.

Principles of Empathetic Communications

Although it is risky and hard, seek first to understand, or diagnose before you prescribe, is a correct principle manifest in many areas of life. It's the mark of all true professionals. It's critical for the physician. You wouldn't have any confidence in a doctor's prescription unless you had confidence in the diagnosis. [2]

If you really seek to understand, without hypocrisy and without guile, there will be times when you will be literally stunned with the pure knowledge and understanding that will flow to you from another human being. It isn't even always necessary to talk in order to empathize. In fact, sometimes words may just get in your way. That's one very important reason why technique alone not work. That kind of understanding transcends technique. Isolated technique only gets in the way. [2]

"Seek first to understand" is a powerful habit of effective interdependence while using empathetic communications.

Empathetic listening takes time, but it doesn't take anywhere near as much time as it takes to back up and correct misunderstandings when you're already miles down the road, to redo, to live with unexpressed and unsolved problems, to deal with the results of not giving people psychological air. A discerning empathetic listener can read what's happening down deep fast, and can show such acceptance, such understanding, that other people feel safe to open up layer after layer until they get to that soft inner core where the problem really lies. People want to be understood. And whatever investment of time it takes to do that will bring much greater returns of time as you work from an accurate understanding of the problems and issues and from the high Emotional Bank Account that results when a person feels deeply understood. Before the problems come up, before you try to evaluate and prescribe, before you try to present your own ideas—seek to understand. When we really, deeply understand each other, we open the door to creative solutions and third alternatives. Our differences are no longer stumbling blocks to communication and progress. Instead, they become the stepping stones to synergy. [2]

Principles of Creative Cooperation

What is synergy? Simply defined, it means that the whole is greater than the sum of its parts. It means that the relationship which the parts have to each other is a part in and of itself. It is not only a part, but the most catalytic, the most empowering, the most unifying, and the most exciting part. The creative process is also the most terrifying part because you don't know exactly what's going to happen or where it is going to lead. You don't know what new dangers and challenges you'll find. It takes an enormous amount of internal security to begin with the spirit of adventure, the spirit of discovery, the spirit of creativity. Without doubt, you have to leave the comfort zone of base camp and confront an entirely new and unknown wilderness. You become a trailblazer, a pathfinder. You open new possibilities, new territories, new continents, so that others can follow. [2]

Synergy is everywhere in nature. If you plant two plants close together, the roots comingle and improve the quality of the soil so that both plants will grow better than if they were separated. The challenge is to apply the principles of creative cooperation, which we learn from nature, in our social interactions. Synergy is exciting. Creativity is exciting. It's phenomenal what openness and communication can produce. The possibilities of truly significant gain, of significant improvement are so real that it's worth the risk such openness entails. [2]

Now that you realize the need to synergize and to develop open communications with business professionals and your program and product development team members, you should learn how to be successful in all your social interactions especially related to your fellow team mates on the system development team. All your social interfaces will greatly improve when you study the wisdom from Dale Carnegie in his book: "How to Win Friends and Influence People" [3]. Dale Carnegie's wisdom will expand your interpersonal skill horizons beyond the seven habits. The seven habits are the starting point, but not the end. Next you need to learn how to influence people to achieve your system safety objectives in designing for safety and get members of your team to work with you with the common goal to improve system safety.

13.4 How to Influence People

We need practical ways to make new friends in your daily work activities and influence people to your way of thinking. You will learn practical examples and methods described by Dale Carnegie in his book 80 years ago. We will discuss the timeless proven methods to develop strong bonds with coworkers and peers that are still applicable today. Actually, these methods will work for all professional business people that you interface with on a regular or intermittent basis, no matter if the personnel work for your organization or they are customers or your suppliers and business partners. Any of these social interactions will benefit by learning the basics from Dale Carnegie [3].

Consider for a moment that a dog makes his or her living by giving you nothing but love.

> When I was five years old, my father bought a little yellow-haired pup for fifty cents. He was the light and joy of my childhood. Every afternoon about four-thirty, he would sit in the front yard with his beautiful eyes staring steadfastly at the path, and as soon as he heard my voice or saw me swinging my dinner pail through the buck brush, he was off like a shot, racing breathlessly up the hill to greet me with leaps of joy and barks of sheer ecstasy. Tippy was my constant companion for five years. Then one tragic night—I shall never forget it—he was killed within ten feet of my head, killed by lightning. Tippy's death was the tragedy of my boyhood. [3]

Dale goes on to explain the valuable lesson he learned from his boyhood best friend. He learned that Tippy knew by some form of instinct that you can make more friends in 2 months by becoming genuinely interested in other people than you can in 2 years by trying to get other people interested in you. Generically speaking, people are not interested in you or me; rather, they are mostly interested in themselves most of the time (morning, noon, and night). Talking about themselves is anyone's favorite subject to talk about. "If we merely try to impress people and get people interested in us, we will never have many true, sincere friends. Friends, real friends, are not made that way" [3].

Dale Carnegie said that there are six ways to make people like you [3]:

1. Be genuinely interested in other people.
2. Smile a lot.
3. Remember people's names.
4. Be a good listener.
5. Learn people's interests.
6. In conversations, sincerely make the other person feel important.

Dale Carnegie said there are 12 ways to win people to your way of thinking [3]:

1. Avoid arguments.
2. Respect a person's opinions and never say, "You're wrong."

3. Admit your errors quickly and emphatically.
4. Begin conversations in a friendly way.
5. Facilitate positive feedback and get a "yes, yes" response immediately.
6. Let the other person do most of the talking.
7. Let the other person feel your great idea is their own.
8. Honestly see things from the other person's point of view (walk in their shoes).
9. Express sympathy for the other person's ideas and desires.
10. Appeal to ethical, noble, and lofty motives.
11. Express passion and dramatize your ideas.
12. Establish a challenging goal to work on together.

If you follow the six ways to make people like you and the 12 ways to win people to your way of thinking, then you will surely be successful in all your social interactions, especially performing as a valued and respected system safety engineer in your development teams. This social success includes convincing people of the difficult systems engineering design decisions that involve safety engineering and the incorporation of system or product design changes for the purpose of the prevention of hazards and safety-critical failures.

What if you have experienced the benefits of the seven habits of highly effective people and you know how to win friends and influence people through repeated practice of your system safety skills and knowledge in your development team, but have not fully realized the benefits with your own career progression and job satisfaction? This is why you should explore the techniques of emotional intelligence (EQ), which will give you additional awareness of yourself and your social surroundings to achieve all your objectives.

13.5 Practice Emotional Intelligence

The daily challenge of dealing effectively with emotions is critical to the human condition because our brains are hard-wired to give emotions the upper hand. Here's how it works: everything you see, smell, hear, taste, and touch travels through your body in the form of electric signals. These signals pass from cell to cell until they reach their ultimate destination, your brain. They enter your brain at the base near the spinal cord, but must travel to your frontal lobe (behind your forehead) before reaching the place where rational, logical thinking takes place. The trouble is, they pass through your limbic system along the way—the place where emotions are produced. This journey ensures you experience things emotionally before your reason can kick into gear. The rational area of the brain (the front of your brain) can't stop the emotion 'felt' by your limbic system, but the two areas do influence each other and maintain constant communication. The communication between your emotional and rational 'brains' is the physical source of emotional intelligence. [4]

When emotional intelligence was first discovered, it served as the missing link in a particular finding: people with the highest levels of intelligence (IQ) outperformed those with average IQs just 20 percent of the time, while people with average IQs outperformed those with high IQs 70 percent of the time. This anomaly threw a massive wrench into what many people had always assumed was the source of success—IQ. Scientists realized there must be another variable that explained success above and beyond one's IQ, and years of research and countless studies pointed to emotional intelligence (EQ) as the critical factor. [4]

Basically, EQ is composed of four skills: self-awareness, self-management, social awareness, and relationship management. Your ability to perform the skills of self-awareness and self-management determines your personal competence. Your ability to perform the skills of social awareness and relationship management determines your social competence. "Personal competence is your ability to stay aware of your emotions and manage your behavior and tendencies. Social competence is your ability to understand other people's moods, behavior, and motives in order to improve the quality of your relationships" [4].

Now, let's drill into the strategies of these four skills.

Self-awareness strategies [4]

- Quit treating your feelings as good or bad.
- Observe the ripple effect from your emotions.
- Lean into your discomfort.
- Feel your emotions physically.
- Know who and what pushes your buttons.
- Watch yourself like a hawk.
- Keep a journal about your emotions.
- Don't be fooled by a bad mood.
- Don't be fooled by a good mood, either.
- Stop and ask yourself why you do the things you do.
- Visit your values.
- Check yourself.
- Spot your emotions in books, movies, and music.
- Seek feedback.
- Get to know yourself under stress.

Self-management strategies [4]

- Breathe right.
- Create an emotion versus reason list.
- Make your goals public.
- Count to ten.
- Sleep on it.
- Talk to a skilled self-manager.
- Smile and laugh more.
- Set aside some time in your day for problem solving.
- Take control of your self-talk.
- Visualize yourself succeeding.
- Clean up your sleep hygiene.
- Focus your attention on your freedoms, rather than your limitations.
- Stay synchronized.
- Speak to someone who is not emotionally invested in your problem.
- Learn a valuable lesson from everyone you encounter.
- Put a mental recharge into your schedule.
- Accept that change is just around the corner.

Social awareness strategies [4]

- Greet people by name.
- Watch body language.
- Make timing everything.
- Develop a back-pocket question.
- Don't take notes at meetings.
- Plan ahead for social gatherings.
- Clear away the clutter.
- Live in the moment.
- Go on a 15-minute tour.
- Watch EQ at the movies.
- Practice the art of listening.
- Go people watching.
- Understand the rules of the culture game.
- Test for accuracy.
- Step into their shoes.
- Seek the whole picture.
- Catch the mood of the room.

We agree with the value of many of these strategies and techniques. Many of these are common sense approaches to social interaction. We realize that many of these are repeats from the "Seven habits" and "How to influence," but when we put these all together, they fill gaps in our social prowess and completely round out our interpersonal development.

We should also point out that we don't necessarily agree with all of these techniques. For instance, we tend to disagree with the fifth strategy for social awareness: "Don't take notes at meetings." We find many times that if we don't take notes at meetings, we will forget what was said or action items that we took and need to respond back to someone at a certain time. Some people in team meetings appreciate when their fellow team members are taking notes based on something important they hear and to remind themselves to follow up on the information to gather more data. We understand what this strategy is on the list since it is difficult to read body language and collect all the data (like nonverbal communication behaviors) from a meeting's participants. You cannot gain a complete awareness of thoughts being communicated in a room if your head is down, stuck in note-taking mode. The trade-off is dependent on the memory abilities of each person. If someone is dedicated as minute taker at the meeting, then this alleviates the need for each meeting participant to take their own notes and can then all focus on reading the chemistry of each person in the meeting.

Relationship management strategies [4]

- Be open and be curious.
- Enhance your natural communication style.
- Avoid giving mixed signals.
- Remember the little things that pack a punch.
- Take feedback well.
- Build trust.
- Have an "open door" policy.
- Only get mad on purpose.
- Don't avoid the inevitable.
- Acknowledge the other person's feelings.
- Complement the person's emotions or situation.
- When you care, show it.
- Explain your decisions, don't just make them.
- Make your feedback direct and constructive.
- Align your intention with your impact.
- Offer a "fix-it" statement during a broken conversation.
- Tackle a tough conversation.

The third strategy of relationship management really resonates with us. "Mixed signals" occur when a person says one thing and does another, or a person says one thing and then a short time later say the exact opposite. It causes confusion when a team is looking for direction on a path to take in approaching future plans or trying to make decisions when multiple options and courses of action exist, and it could also extend beyond team confusion to an erosion in the team's confidence and trust in a team leader's abilities. Team leaders could send mixed signals when designating team members with certain roles without demonstrating confidence and trust in that team member. As an example, if the team leader places responsibility to accomplish an engineering analysis task on a particular team member, and the team leader goes behind this person's back and assigns the same task to someone else who the team leader has already established a certain level of trust with, this undermines the team's trust in the team leader. Team leaders must strive to build trust in their team and follow the sixth strategy of relationship management.

If you believe passionately about a design change that must take place to eliminate the risk of a catastrophic effect from a product failure or hazard in order to save lives when a product is used in a customer application, then you should do everything in your power to make the design change a reality and fulfill your right to become an unsung hero. You should practice the techniques discussed in this chapter to interface and integrate with all the functional disciplines that exist in your organization. Practice Dr. Stephen Covey's seven habits, use Dale Carnegie's 12 ways to win people to your way of thinking, and exercise your EQ to achieve the objectives that you seek. If you follow Dale Carnegie's 12 ways to win people over, step by step, and you flawlessly perform the EQ relationship management strategies and still are unsuccessful in implementing a design change, keep trying and be persistent and committed to the objective you seek. If you have tried and tried and tried to make the effort using these techniques to enact the design change, but have repeatedly been unsuccessful, we recommend you consider another approach, the ultimate approach to influencing people. This approach is called positive deviance. You tap your personal and professional passions to influence vital behaviors.

13.6 Practice Positive Deviance to Influence People

To learn how to develop one of the most important of all influence methods, we travel to Atlanta, Georgia, and meet Dr. Donald Hopkins and his staff at The Carter Center. Their work across Africa and Asia teaches us how to identify a handful of vital behaviors that help change the habits of millions of people. In this case, he and his colleagues help change the dangerous water-drinking habits of millions of remote villagers. Hopkins's work on applying principles of 'positive deviance' helps us all understand what it takes to discover a handful of high-leverage behaviors that drive virtually every change effort we'll ever undertake. [5]

Since 1986, Dr. Hopkins and his team at The Carter Center in Atlanta have focused on the eradication of the Guinea worm disease. The Guinea worm is one of the largest human parasites (it can grow to three feet long), and it has caused incalculable pain and suffering in millions of people. When West Asian and sub-Saharan villagers drink stagnant and unfiltered water, they take in the larvae of Guinea worms, which then burrow into abdominal tissues and slowly grow into enormous worms. Eventually the worms begin to excrete an acid-like substance that helps carve a path out of the host human's body. Once the worm approaches the skin's surface, the acid causes painful blisters. To ease the horrific pain, victims rush to the local water source and plunge their worm-infected limbs into the pond for cooling relief. This gives the worm what it wanted—access to water in which to lay hundreds of thousands of eggs, thus continuing the tragic cycle. [5]

Sufferers cannot work their crops for many weeks. When parents are afflicted, their children may drop out of school to help out with chores. Crops cannot be cultivated. The harvest is lost. Starvation ensues. The cycle of illiteracy and poverty consumes the next generation. For over 3500 years the Guinea worm has been a major barrier to economic and social progress in dozens of nations. Hopkins was interested in this particular disease because he knew that if 120 million people in 23,000 villages would change just a few vital behaviors for just one year, there would never be another case of the infection. But imagine the audacity of intending to influence such a scattered population in so many countries—frequently faced with corrupt or nonexistent health systems or fragile political stability. And yet this is exactly what Hopkins's team has done. Soon he and his colleagues will have laid claim to something never before accomplished in human history. They will have eradicated a global disease without finding a cure. Despite this enormous disadvantage, Hopkins and his small band of intrepid change agents will have beaten the disease with nothing more than the ability to influence human thought and action. The implications of Hopkins's work for individuals, businesses, and communities are enormous. [5]

Positive deviance can be extremely helpful in discovering the handful of vital behaviors that will help solve the problem you're attacking. That is, first dive into the center of the actual community, family, or organization you want to change. Second, discover and study settings where the targeted problem should exist but doesn't. Third, identify the unique behaviors of the group that succeeds. When members of The Carter Center team began their assault on Guinea worm disease, they used this exact methodology. They flew into sub-Saharan Africa and searched for villages that should have Guinea worm disease, but didn't. They were particularly interested in studying villages that were immediate neighbors to locations that were rife with Guinea worm disease. Eventually, the team discovered its deviant village. It was a place where people rarely suffered from the awful scourge despite the fact that the villagers drank from the same water supply as a nearby highly infected village. It didn't take long to discover the vital behaviors. Members of the team knew that behaviors related to the fetching and handling of water would be particularly crucial, so they zeroed in on those. In the worm-free village, the women fetched water exactly as their neighbors did, but they did something different when they returned home. They took a second water pot, covered it with their skirts, and poured the water through their skirts into the pot, effectively straining out the problem-causing larvae. That was a vital behavior. The successful villagers had invented their own eminently practical solution. [5]

The six sources of influence that came into play with the Guinea worm project [5]:

1. Personal motivation—Make the undesirable desirable
2. Personal ability—Surpass your limits
3. Social motivation—Harness peer pressure
4. Social ability—Find strength in numbers
5. Structural motivation—Design rewards and demand accountability
6. Structural ability—Change the environment

All of these six sources of influence can be applied to the system safety engineering role to realize big problems that need to be solved and take drastic action to make a big impact for the benefits of your team, your company, and your customers. Find the Guinea worm in your daily development activities that must be eradicated to prevent hazards and potential disasters.

Assume you have realized the benefits of all these strategies and techniques and wonder what more could exist, that is, when you turn and look around you to find someone who needs to know what you know, to ramp up their development faster than you and "pay it forward."

13.7 Practice "Pay It Forward"

The goal of true leadership is to help others—teammates, employees, and colleagues—become more capable, confident, and accomplished than their leaders. Through the actions of a forward-thinking and extraordinarily successful CEO, Farber reveals the three keys to achieving what he calls Greater Than Yourself (GTY): Expand Yourself, Give Yourself, and Replicate Yourself. [6]

Your job is to extend and offer yourself to another, with the expressed purpose of elevating that person above yourself.—Charles Roland [6]

A vow is not something to take lightly. But with Greater Than Yourself, there is no preaching without the practice; no saying without doing. Here is a framework to help you commit to a Greater Than Yourself vow of your own. [6]

Expand Yourself

"Self-expansion is a perpetual enterprise. And because it's the foundation of whatever you do for others, expanding yourself is the furthest thing from selfishness. You expand yourself in order to give yourself to others."—Plumeria Maple [6]

I will create a deep and expansive sense of who I am (so that eventually I can give it all away) by doing the following [6]:

- Shift my perception about myself:
  o Isolation ≫ Connection
  o Alone ≫ Interdependence
  o Me ≫ Us
- Taking personal inventory of:
  o Things I do well
  o Meaningful experiences I have had
  o Life Lessons I have learned
  o People I know
  o My admirable qualities
  o My personal values
- Asking Myself:
  o Is what I am currently doing helping me to:
    i. Expand on the items already in my inventory?
    ii. Add to my inventory?
    iii. Deepen my mastery and wisdom?
  o What more can I do to improve the quality and depth of my experience and knowledge?
- Choosing my GTY project wisely
  o Focus on someone
    i.   Whom I trust
    ii. Whom I believe in
    iii. Who can benefit from and improve upon my gifts
    iv. Who has the drive, energy, heart, and desire to take full advantage of what I give them
    v.   Whose values are congruent with my own
    vi. Who has qualities and abilities I admire
    vii. Whom I love or care deeply about

Give Yourself [6]:
- Tithe your time
- Make a difference by promoting their welfare, fortunes, success, and capacity for achievement
- Invest in the GTY relationship
- Give it all. Give everything in your inventory including your knowledge, connections, experience, insights, advice and counsel, life lessons, confidence, encouragement, and honest feedback

Replicate Yourself [6]:
- Establish the one condition with your GTY recipient: that you expect nothing in return except they commit to taking on GTY projects of their own, and their GTY will, in turn, commit to taking on GTY projects of their own.
- Challenge everyone you know to do the same

Imagine the benefits to your teams and your company if your vast system safety engineering skills and DfS knowledge gained over years of practical experience could be passed on to several individuals when you near the end of your career. These early or mid-career employees that you selected and took the time to mentor over several months or years could become even more successful than yourself with your coaching, guidance, and direction. The benefits are enormous to your company and development teams, and to you personally, providing internal satisfaction that transcends past experiences, and would be beyond words, difficult to fully express. These benefits don't need to be confined only to your employees and team members. They could be applied to your customers and your suppliers.

13.8 Interfaces with Customers

The main purpose of interfacing with the customer organizations is to synergize with your customers who have expressed their needs and objectives for a safe design and to establish a business and technical rapport that builds trust and confidence between your customer counterparts and yourself. This trust and confidence may begin with scheduling regular coordination meetings, creating working groups, holding training sessions, and providing periodic status reports. Listen to your customers, being very careful to pay particular attention to their safety concerns. You should use the "voice of the customer" to influence the system or product designers to drive design enhancements to improve safety features and remedy safety concerns.

Programs awarded to defense contractors by military customers will have system safety engineering tasked to meet requirements specified under the contract Statements of Work (SOWs). System safety engineering will be the primary Point of Contact (POC) to the military customer system safety counterpart. It is the responsibility of the program system safety engineer to represent the system IPT at meetings with the customer's system safety engineer representative. There will not be a system safety engineer representing the customer for commercial programs. Government regulatory agencies may be involved during commercial system or product development to ensure they are safe for potential customers. Examples of government regulatory agencies are the Federal Aviation Administration (FAA), the Federal Communication Commission (FCC), the Nuclear Regulatory Commission (NRC), and the Food and Drug Administration (FDA).

13.9 Interfaces with Suppliers

Safety engineering should be familiar with those suppliers that are considered strategic suppliers and key to the smooth and continuous performance of development and production operations. Strategic suppliers with essential components, materials, products, and services require strong business relationships to ensure program success for delivery of systems or products that consistently satisfy performance requirements, minimize risks of hazards, and achieve price and cost targets. The main purpose of interfacing with the key supplier organizations is to establish a technical and business rapport that builds trust and confidence between your supplier counterparts and yourself, to assess risks of the entire supply chain, integrating the supply chain vertically and horizontally, to evaluate your suppliers' supply chain down to the level where raw materials and chemicals are procured, and to discover potential hazards and opportunities to mitigate risks of hazards. Trust and confidence occurs with constant communication, scheduling progress reviews, conducting regular coordination meetings, creating working groups at all levels of the supply chain, holding training sessions, and providing periodic status reports.

13.10 Five Hats for Multi-Disciplined Engineers (A Path Forward)

We have shared the benefits of interpersonal actions, habits, influences, and EQ related to your success as a system safety engineering team member. But what if you could perform in several engineering or functional roles for a team, basically wearing more than one hat? You would greatly increase your value to a team by being a multidisciplinary engineering expert for the team to draw knowledge from, without adding to headcount, and ultimately lowering development costs and affecting the corporate bottom line. Instead of interfacing with a reliability engineer in your team, for instance, you become a reliability engineer. You could also take on other roles for the team, either engineering or non-engineering roles.

Just imagine how valuable you would become by assuming multidisciplinary roles on a development team. Did you ever wonder how certain people get promoted? This is one of those ways that separates you from the others.

In the article viewable to the public on the internet titled "Five Hats for Reliability Engineers" [7], the author states that in a meeting that he attended with new engineers stepping into their new roles, they asked about the expectation in their role. The author described five hats that they should wear on a regular basis. These five hats are:

1. The technical hat
2. The trainer hat
3. The coach hat
4. The sales and marketing hat
5. The meeting hat

To focus on the fundamentals for wearing these five hats, the primary notion is to consider multiple engineering disciplines at the same time to be effective and affordable on the job. This aligns with the popular phrase "Doing more with less," which was coined by the Government in the 1990s. Even though we consider the five hats as stated previously very important, it requires a multidimensional view of the different roles or hats. Five hats is the initial way to think outside a singular unidimensional role. When wearing the technical hat, it is not enough to be concerned with only one engineering discipline, such as system safety, but rather concern yourself with multiple disciplines at once. Get additional training in multidisciplinary engineering tasks. Get assigned to cover the multiple roles in technical meetings. Once you are trained in the role, and have several years of experience performing these functions on one or more jobs, become a trainer to persist the ideology of five hats and demonstrate the value in the jobs with noteworthy accomplishments. To summarize a path forward, beyond the five hats to become even more valuable to your development organizations:

- Accomplish the same engineering tasks (e.g., FMEA, FTA) from different perspectives, such as safety engineering, security engineering, and reliability engineering, and be influential with your conclusions and recommendations.
- Perform tasks once involving engineering experts in security, safety, reliability, quality, and designers, as a minimum, and peer review the task results with them.
- Train and develop engineers to perform one task that represents the care-abouts of multiple functional groups and make some of these engineers your GTY project.
- Become a multidisciplinary engineer who can be assigned to a development team to perform multiple functions at once.

Become a really good system safety engineer first, then seek to understand other functional disciplines to become a really good multidisciplinary engineer, with at least four additional fields of expertise or professional roles that the team needs. Imagine performing five roles in a job, wearing five hats including system safety engineer, and then realize that these five roles were previously accomplished by five people before you took these roles over. Think in terms of value-added benefits to the company. The thought is mind-bending.

13.11 Conclusions

In this chapter, we covered several ways of integrating safety with other engineering and functional disciplines. We showed you the many key interfaces to SSE and defined the cross-functional teams. We touched on modern decision-making in a digital world and knowing who are your friends and your foes. We created an analogy of the importance of constant communication to within your teams to safety-critical systems like TCAS and IFF to demonstrate a correlation between machine interaction and human interaction. We touched on Raytheon's code of conduct and values. We introduced you to the paradigms of the seven habits of highly effective people, how to win friends and influence people, EQ, influencer— the power to change anything, greater than yourself, and the five hats for reliability engineers. We hope this will become paradigms that are practiced repeatedly in your careers so that you ramp up your career successes, teach mentees to become greater than yourself, and help your company improve their profit margins, cash flow, and financial balance sheet bottom line.

References

[1] Raytheon's Code of Conduct, Company Policy 74-RP, Doc No. RP-OGC-ETH-001.

[2] Covey, S. R., *The 7 Habits of Highly Effective People*, 1989, Fireside, Simon and Schuster, Inc., New York.

[3] Carnegie, D., *How to Win Friends and Influence People*, 1936, Pocket Books, Simon and Schuster, Inc., New York.

[4] Bradberry, T. and Greaves, J., *Emotional Intelligence 2.0*, 2009, TalentSmart, Inc., San Diego, CA.

[5] Patterson, K., *Influencer: The Power to Change Anything*, 2008, McGraw-Hill, New York.

[6] Farber, S., *Greater Than Yourself*, 2009, Doubleday, New York.

[7] Isenhour, S., Five Hats for Reliability Engineers, 2012, http://www.ReliabilityNow. Net (Accessed on August 4, 2017).

14

Design for Reliability Integrated with System Safety

Louis J. Gullo

14.1 Introduction

In the previous chapter, we discussed the ways system safety engineers should integrate and interface with other types of engineers and functional disciplines. One of those engineering discipline interfaces, which is key to the success of a system safety engineer, is reliability. System safety benefits greatly from a well-established and reinforced interface to reliability engineering. The integration with all functional disciplines is very important for effectively and efficiently practicing system safety engineering, but the most important of these functional discipline interfaces is the interface to reliability engineering. The system safety engineer should approach this chapter building on and applying the lessons from Chapter 13 to establish a key interface with reliability engineering.

Much of what system safety engineering does is beneficial to reliability engineering, and system safety engineering benefits greatly from reliability engineering. The two disciplines should work together to gain mutual benefits from one another. It is imperative that system safety and reliability work closely together on any system development team. This is necessary to allow for engineering productivity and efficiency in task accomplishments. Both disciplines should

Design for Safety, First Edition. Edited by Louis J. Gullo and Jack Dixon.
© 2018 John Wiley & Sons Ltd. Published 2018 by John Wiley & Sons Ltd.

divide up analysis efforts realizing the data inputs and outputs that each one needs or provides between the two disciplines. By agreeing early in development planning who will accomplish what tasks and what the inputs and outputs that result from those tasks will be, a smooth hand-off and data exchange will occur within their engineering interface. An even better situation that ensures a perfect interface between the two disciplines is if the two roles were assigned to the same engineer. We discussed the possibility of a systems engineer wearing multiple hats as a multidisciplined engineer at the end of Chapter 13. Assigning one engineer to perform both system safety and reliability is preferred by any program manager or systems engineering manager, in terms of lowering the cost of development. But this type of engineer is not easily to find or develop. For purposes of this chapter, we will assume the two roles are carried out by two different engineers.

Certain paradigms introduced in Chapter 1 are areas where commonality can be found between reliability and system safety engineering. Four of these paradigms are listed here:

Paradigm 1: Always Aim for Zero Accidents.
Paradigm 3: Spend Significant Effort on Systems Requirements Analysis.
Paradigm 4: Prevent Accidents from Single as well as Multiple Causes.
Paradigm 6: Design for Prognostics and Health Monitoring (PHM) to Minimize the Number of Surprise Disastrous Events or Preventable Mishaps.

Even though there are overlaps and commonality in the methodologies of reliability and system safety engineering, there are unique differences in approaches between the two roles. Therefore, most of this chapter will address the integration of system safety and reliability disciplines as different people assigned on a team working different roles but finding areas that are common between them and avoiding duplication of effort by applying similar methodologies to perform their tasks. In order to address this integration, the system safety engineer will be exposed to standards and technical material important to the reliability engineering discipline but will be covered in a cursory manner within these pages without delving too deeply into the subject of reliability engineering. Much of the reliability engineering content introduced here includes those definitions and methodologies used by reliability engineering, which are important to know in order to develop a strong interface and to enjoy a solid technical data exchange.

14.2 What Is Reliability?

Reliability is the ability of an item to perform a required function under stated conditions for a stated period of time, as defined by IEEE Std 1413. IEEE Std 1413 provides a standard framework for reliability data collection, analyses, assessments, and predictions of hardware. This standard provides a framework for the disclosure of reliability analysis methods and for all necessary information within a reliability prediction report: "This standard can be used by the developer of the prediction for planning (e.g., gathering input information) and performing predictions, and by the user of the prediction to assess the value of the predictions. The usefulness of a reliability prediction is dependent on the accuracy and completeness of the information utilized as input to the prediction, and the analysis method(s) used to create the prediction" [1].

IEEE Std 1413 was written as a standard framework for hardware reliability analyses. For software reliability, there is IEEE Std 1633. IEEE Std 1633 is the recommended practice on software reliability.

The scope of this recommended practice is to address Software Reliability (SR) methodologies and tools. This recommended practice does not address systems reliability, software safety, nor software security. This recommended practice provides a common baseline for discussion and prescribes methods for assessing and predicting the reliability of software. The recommended practice is intended to be used in support of designing, developing, and testing software and to provide a foundation on which practitioners and researchers can build consistent methods for assessing the reliability of software. This recommended practice contains information necessary for the application of SR measurement to a project. This includes SR activities throughout the Software Life Cycle (SLC) starting at requirements generation by identifying the application, specifying requirements, and analyzing requirements and continuing into the implementation phases. It also serves as a reference for research on the subject of SR. [2]

IEEE Std 1332 provides an IEEE standard for reliability programs for the development and production of electronic products.

This document provides a standard set of reliability program objectives for use between customers and producers, or within product development teams, to express reliability program requirements early in the development phase of electronic products and systems. The purpose of this document is to establish a standard set of objectives which provide an effective structure for the life-cycle activities needed to design, manufacture and utilize reliable electronic products, and systems across the supply chain. The objectives of a reliability program apply to products, systems, or systems of systems, which are referred to as 'items' for the purposes of this standard. An electronic system refers to both hardware and software. The reliability program has three objectives. The reliability program is integrated to the engineering life cycle, and the objectives of the reliability program are included in the overall engineering goals during the entire engineering life cycle for the item. The three objectives provide a flexible structure to promote necessary and sufficient interaction between the customer and supplier. The 'supplier' is an entity that designs or produces electronic items. For the purposes of this standard, the supplier represents the community of developers and producers for the particular item. The 'customer' is the entity that either purchases the electronic item from the supplier or is the recipient of the item from within the same organization. The customer represents the community of users who will actually use the item. In the case of Commercial Off-the-Shelf (COTS) items, for which requirements are not communicated directly from a customer to the suppler, any references in this standard to the customer's reliability requirements refer to the supplier's reliability specifications for the item. Any of these requirements can be used to define the item's reliability requirements. [3]

The design for reliability process is an integrated, multidisciplinary endeavor. The supplier shall establish, document, and implement a design for reliability process that includes characterizing the extent and rate of item degradation, and thus reliability, based on the nature, magnitude, and duration of exposure to the loads. The design for reliability process should document the trade studies performed to evaluate the item's performance, use, environmental, and physical features against its safety, reliability, and quality assurance features to define an optimal design, with regard to item robustness versus item cost. For example, circuit-card design features that mitigate vibration and shock-induced damage that may dominate early failures during shipment and installation are balanced against circuit-card design features that mitigate thermal cycling induced damage that may lead to solder joint fatigue after many years of operation to define a design that optimizes reliability life versus unit cost. The design for reliability process should reflect the inclusion of early or out-of-box failures, random occurrence failures, and wear out, as each contributes to the overall item reliability performance. [3]

The supplier establishes a supply chain that is capable of meeting the customer performance and reliability requirements. The supply chain shall be assessed for its capability prior to placing a purchase order or establishing a contract. The selection of a supplier is often based on factors that do not explicitly address reliability, such as technical capabilities, production capacity, geographic location, support facilities, and financial and contractual factors. A selection process that takes into account the ability of suppliers to meet reliability objectives during manufacturing, testing, and support can improve reliability of the final item throughout its lifecycle and provide valuable competitive advantages. Reliability capability assessment quantifies the effectiveness of reliability activities within the supplier's organization. Reliability capability indicates whether the key reliability practices within an organization are monitored and improved. One way to ensure that an organization's supply chain demonstrates reliability capability is by executing supplier self-assessments. Supplier organizational reliability capability assessment score should be an output from any supplier assessment conducted by a supply chain management organization, especially when the system requires highly reliable performance over the system life cycle. [3]

As stated by Brian Moriarty in chapter 17 of *Design for Reliability* [4], "Design for Safety is a major effort to assure that the product being designed has considered all known hazards and that the hazards are mitigated or acceptable to operate the system. Reliability is a key feature in obtaining the total product base to examine hazards."

Safety engineering utilizes reliability information collected from the system level through the subsystem level to the parts level of all the areas of the product design in the process of analyzing the system design to leverage the reliability data and obtain the Risk Assessment Codes (RAC) for each hazard identified. Brian Moriarty emphasized the importance for the system safety engineer to know the reliability of the system, its subsystems, boxes, assemblies, circuit cards, and all the way down the system hierarchy to the piece part level for all the components, off-the-shelf products, and materials designed into the system. The Design for Safety (DfS) activities and the Design for Reliability (DfR) activities should be inseparable: "The design project team must work together in the direct compatible use of the information to reach the conclusion that the design can be reliably used with the hazards defined for product that are 'acceptable hazards'" [4].

The goals of system safety are to have complete knowledge of the system in its original generic design to the final design with examination of the hazards that exists with the product including human errors that lead to unsafe consequences. Documenting importance level of each and every design analysis for safety is a key factor for the history of building the design that is retained in records at each step of the building process in which safety has been analyzed. The use of previous 'Hierarchy System' from which a revised product is made begins with examination of the previous product and the safety status of that system. Previous systems, having gone through acquisition to final operational field use, assist greatly in a knowledge base of the enhanced system that will be a revision update to the hierarchy system. [4]

The interfacing with personnel who use similar systems in field operating environments, such as user, operator, or maintenance personnel following operation and maintenance procedures, will have valuable experiences and knowledge on how the system does or does not perform reliably and safely. These personnel should be sought out as well as the reliability engineer for the collection of valuable data to improve the DfS and DfR capabilities for the newly developed systems to ultimately replace the predecessor systems out in the customer applications in the field environments. One factor that is critically important is related to the environment, which may degrade the components and materials in the system design with high or low temperature extremes above specifications, or mechanical shock and vibration profiles, that can adversely affect and degrade critical functions and that can prematurely wear out and age portions of the design, leading to accelerated failure mechanisms and hazards. These environments and interfaces are key parts of the safety evaluation that is required as the system is developed.

Now, let us examine the types of data that reliability engineering provide the system safety engineer, which decreases the system safety engineer's workload and has been proven to be very valuable in successful achievement of many types of system safety analyses.

14.3 System Safety Design with Reliability Data

To determine a safe world for our products and systems, we rely on reliability data, predictive analytics, PHM, and Machine Learning (ML) capabilities. Reliability data may be raw data inputs from various test and customer application sources, or the data may be refined and processed online using PHM sensors with prognostic reasoners and ML techniques or offline using human-in-the-loop predictive analytics to assess preventive maintenance actions. This is a major input into the safety engineer's development of a hazard analysis. This should be researched from multiple sources to provide influence to the system safety design. This data related to performance of a product measured over a period of time with predictive algorithms and reliability models to predict future failure or hazard occurrences is useful to calculate risk levels and to recommend enhancements to a product's design features for safe operation. Likelihood of hazards or failures must be evaluated in terms of their probability of occurrence of known and defined failures or hazards. Probability of occurrence of failures or hazards is the primary numerical input provided by reliability engineers to the system safety process in conducting risk assessments and defining the risk levels of hazards. The probability of occurrence of a safety-critical hazard cause may be a one-to-one correlation to a mission-critical reliability failure cause. The safety assessment needs to determine this possibility. If the safety assessment concludes that the safety-critical hazard cause correlates to a mission-critical failure cause, then the hazard cause will have a corresponding failure mode with reliability data. This reliability data substantiates the hazard cause probability of occurrence and provides evidence to support assignment of hazard risk levels for each hazard that is tracked [4].

System and product developers establish a closed-loop root cause analysis process to collect data on failed items. Sources of this data may include developmental units in test, prototype tests, preproduction unit tests, qualification unit test, operational production unit tests, and units operating and tested in the field in customer applications, which are assessed as customer returns due to failures or repaired in the customer applications. This closed-loop root cause analysis process enables the system or product developer to continuously assess and improve reliability. This process may be a joint customer–system/product developer responsibility to maintain the system or product and determine the root causes of failure modes that may impact item performance, supportability, reliability, or availability. The end result of such closed-loop root cause analysis process efforts is to determine the corrective action(s) necessary to reduce failure risks to acceptable or specified levels. An output from the closed-loop failure analysis process is data feedback provided to the system or product developer's design and manufacturing functions in each stage of the item's life cycle. Failure data obtained from manufacturing, maintenance, operation, and field returns should be used to improve the DfR of future items and sustain the item in the field in a more cost-effective manner. A formal Failure Reporting, Analysis, and Corrective Action System (FRACAS) provides a framework for controlling the closed-loop root cause analysis and corrective action processes [3].

Reliability engineers collect, analyze, develop, and retain various types of reliability data that benefits the system safety engineer. Figure 14.1 illustrates the interchange of some of the data between system safety and reliability engineering to complete engineering analyses, such as Failure Modes and Effects Analysis (FMEAs). The various types of reliability data include qualitative and quantitative data, deterministic and nondeterministic probabilistic data, and FRACAS data:

- Failure or hazard probability of occurrence data
- Reliability or the "probability of success"
- Time-to-Failure (TTF) data
- Mean-Time-Between-Failures (MTBFs) and Mean-Time-To-Failures (MTTFs)
- Mission-critical failure data including failure rates, hazard rates, and intensity functions
- Root cause analysis and failure effect results from FRACAS reports
- Operational mission times and test run times

An example of reliability data collection methodology and sample data is given in Chapter 12 for system safety testing. System safety engineers can leverage

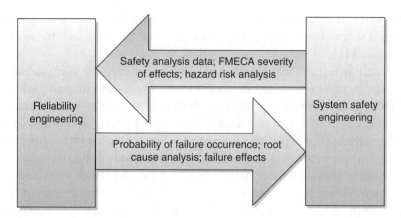

Figure 14.1 Reliability and system safety data interfaces

reliability data as shown in the aforementioned list and Figure 14.1, from other sources besides testing. These types of reliability data are collected from multiple sources. Some of these sources are as follows:

- Field data records
- Customer use applications
- Development, production, and operational test events
- System qualification tests
- Development models and simulation (such as damage, stress, and accumulated fatigue models and simulations)
- Finite Element Analysis (FEA)
- Supplier test and empirical data
- Historical data sources:
  - Company proprietary data
  - Physics of Failure (PoF) models (proprietary or public models)
  - Industry consortium proprietary data
  - Publicly available commercial and military handbook methodologies

The system safety engineers should understand the different ways probability of failure occurrence data is collected and generated by reliability engineering and used in assessing hazards in the product being developed. Reliability engineering should be able to provide reliability metrics that support their probability of failure occurrence data. Sometimes, reliability probability data is not available, so qualitative data will be provided instead of quantitative data. Each hazard identified and documented in the hazard analysis must be assigned a hazard probability level as defined in Table 14.1. Probability level A (frequent probability)

Table 14.1 Hazard probability levels (excerpt from Table 7.1)

Probability level	Qualitative probability of occurrence	Likelihood of occurrence applicable to the item
A	Frequent	Likely to occur often in the life of an item
B	Probable	Will occur several times in the life of an item
C	Occasional	Likely to occur sometime in the life of an item
D	Remote	Unlikely, but possible to occur in the life of an item
E	Improbable	So unlikely, it can be assumed occurrence may not be experienced
F	Eliminated	Incapable of occurrence. This level is used when potential hazards are identified and later eliminated

is a qualitative definition that equates to a quantitative metric associated with the specific individual item's hazard or failure probability. The quantitative metric associated with the qualitative definition is provided in Table 14.2.

The hazard or failure probability of occurrence is combined with the severity level creating an RAC, which represents the risk level associated with the hazard being examined. The detailed list of RACs is shown in Table 14.3 (referenced from Table 7.3).

Each of these probability levels equate to a numeric probability or range of probabilities as listed in Table 14.2. For instance, probability level A equates to a specific individual item's hazard or failure probability of occurrence greater than 0.10. Probability level F equates to a specific individual item's hazard or failure probability of occurrence of zero.

Table 14.2 Quantitative probability of occurrence ranges

Probability level	Quantitative probability of occurrence
A	Probability $(P) > 0.10$
B	$0.100 > P > 0.010$
C	$0.0100 > P > 0.0010$
D	$0.0010000 > P > 0.0000010$
E	$P < 0.0000010$
F	0

Table 14.3 Risk assessment matrix

Frequency of occurrence	Hazard severity categories			
	1 Catastrophic	2 Critical	3 Marginal	4 Negligible
A—Frequent	High	High	Serious	Medium
B—Probable	High	High	Serious	Medium
C—Occasional	High	Serious	Medium	Low
D—Remote	Serious	Medium	Medium	Low
E—Improbable	Medium	Medium	Medium	Low
F—Eliminated	Eliminated			

14.4 How Is Reliability Data Translated to Probability of Occurrence?

System safety engineers and other types of engineers should understand that the goal of the reliability engineer is to use processes and tools to detect and uncover design weaknesses. Part of the data collected by reliability engineering is called root cause failure analysis data. From this data, engineers determine why an item fails and recommend design change corrective actions to prevent or eliminate future failure occurrences. This failure analysis data are usually entered into an FRACAS database for ease of retrieval in the future. By conducting failure analysis to determine a root cause failure mechanism and incorporating design changes to prevent recurrence of failure mechanisms, DfR improves, such that:

- Mission-stopping failures are minimized or reduced over the anticipated life.
- Unscheduled downtime is minimized or reduced.
- ROI must be more than the DfR investment to provide zero net cost.

When you have collected enough statistical data from enough samples over a period of time, you calculate reliability using standard mathematical equations or distribution functions. The exponential distribution is a commonly used reliability function. The exponential distribution represents the failure probability density function, $f(t)$, and the reliability function, $R(t)$, when the failure rate, λ, is a constant. The mission time, t, and failure rate are inputs to the reliability function, $R(t)$, and $f(t)$ probability density function as shown:

$$f(t) = \lambda e^{-\lambda t}$$
$$R(t) = e^{-\lambda t}$$

In general terms, density means the quantity of something per unit of measure. Density refers to the number of occurrences over a unit area, length, volume, or period of time. It can be used as a measure of compactness, such as mass per unit volume. In probability theory, a Probability Density Function (PDF), or density of a continuous random variable, is a function whose value at any given point in the sample space can be interpreted as providing a relative likelihood that the value of the random variable would equal that sample. The sample space refers to the set of possible values defined by and used for the random variable. The PDF is used to specify the probability of the random variable falling within a particular range of values, as opposed to taking on any one value. This probability is given by the integral of this variable's PDF over that range. The probability is given by the area under the density function curve above the horizontal axis and between the lowest and greatest values over the entire range of the curve. The PDF is always nonnegative, with a value between 0 and 1 at any point on the curve, and the integral of the PDF over the entire range of the curve is equal to one. The probability distribution function and probability function may be used to denote the probability density function. It should be understood that this use of the terms as synonyms is not standard among those in the mathematical probability and statistics profession. PDF referring to the probability distribution function may be used when the probability distribution is defined as a function over general sets of values, or it may refer to the Cumulative Distribution Function (CDF). PDF and CDF are discussed further in Chapter 12, "System Safety Testing," and data collection methods.

The graph shown in Figure 14.2 illustrates the reliability function for an exponential distribution. The curve shows how reliability of hardware begins its life

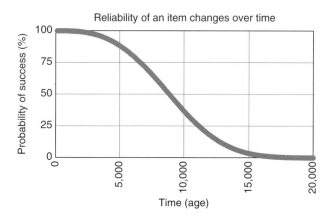

Figure 14.2 Graph of reliability over time

near 100% and degrades over time due to accumulated fatigue and stresses that cause wear-out time-limiting failure mechanisms.

Many times, the exponential distribution does not fit the reliability data collected. Reliability engineering may assume an exponential distribution to assess reliability of items, since the calculations for reliability metrics such as failure rate and MTBF are simpler than the calculations for other types of probability distributions that may fit reliability data better than an exponential. Some of these other types of probability distributions are shown in Table 14.4. Some of these distributions are used for specific purposes with specific types of data. For instance, the binomial distribution is used with the reliability data collected that are test pass or fail data. The binomial calculates reliability using total number of tests or trials and the number of successful tests or trials. This

Table 14.4 Example types of distributions used in reliability

Distribution name	Density function	Where used
Exponential	$\lambda \exp(-\lambda t)$, where λ is the constant failure rate and the inverse of the mean life or MTBF	Most common reliability distribution; applies to middle section of bathtub curve
Weibull	$\alpha \beta t^{\beta-1} \exp(-\alpha t^\beta)$, where α is the scale parameter and β is the shape parameter	Infant mortality (shape parameter <1); wear-out (shape parameter >1); constant failure rate (shape parameter $=1$)
Gamma	$(1/\alpha!\beta^{\alpha+1})t^\alpha \exp(-t/\beta)$, where α is the scale parameter and β is the shape parameter	Time to multiple failures for redundant equipment where α is the number of failures before product failure
Normal or Gaussian	$(2\pi\sigma^2)^{-1/2} \exp(-(t-\mu)^2/2\sigma^2)$, where μ is the mean and σ is the standard deviation	Wear-out
Lognormal	$(2\pi t^2\sigma^2)^{-1/2} \exp(-\ln(t-\mu)/2\sigma)^2$, where μ is the mean and σ is the standard deviation	Fatigue life and maintainability (repair times)
Binomial	$\text{Comb}\{n;x\}p^x q^{n-x}$, where n is the number of trials, x ranges from 0 to n, p is the probability of success, and q is $1-p$	Reliability of "one-shot" devices
Poisson	$(\lambda t)^x \exp(-\lambda t)/x!$, where x is the number of failures and λ is the constant failure rate	Number of failures in a time period

method of probability calculations is useful for assessing the reliability of one-shot devices, such as projectiles or missiles.

Reliability Probability Concept

To better understand the data that reliability engineering uses to calculate probabilities of occurrence for failures and hazards, certain concepts and equations should be understood.

Let $f(t)$ be a continuous function defined such that

$$f(t) \geq 0$$

$$\int_{-\infty}^{\infty} f(t) dt = 1$$

$$P(a < t < b) = \int_{a}^{b} f(t) dt$$

Then $f(t)$ is a probability density function for the continuous random variable t. The reliability function is also known as a probability of success.

Figure 14.3 shows a probability of success example for a set of four input parameters, each with a numeric probability value between 0 and 1. These four parameters are funds, resources, time, and skills. These four parameters are assigned values that are input to an AND gate. The output of the AND gate is a parameter called success. If all four inputs have a probability of "1," then the output is a probability of "1."

Figure 14.3 Probability of success example

Notation for Reliability Distribution Functions

Probability density function:

$$f(t) = \frac{dF(t)}{dt}$$

Cumulative failure distribution function:

$$f(t) = \int_0^\infty f(t)dt$$

Reliability function:

$$R(t) = 1 - F(t)$$

Failure rate (FR = failures/time):

$$\lambda = \frac{f(t)}{R(t)}$$

These reliability functions product reliability metrics that are used for various types of safety analyses such as Fault Tree Analysis (FTA) and FMEA as follows:

- For non-repairable systems, failure rate is the hazard rate function.
- For repairable systems, failure rate is the recurrence rate function, also known as Rate of Occurrence of Failure (ROCOF).

As discussed earlier, system safety engineers and other types of engineers should understand that the goal of the reliability engineer is to detect and uncover design weaknesses. Part of the data collected by reliability engineering is root causes analysis data including failure mechanisms and PoF data and models used for reliability predictions and assessments. This data is usually entered into an FRACAS database for ease of retrieval in the future. This data may also be used in FMEAs and FTAs.

Safety engineers provide failure effect severity to reliability engineers for FMEAs. Testability engineers provide probability of fault detection and probability of fault isolation data to reliability engineers for FMEAs.

System safety engineers should work together with reliability engineers to develop FMEAs and FTAs and provide justification for design improvements that benefit design safety and design reliability.

As discussed in Chapter 13, system safety and reliability can be performed by the same person, wearing multiple hats as a multidisciplined engineer.

Definitions of severity of the hazard are assigned with the criteria shown in Table 14.5. The safety designer must evaluate about the result of the hazard in examination of the product.

Table 14.5 Hazard severity levels (from Table 7.2)

Severity	Category	Mishap definition
Catastrophic	1	Death, system loss, or severe environmental damage
Critical	2	Permanent partial disability, injuries, or occupational illness that may result in hospitalization of at least three personnel; reversible significant environmental impact
Marginal	3	Injury or occupational illness resulting in one or more lost work days; reversible moderate environmental impact
Negligible	4	Injury or occupational illness not resulting in a lost work day; minimal environmental impact

These definitions could be modified for program-specific purposes, as shown in Table 14.6.

Table 14.6 Severity definitions

Description	Category	Environmental, safety, and health result criteria
Catastrophic	1	Could result in death, permanent total disability, loss exceeding $1M, or irreversible severe environmental damage that violates law or regulation
Critical	2	Could result in permanent partial disability, injuries, or occupational illness that may result in hospitalization of at least three personnel, loss exceeding $200K but less than $1M, or reversible environmental damage causing a violation of law or regulation
Marginal	3	Could result in injury or occupational illness resulting in one or more lost work days(s), loss exceeding $20K but less than $200K, or mitigation environmental damage without violation of law or regulation where restoration activities can be accomplished
Negligible	4	Could result in injury or illness not resulting in a lost work day, loss exceeding $2K but less than $10K, or minimal environmental damage not violating law or regulation

14.5 Verification of Design for Safety Including Reliability Results

As stated in Chapter 12, "System Safety Testing," DfS is verified and validated by demonstration and testing:

> The purpose of verification is to confirm that the design meets the item reliability requirements. Validation confirms that the item will fulfill its intended use. These practices assure the reliability assertions made at the design stage by comparing them with the tracked and observed reliability during testing and usage. Item field reliability data (when available) should be compared with reliability estimates, reliability test conditions, and warranty return estimates. Verification and validation activities encompass comparison of identified potential problems against those experienced in the field. These activities include the comparison of expected and field failure modes and mechanisms and the comparison of reliability prediction models for an item against field failure distributions. [3]

In the system development schedule, system demonstration and testing will be required to prove that the system safety control requirements are properly met. Testing programs require a test plan, test procedures, and a test report to record all results of the test. As the safety control requirements are defined to lower subsystem levels from the functional specification documents, this list must be assembled in order to assure complete testing is performed. The system safety engineer must be involved in performing or witnessing the tests where safety requirements are validated and reviewing the test data and results when the tests are completed. There may be Problem Reports (PRs) that are written against the safety requirements that fail their test. Reliability will be involved in conducting root cause analysis of PRs and recommending corrective actions to resolve the PRs. Corrective actions to resolve the PRs must be performed by the safety engineers and other relevant engineers to reach a conclusion on the changes required. Following the incorporation of changes, retest or regression testing is to be performed. The test is completed when a "passed" decision is reached for the testing that verifies the safety control requirement [4].

The test performed must show proof of the integrity of the hazard RAC that was found in the design of the product. If the test results in the product not meeting its RAC established hazard risk level, reexamination of the RAC must be performed. In the situation where the RAC is a lower level of risk than previously found in the analysis, the test results will probably be accepted. However when the RAC change results in a higher level of hazard definition, action must be taken to properly reexamine the product for a change in order to bring the product to an acceptable hazard level. The safety engineer is responsible to assure that action is taken and resolution of change is reached before the change can be made to the product and a retest can be performed [4].

14.6 Examples of Design for Safety with Reliability Data

There are different types of reliability data that may be provided to the system safety engineer [4]. Some of these include failure cause data from root cause failure analysis and probability of occurrence data from test and customer application failure rate sources. The following three examples are provided using an FMEA worksheet to show a type of hazard analysis methodology performed for three specific product types using different forms of reliability data. The product for each example is described followed by the purpose of the product and the basic functional requirements of the product. The FMEA worksheet requires 11 hazard data elements be listed, collected, and analyzed for the following categories of data:

1. Hazard number
2. Hazard description
3. Hazard or failure causes
4. System state
5. Possible hazard or failure effects
6. Severity/rationale
7. Existing controls or requirements
8. Probability of occurrence likelihood/rationale
9. Current or initial risk
10. Recommended safety controls or requirements
11. Predicted residual risk

The following are the three examples for hazard analysis method using an FMEA worksheet:

1. Use of new infant crib without past historical data on previous product designs
2. Development of refrigeration product that has large number of interfaces and failure causes
3. Development of power supply design using past historical reliability data

1. Use of New Infant Crib without Past Historical Data on Previous Product Designs

Product: Infant crib (age birth to one year)

Purpose of Product: Provides area for infant to be in horizontal position (either awake or sleeping)

Basic Functional Requirements: Infant crib has several functions: (1) rectangle infant crib design with two adjustable wooden fences and two solid wooden barriers, (2) locking gates in wooden fence with vertical bars on two sides of crib and fixed position fence at the head and foot of infant crib, (3) gate moving up and down vertically when parent uses footswitch, and (4) low position of mattress in the crib for infant to lie comfortably in horizontal position

(1) Hazard #	(2) Hazard description	(3) Hazard or failure causes	(4) System state	(5) Possible hazard or failure effects	(6) Severity/ rationale	(7) Existing controls or requirements	(8) Likelihood/ rationale	(9) Current or initial risk	(10) Recommended safety controls or requirements	(11) Predicted residual risk
Crib-1	Infant pulls mattress over his head	Infant is able to pull mattress because no tie down of mattress to crib is present	Use of crib for infant anytime during the 24 hour day	Anytime infant can pull mattress over his/her head potential effect on breathing of infant can be unacceptable	1 Catastrophic Fatality due to suffocation of infant	1. Mattress with no tie down to crib 2. No other method to prevent mattress from changing position	D Extremely remote Difficulty for infant to pull mattress over head	1D High-level hazard Unacceptable	1. Place tie down lines on mattress and connect them to crib 2. Develop procedure in crib manual to alert crib owner of necessity to tie does mattress 3. Add alarm warning system to trip on, if mattress tie down lines are opened	3B Low-level hazard Acceptable
		(Customer application reliability data provided from FRACAS database)		(Customer application reliability data provided from FRACAS database)	(Customer application reliability data provided from FRACAS database)					

2. Development of Refrigeration Product That Has Large Number of Interfaces and Failure Causes

Product: Refrigerator

Purpose of Product: Preserve food in temperature-controlled refrigerator

Basic Functional Requirements: Refrigerator provides controlled lower temperature for enclosed food in its enclosed framed box. Product has multiple electrical and fluid interfaces. House power maintains refrigerator with controlled temperature to regulate temperature with Freon tubing. Cold water provided to refrigerator with water tubing. Separate power outlet in circuit breaker box provided for refrigerator

(1) Hazard #	(2) Hazard description	(3) Hazard or failure causes	(4) System state	(5) Possible hazard or failure effects	(6) Severity / rationale	(7) Existing controls or requirements	(8) Likelihood/ rationale	(9) Current or initial risk	(10) Recommended safety controls or requirements	(11) Predicted residual risk
Reg-1	Refrigerator fails to operate	House power fails; Temperature control switch fails to operate; Freon liquid tube fails, pressure leak; Power fails; Water tubing breaks and shorts power, causing circuit breaker to open	Use of refrigerator cannot keep food at designated temperature	Contaminated food from unregulated temperature	3; Moderate (no possibility the failure will cause equipment damage or harm to customers)	Provide additional separate house power outlet to reconnect refrigerator to other operating power outlet; Reconnect circuit breaker if power disconnects from power short	C; Low (factory test and customer application reliability data supports low probability of occurrence)	3C; Moderate-level hazard	Add backup power for refrigerator; Place additional instructions in operator's manuals for user to know more information on refrigerator hazards; Additional sensors for diagnostics to isolate failure causes for warranty maintenance	4E; Low-level hazard; Acceptable

3. Development of Power Supply Design Using Past Historical Reliability Data

Product: Uninterrupted Power Supply (UPS) for circuit card assembly in nuclear power plant electronics

Purpose of Product: Provide an additional power supply to backup primary power supply in case of a loss of primary power output

Basic Functional Requirements: UPS is wired in parallel with double side switch to provide power when primary power supply fails

(1) Hazard #	(2) Hazard description	(3) Hazard or failure causes	(4) System state	(5) Possible hazard or Failure Effects	(6) Severity / rationale	(7) Existing controls or requirements	(8) Likelihood/ rationale	(9) Current or initial risk	(10) Recommended safety controls or requirements	(11) Predicted residual risk
UPS-1	UPS cannot function when required to operate	Switch fails to provide power when Primary Power fails Power switch is not able to change position, stuck in one position UPS is not operational due to incorrect wiring UPS has failed starter	Use of UPS cannot provide power to circuit card assembly function	Circuit card assembly in nuclear power plant fails causing loss of power to control water temperature. Water temperature rises, causing overheating condition that could lead to a nuclear power facility damage and loss of life	1 Catastrophic	UPS power switch is checked daily Switch is wired to the Circuit Card Assembly	D Extremely remote Probability of occurrence (0.000001) based on reliability data from customer application data and factory testing	1D Unacceptable risk High-level hazard	Add signal generator for backup to UPS if UPS fails to take over basic power Add a health monitoring sensor to detect power signal degradation and to alert operator with an alarm when imminent failure is possible Add continuous check of UPS switch operability Add training to nuclear power plant personnel	1E Medium risk May be deemed Acceptable

14.7 Conclusions

The primary key to the success of a system safety engineer is a well-established interface to reliability engineering. Much of what system safety engineering does benefits reliability engineering, and vice versa. It is imperative that system safety and reliability work closely together on any system development team. An even better situation is if the two roles were handled by a single engineer. But this type of engineer is not easy to find or to develop. Even though there are overlaps in methodology, there are unique differences in approaches between the two roles. Therefore, most of this chapter has addressed the integration of system safety and reliability disciplines as different people assignments on a team. In order to address this integration, the system safety engineer has been exposed to the reliability engineering discipline in a cursory manner. Most of the content introduced the reader to the definitions and methodologies used by reliability engineering. It is highly important for system safety engineering to build the interface to reliability engineering to leverage the work output from reliability analysis to avoid duplication of effort and minimize cost in development.

The Safety Designer must always consider the reliability of the product he is responsible for. Probability information is a key to the hazard analysis that must be performed on the product. Knowledge of the hazards associated with the product must be performed so that the Risk of the hazard can be properly evaluated. It is recognized that there will always be hazards in a product. Electronic equipment has power which must be evaluated for the use of the Operator/ Maintainer/User. The effort in the Design for Safety is to recognize any hazards including human errors, properly evaluate it and assure that the hazard is an acceptable hazard. [4]

Safety and reliability are inseparable. The standard Risk Assessment Code (RAC) defines the combination of the severity and likelihood to a RAC number that places the evaluation in the Low, Medium or High hazard level. Based on this evaluation the final step is to determine if additional requirements are needed to mitigate the hazard to an acceptable hazard level so that the product can be used. Without the reliability data the total evaluation of the hazard cannot be completed. The hazard level associated with the failure resulting in the hazard occurrence must be defined in the evaluation the safety designer performs on the product. [4]

The examples provided illustrate the method of developing a hazard worksheet that will document the product and the data needed for total evaluation of product hazards. Detail in listing the hazard data is a major factor in this evaluation. Peer review of the hazard data is normally performed as an independent review to assure that the data is correct and the hazard risk level is correctly displayed. Definitions of the worst credible event occurring from the hazard must be shown with the associated data for the exact hazard level information in the Design for Safety. [4]

Acknowledgment

The author would like to recognize the contributions of Brian Moriarty from chapter 17 of the *Design for Reliability*, published by John Wiley & Sons in 2012.

References

[1] IEEE Std 1413 (2010) *IEEE Standard Framework for Reliability Prediction of Hardware*, IEEE, New York.
[2] IEEE Std 1633 (2008) *IEEE Recommended Practice on Software Reliability*, IEEE, New York.
[3] IEEE Std 1332 (2012) *IEEE Standard Reliability Program for the Development and Production of Electronic Products*, IEEE, New York.
[4] Raheja, D. and Gullo, L.J. (2012) *Design for Reliability*, John Wiley & Sons, Inc., Hoboken, NJ.

15

Design for Human Factors Integrated with System Safety

Jack Dixon and Louis J. Gullo

15.1 Introduction

In starting this chapter, we refer back to the previous two chapters where we discussed the ways system safety engineers should integrate and interface with other types of engineers and functional disciplines. One of those engineering discipline interfaces, which is key to the success of a system safety engineer, is reliability. Another important engineering interface for a system safety engineering is Human Factors Engineering (HFE). System safety benefits greatly from a well-established and reinforced interface to HFE, same as with reliability engineering. The integration with all engineering and functional disciplines is very important for effectively and efficiently practicing system safety engineering. The system safety engineer should approach this chapter building on and applying the lessons from Chapters 13 and 14 to establish a key interface with HFE. As stated in Chapter 14, for purposes of this chapter, we will assume that the two roles (HFE and system safety) are carried out by two different engineers. It would be wise for any safety engineer to learn the role of the Human Factors Engineer (HFE) to be able to provide more value to the system/product development team

Design for Safety, First Edition. Edited by Louis J. Gullo and Jack Dixon.
© 2018 John Wiley & Sons Ltd. Published 2018 by John Wiley & Sons Ltd.

to wear multiple hats and contribute in a greater way as a multidisciplined engineer. Just as there are certification procedures for system safety and reliability, there are also certifications for HFE. This might be a great way to stand out on a team, holding these multiple certifications from credible organizations.

The design team must recognize the role that the human will play in the product or system being designed. Human factors engineers, safety engineers, designers, and systems engineers must work together to develop a design in which the human system interfaces are compatible with the user. HFE must be included in the design process early and must be embraced in all phases of system development to include requirements specification, design, and testing. As in all good systems engineering, the requirements development phase is critical to success. HFE activities during requirements specification include evaluating legacy systems and operator tasks, analyzing user needs, analyzing and allocating functions between man and machine, and analyzing tasks and associated workload. During the design phase HFE activities include participating in the evaluation of alternative designs, evaluating software by performing usability testing, refining analyses of tasks and workload, and using modeling tools such as human figure models to evaluate crew station and workplace design and operator procedures. During the testing phase HFE activities include confirming that the design meets HFE specification requirements, measuring operator task performance, and identifying any undesirable features. This emphasis on integrating the human with the system will help to prevent human-induced accidents by making the design less sensitive to human errors. It will also help to reduce the risk to the human that use or operate the product or system.

Since all products or systems are a collection of components that interact with each other and the external environment to achieve a common goal, the designer must be concerned about all the system/product components and their interactions. Some of these components include humans. A human may be part of a system, a user of the system, or a controller of the system. Accidents with products or systems are often blamed on human error. Quite often, human error is a symptom of something wrong with the design of the product or system. This may indicate a lack of consideration of the human aspect during the design phase of a product. Ideally, we want designers to design out all possibility of human errors. This is achievable with human-centered design practices.

The reliability, safety, and usability of products and systems will be enhanced substantially if they are designed with people in mind. Safe and efficient operation of products and systems depends on properly designed and engineered interactions between the human and the machine. Let us get started with an understanding of the role of the HFE.

15.2 Human Factors Engineering

"Human Factors Engineering (HFE) is a specialty engineering discipline that focuses on designing products and systems with the user in mind" [1]. The goal of HFE is to maximize the ability of a person or a team to correctly operate and maintain a product or a system, to eliminate design-induced impediments and errors, and to improve system safety and reliability. It is important that any design engineer recognizes the capabilities and preferences of the user of the system/ product, as well as the limitations of the system/product that a user must be aware of. A system/product that has been designed from these perspectives will ensure customer satisfaction, require less training to operate and maintain, result in fewer human errors and system/product failures, and be safer to use [1].

HFE is defined in MIL-STD-1908 [2] as

the application of knowledge about human capabilities and limitations to system or equipment design and development to achieve efficient, effective, and safe system performance at minimum cost and manpower, skill, and training demands. Human engineering assures that the system or equipment design, required human tasks, and work environment are compatible with the sensory, perceptual, mental, and physical attributes of the personnel who will operate, maintain, control and support it.

Human factors engineering as we know it today has its origin in World War II due to military needs to design and operate aircraft safely. Prior to that time, humans were typically screened to fit the job or equipment rather than having the equipment designed with the human in mind. While human factors engineering grew initially in defense and aerospace industries, it has since spread to all industries including nuclear, space, health care, transportation, and even furniture design. [1]

System safety and human factors engineering go hand in hand to create safe and usable product designs.

15.3 Human-Centered Design

Human-centered design encompasses a wide variety of concerns. As the name implies, human-centered design places people at the center of design considerations rather than making people conform to the design. The design must accommodate humans' physical characteristics, such as strength, vision, and limits, and mental processes, such as perception and cognition. It must also take into account the environment in which a person must operate and the characteristics of the equipment, product, or system. The user also operates within an organizational framework that must also be considered. Other factors that influence human behavior, and should therefore be considered in the design process, include the technology being used, the management systems that are in place, and the procedures and processes under which the user will operate.

15.4 Role of Human Factors in Design

The product design should minimize human error and maximize human performance. Design considerations must include human capabilities, human limitations, human performance, usability, human error, stress, and the operational environment. Some typical topics that should be taken into account during the design process are shown in Table 15.1. This table is not meant to be all-inclusive; it is provided to highlight some major topics and to stimulate thought about what should be considered in a design.

With products and systems becoming more complex, organizations must be committed to human-centered design. Product development efforts must fully integrate systems engineering with all types of specialty engineering, including human factors engineering; must develop system requirements to include user requirements; must test the product being developed with real users, assess the usability, and fix any shortcomings identified; and must use the tools and techniques that will facilitate this integrated design approach.

15.4.1 Hardware

Designers must consider the human being during the design process as related to hardware. Typical hardware-related considerations include weight, layout, access, anthropometrics, and ergonomics. Many examples of bad hardware design can easily be found in everyday products:

- Remember the Ford Pinto that required that the engine be dropped in order to change the spark plugs?
- Have you ever rented a car, pulled into a gas station to refuel it, and didn't know where the gas cap was? It could be on the right side, on the left side, or under the license plate. Inevitably, it is always on the side opposite the one nearest the pump when you pull in. Why aren't they always on the same side? Why aren't they always on the driver's side, as that's probably who will fill up the car?
- Have you ever looked at a stove and wondered which knob controls which burner? Confusing? Turning on the wrong one could result in a fire hazard or cause someone to be burned.

Table 15.1   Considerations in human-centered design

Topic	Considerations
Displays	Can they be seen and understood by the expected user audience
	Are they grouped in an intuitive way that makes sense to the user
	Are they properly integrated with the controls
	Is the quality, relevance, and quantity of information presented appropriate
	Are audio alerts appropriate for their intended function
	Are warnings provided for anomalous events
Controls	Can they be operated by the expected user audience
	Are they grouped for ease of operation
	Are they labeled for easy understanding
	Are they properly integrated with the displays
	Is accidental activation prevented
	Are emergency controls clearly identified
Workspace design	Will the human fit comfortably
	Can the person perform the assigned operations using the workspace provided
	Is sufficient space provided
	Is temperature, humidity, and ventilation adequately controlled
	Is the lighting level correct for the functions to be performed
	Are viewing angles correct
	Is the workstation eye height at the proper level
	Does the workspace have adjustability
Workplace environment	Will the human be safe as well as comfortable in the environment
	Is lighting adequate
	Is the temperature and humidity properly controlled
	Is the background sound level suitable for the function being performed
	Are the sound levels safe (i.e., below thresholds that damage hearing)
	Is ventilation sufficient
Maintenance	Has the product or system been designed with the maintainer in mind
	Is the product modularized
	Has proper access been provided
	Are test points provided
	Is diagnosis easy to perform
	Are personnel protected from dangerous voltages

All these design flaws could have been eliminated if the designers had taken into account that essentially all products and systems have users who need to be considered and the design needs to accommodate them.

15.4.2 Software

Similarly, software design should consider the user. How many times have you installed a new computer program, opened it up, and discovered that you have no idea how to do anything with it—the icons are indecipherable, the order in which things need to be done is a puzzle, the color scheme makes the fonts unreadable, and no useful instructions are provided. What if this computer program controlled some hazardous operation; could it be operated safely if it contained those problems?

What about Microsoft Windows' non-intuitive design? Why would anyone design a system where one must go to the start button to shut off the machine? In the newest versions, the design seems to have gotten worse. When Windows 8 was released, there was no start button; therefore since the shutoff has always been available only after you press the start button, there was no way to shut off the computer. The resourceful user had to find and install a third-party fix that installed a script with an icon button to shut down the machine. In an update of Windows 8, a fix was provided to add the start button back in. Why not fix it right and add a shutdown button? Later, Windows 10 was introduced and the design continued to get worse. Now the user must press the start button, which is no longer called "Start," then press the power icon, and then choose the shutdown option to turn off the computer. Just think how safe would it be if these same people provided software to control a hazardous process. The operator would have to go through three button pressings to shut down the process during an emergency while operating under the stress of an emergency situation—not a good design.

Traditionally, human factors considerations in the design process have been focused on hardware aspects of the product or system. In the last couple of decades, more and more products have become more software intensive. The proportion of software to hardware in products has been increasing steadily. This trend has made development and change more rapid, and the greater use of software imparts a greater risk in both operations and maintenance. Software requires different skills; it can add risk to the system by introducing new hazards; it makes diagnosis more challenging and generally imposes new considerations for the designer to think about during the design process.

Usability is a term used to describe the ease with which a person can employ a product or system. ISO 9241 [3] defines usability as the "extent to which a product can be used by specified users to achieve specified goals with effectiveness, efficiency and satisfaction in a specified context of use." Although the term can be applied to both hardware and software aspects, it is most often used in relation to software. As complex computer systems find their way into our everyday life, usability has become more popular and more widely utilized in recent years. Designers have seen the benefits of developing products with a user orientation. By understanding the interaction between a user and a product, the designer can produce a better, safer, and more widely accepted product. Desirable functionality or design flaws may be identified that may not have been obvious if human factors had not been considered. Implementing this human-centered design paradigm, the intended users of the product are kept in mind at all times. Maybe, then, the user won't have to go to the start button to stop the machine!

Paradigm 3: Spend Significant Effort on Systems Requirements Analysis

As with all good systems engineering, the most important part of the process is the up-front definition of requirements. It is critically important that complete user interface requirements be identified early in the development process. There are numerous guidelines and style guides for user interfaces. An example of a guideline is ISO 9241, one of a series of guidelines for various aspects of computer–user interfaces [3]. These provide a starting place for user interface requirements generation, but they must be customized and tailored to fit the application.

Also important to implementing user interfaces successfully is design evaluation. This can be done using mock-ups or prototyping the user interface and testing with actual users. Requirements and implementations can then be adjusted in the design process when it is still cost effective to make changes. Continuing user evaluation as the design evolves will ensure the best usability of the end product.

Human factors specialists do more than design-friendly icons. They bring two important types of knowledge to bear on system development: (1) human abilities and limitations and (2) empirical methods for collecting and interpreting data from people. They define criteria for ease of use, ease of learning, and user acceptance in measurable terms. New technology demands much thought about the role of the tool. The distress and aversion that many people manifest toward computerization is perfectly rational in the presence of ill-conceived designs. Too many of our systems confuse the operator.

One example of a human capability is time perception, which can be shown to affect the functional requirements of a software design. Table 15.2 shows time perception across various tasks and media. Transaction interactions should be without a perceived wait, and the standard deviation of all transactions should be less than 50% of the mean.

Table 15.2 Perceptions across media

Human perception	Transaction time	Application	Preferred physical architecture
"Instantaneous"	Less than 1/3 second	Software development	Personal computer or workstation
"Fast"	Between 1/3 and 1 second	Simple query	Client/server
"Pause"	Between 1 and 5 seconds	Complex query and application launch	Thin client
"Wait"	Greater than 5 seconds	Action request	Background batch

15.4.3 Human–Machine Interface

The interface between a person and a machine is of prime importance. The designer must be concerned with any and all parts of the product where a person must interact with the equipment. The human–machine interface becomes a critical component to be considered. This interface is defined as the plane of interaction between the person and the machine. It is across this plane that information and/or energy flows. The information is transferred across this interface from the machine to the person via displays and from the person to the machine via controls. Therefore, displays and controls are of major importance in product or system design. Effective displays help to determine the proper action needed. Ineffective displays and controls contribute to errors that may lead to accidents.

The designer must also be concerned with the proper allocation of functions between a person and machine or user and equipment. Machines are consistent; people are flexible. Machines are more capable of performing repetitive and physically demanding functions; people are more capable of performing functions that require reasoning. These are of prime significance when allocating functions to the machine or the person. This allocation of functions must ensure that the tasks assigned to each take into account what they do best, what their capabilities are, and what limitations they each have. Trade-offs between human and machine must be made regarding speed, memory, safety, complex activities, reasoning, overload, and so on. In the next three sections, we describe some of these considerations in greater detail.

15.4.4 Manpower Requirements

The staff required to operate and maintain the system must be a design consideration. How many and what type of people will be needed to operate and maintain the system or product? Are properly qualified people available? Will training be required? If so, how much? Will these types of people be available in the future to support the entire life cycle of the system? Often, there are trade-offs that can be made during the design process that can reduce the number of people needed or the amount of training that will be necessary. For example, if the graphical user interface is intuitive to use, the time for the operator to learn to use the system can be greatly reduced.

15.4.5 Workload

The workload the system or product imposes on a user is a related concern. The tasks that a person must perform must be delineated. An assessment must be made of the amount of effort that each task will take both physically and intellectually. A matchup of the capabilities of the user to the tasks at hand must be ensured; otherwise, the user will become quickly dissatisfied with the product or, worse, will make errors as a result of task overload, which could lead to substantial undesirable consequences including accidents.

15.4.6 Personnel Selection and Training

Another important factor to consider during system design is who will be needed to operate and maintain the system. The proper people with sufficient skills and knowledge must be selected for the best fit to operate and/or maintain a system. Once selected, these people must be properly trained to do the job functions that have been allocated to them as a result of the system design process. Again, good product design can help to reduce these demands on the human operator or maintainer. This role reflects back to paradigm 8 from Chapter 1.

 Paradigm 8: Develop a Comprehensive Safety Training Program to Include Handling of Systems by Operators and Maintainers

15.5 Human Factors Analysis Process

Almost any technique used in system analysis can be applied to address the human element. However, there are numerous human factors-specific analysis techniques from which a designer can choose. Many of the system safety analysis techniques covered in this book will include the human element. Likewise many Human Factors Analyses (HFA) techniques will consider the safety.

15.5.1 Purpose of Human Factors Analysis

The overarching purpose of HFA is the inclusion of human considerations (e.g., man-in-the-loop, man–machine integration) in system safety analyses to develop a reliable, usable, and safe product or system. Various HFA are conducted at different times in the development process and for different reasons. Analysis of human factors assists in the development of requirements, which is a critical step in the design process. As the product or system evolves, different analyses are conducted to define the human role in the system to ensure that the human needs and limitations are being considered, to determine the usability of the product, and to confirm the ultimate safety of the system. Other analyses can be conducted to help guarantee customer acceptance of the product or system.

While the human factors engineer may be the lead for conducting the HFA, it should be a team effort. The team may vary depending on the stage of development or the particular analysis being conducted, but the results will always be better if it is a joint effort by a cross-functional team. Participants in the HFA will always include the design engineer. The team may include other specialty engineers, such as system safety, reliability, software, and manufacturing engineers. Often, team participants may include management, marketing, sales, and service personnel. The team should be tailored to enhance the particular analysis being conducted.

15.5.2 Methods of Human Factors Analysis

As our products and systems have become more complex, it has become imperative that the old approach—either totally ignoring human considerations or making "educated guesses" based on intuition of how best to accommodate the human—be replaced by systematic analytical techniques to better match the human being and the machine. Although it is beyond the scope of this book to elaborate on all the various HFA, Table 15.3 presents a sampling of the analyses available to a design team. The reader can find more detailed coverage of these and many other techniques in the books of Raheja and Allocco [5] and Booher [6].

15.6 Human Factors and Risk

Risk is inherent in all systems, and people add an additional dimension to the risk concern. Designers must be aware of both the risk posed by humans and the risk imposed on humans.

15.6.1 Risk-Based Approach to Human Systems Integration

The goal of all good systems engineering design, as well as good system safety practice, is to reduce risk throughout the development cycle. This is accomplished by applying all relevant disciplines. In the past, however, the risks associated with human systems integration have often been ignored. Engineering risks are noticed at various times during the development due to implementation problems or cost overruns.

Table 15.3 Human factors analysis tools

Technique	Purpose	Description	Cost/difficulty	Pros	Cons
Prototyping	Addresses design and layout issues	Mock-up of user version	Cheap if done early; more expensive as design progresses	Quick way to show possible design before investing time and money on detailed development	Can be expensive for complex systems or if the prototypes have to be updated each time the product or system changes
Improved Performance Research Integration Tool (IMPRINT) (Reference "MANPRINT in Acquisition: A Handbook" [4])	IMPRINT is appropriate for use as both a system design and acquisition tool and a research tool IMPRINT can be used to help set realistic system requirements, to identify user-driven constraints on system design, and to evaluate the capability of available manpower and personnel to effectively operate and maintain a system under environmental stressors. It incorporates task analysis, workload modeling, performance-shaping and degradation functions and stressors, a personnel projection model, and embedded personnel characteristics data	IMPRINT, developed by the Human Research and Engineering Directorate of the US Army Research Laboratory, is a stochastic network modeling tool designed to help assess the interaction of soldier and system performance throughout the system life cycle—from concept and design to field testing and system upgrades. IMPRINT is the integrated Windows follow-on to the Hardware versus Manpower III (HARDMAN III) suite of nine separate tools	Moderately time consuming. Requires training. Inputs can be extensive	Generates manpower estimates and can be used to estimate life cycle costs. Produces many types of reports	Time consuming and a lot of data are needed

(Continued)

Table 15.3 (Continued)

Technique	Purpose	Description	Cost/difficulty	Pros	Cons
Task analysis	Analyzing tasks and task flows that must be performed to complete a job	A job/task is broken down into increasingly detailed actions required to perform the job/task. Other data such as time, sequence, skills, etc. are included. This analysis is used in conjunction with other HFE analyses such as functional allocation, workload analysis, training needs analysis, etc.	Moderate but can get expensive for large, complex systems	Relatively easy to learn and perform	Task analysis was first developed for factory assembly line jobs that were relatively simple, repetitive and physical, and easy to define and quantify. It is much more difficult to apply to complex or highly decision-based tasks
Human Reliability Analysis (HRA)	Used to obtain an accurate assessment of product or system reliability, including the contribution of human error	Considers the factors that influence how humans perform various functions. It may include operators, maintainers, etc. The analysis is conducted using a framework of task analysis. First, the relevant tasks to be performed must be identified. Next, each task is broken down into subtasks, and the interactions with the product or system are identified, and the possibility for errors is identified for each task, subtask, or operation. An assessment of the impact of the human actions identified is made. The next step is to quantify the analysis using historical data to assess the probability of success or failure of the various actions being taken Also, any performance-shaping factors, such as training, stress, environment, etc., are taken into account. These factors may have an effect on the human error rate, which can be either positive or negative, but is usually negative. These might include heat, noise, stress, distractions, vibration, motivation, fatigue, boredom, etc.	Expensive and time consuming on large systems	Thorough analysis of human errors and human–machine interactions	Requires extensive training and experience in several disciplines

Method	Use	Description	Cost/Ease	Advantages	Disadvantages
Job safety analysis	Used to assess the various ways a task may be performed so that the most efficient and safest way to do that task may be selected	Each job or process is analyzed element by element to identify the hazards associated with each element. This is usually done by a team consisting of the worker, supervisor, and safety engineer. Expected hazards are identified, and a matrix is created to analyze the controls that address each hazard	Easy for simple jobs; harder for complex jobs	Good for structured jobs	Difficult if there is much variation in the job
Technique for Human Error Rate Prediction (THERP)	Used to provide a quantitative measure of human operator error in a process	THERP was initially developed during the 1960s for use in nuclear industry for probabilistic risk assessment. It is a means of quantitatively estimating the probability of an accident being caused by a procedural error The method involves defining the tasks, breaking them into steps, identifying the errors, estimating the probability of success/failure for each step, and calculating the probability of each task	Can become expensive for processes with a lot of tasks	Can be very thorough	Obtaining good probability data
Link analysis	Used to evaluate the transmission of information between people and/or machines. It is focused on efficiency and is used to optimize workspace layouts and man–machine interfaces	Links are identified between any elements of the system. The frequency of use of each link is determined. The importance of each link is then established. A link value is calculated based on frequency, time, and importance. System elements are then arranged so that the highest value links have the shortest length	Moderate	Operations and training can be enhanced. Safety critical areas can be identified	Can become cumbersome on large systems

(Continued)

Table 15.3 (Continued)

Technique	Purpose	Description	Cost/difficulty	Pros	Cons
Hazard analysis	To identify hazards and their mitigations during various phases of product/system development	There are numerous types of hazard analyses, each with its own focus and each conducted at specific times during the design effort. Preliminary Hazard Analysis (PHA) is done early in the concept phase and helps provide information for conducting trade studies and/or developing requirements. Later, as the product development progresses, more detailed analyses are conducted—Subsystem Hazard Analysis (SSHA), System Hazard Analysis (SHA), and Operating & Support Hazard Analysis (O&SHA). Most often all of these analysis techniques use a matrix format. This matrix (or sometimes individual hazard tracking sheets) typically contain the following information: hazard description, effect, operational phase, recommended controls, and risk assessments (both before and after controls are implemented). These hazard analysis techniques are described in greater detail in Chapter 7	Moderate depending on the size of the system being analyzed	Provides a systematic way to determine hazards, assesses risks, and presents recommendations to mitigate or control	Effectiveness and usefulness is dependent upon knowledge (of system and of hazards) of the team performing the analysis

| Fault Tree Analysis (FTA) | To evaluate the likelihood of an undesirable event happening and identifying the possible combination of events that could lead to it | FTA is a deductive top-down analysis. It identifies an undesirable event (top level) and determines the contributing elements (faults/conditions/human errors) that would precipitate it. The contributors are interconnected with the undesirable event using in a tree structure through Boolean logic gates. This tree construction continues to lower levels until basic events (those that cannot be decomposed further) are reached. Combinations of these bottom level basic events that can cause the top-level, undesirable event to occur can be calculated. They are called cut sets. If failure rate data is provided, the probability of occurrence of the undesirable event can be determined. FTA is described in greater detail in Chapter 9 | Can get very expensive and time consuming for large systems | FTA provides a graphical representation that aids in the understanding of complex operations and the interrelationships between subsystems and components. Can be used either qualitatively or quantitatively | Significant training and experience are required |

Risks associated with human systems integration, however, are usually noticed only after a product or system is delivered to the customer. These end-state problems may lead to customer dissatisfaction and rejection of the product, due to it being too difficult or inefficient to use, or, worse, they may lead to human error in the use of the product, which could lead to an accident having catastrophic consequences.

These operating risks can be traced back to failure to properly integrate the human needs, capabilities, and limitations at an early stage of the design process. Like all risk reduction efforts, risk reduction in the area of human systems integration must be started early and continue throughout the development of a product or system. This will ensure that requirements based on human factors are incorporated. It will allow design trade-offs to consider the use of the product. This will produce a high level of confidence that the product or system will be accepted, usable, and safe for the consumer to use.

 Paradigm 3: Spend Significant Effort on Systems Requirements Analysis

As has been emphasized earlier in this book, the key to design success rests in the development of requirements. Requirements involving human factors are no different. A thorough and complete effort to specify human factors requirements in the earliest stages of product or system development will reap large rewards later by resulting in a product or system that is usable and safe to use. During development, other approaches to mitigating risk in the area of human systems integration include using task analysis to refine the requirements and conducting trade studies, prototyping, simulations, and user evaluations.

15.6.2 Human Error

It is easy to blame accidents and failures on human error, but human error should be viewed as a symptom that something is wrong within the system. Generally speaking, humans try to do a good job. When accidents or failures occur, they are doing what makes sense to them for the circumstances at hand. So to understand human error, one must understand what makes sense to people. What "reasonable" things are they doing, or going to do, given the complexities, dilemmas, trade-offs, and uncertainties surrounding them? The designer must make the product or system make sense to users, or they will be dissatisfied with it, failures will occur, or they will have accidents using it.

Designers often think that adding more technology can solve all human error problems. Quite often, however, adding more technology does not remove the potential for human error. It changes it and may cause additional problems. Make sure that you know whether you are really solving the problem of human error by adding technology or are merely causing different problems. Take the simple addition of a warning light. What is the light for? How is the user supposed to respond to it? How does the user make it go away? If it lit up before and nothing bad happened, why should the user respond to it now? What if the light fails just as it is needed? Or, for another example, think of all the new technology being added to automobiles today—Bluetooth, screens with multiple menus and multiple functions, GPS displays, etc. Are these helping automobile safety, or are they adding to the already existing problem of the distracted driver?

Many error-producing conditions that may cause a failure or an accident can be added to a product or system inadvertently. The designer must remain cognizant of the many things that can lead to human error, including environmental factors such as heat, noise, and lighting; confusing controls, inadequate labels, and poor training; difficult-to-understand manuals or procedures; fatigue; boredom; and stress. The goal is to eliminate those things that can contribute to human error.

15.6.3 Types of Human Error

Human error affects system/product reliability and safety. There are many ways in which people can make errors. They can commit errors in calculations, they can choose the wrong data, they can produce products with poor-quality workmanship, they can use the wrong material, they can make poor judgments, and they can miscommunicate (just to name a few). Borrowing from Dhillon [7], we categorize human error using the design life cycle perspective:

- **Design errors:** These types of human errors are caused by inadequate design and design processes. They can be caused by misallocation of functions between a person and a machine, by not recognizing human needs and limitations, or by poorly designed human–machine interfaces.
- **Operator errors:** These errors are due to mistakes made by the operator or user of the equipment design and to the conditions that lead up to the error being made. Operator errors may be caused by improper procedures, overly complex tasks, unqualified personnel, inadequately trained users, lack of attention to detail, or nonideal working conditions or environment.
- **Assembly errors:** These errors are made by humans during the assembly process. These types of errors may be caused by poor work layout design, distracting environment (improper lighting, noise level, inadequate ventilation, and other stress-inducing factors), poor documentation or procedures, or poor communication.
- **Inspection errors:** These errors are caused by inspections being less than 100% accurate. Causes may include poor inspection procedures, poor training of inspectors, or a design being difficult to inspect.

- **Maintenance errors:** These are errors made by maintenance personnel or the owner after a product is placed in use. These errors may be caused by improper calibration, failure to lubricate, improper adjustment, inadequate maintenance procedures, or designs that make maintenance difficult or impossible.
- **Installation errors:** These errors can occur because of poor instructions or documentation, failure to follow the manufacturer's instructions, or inadequate training of the installer.
- **Handling errors:** These errors occur during storage, handling, or transportation. They can be the result of inadequate material handling equipment, improper storage conditions, improper mode of transportation, or inadequate specification by the manufacturer of the proper handling, storage, and transportation requirements.

When most people think of human error, the natural tendency is to think of an error by the operator or user. However, as can be seen from the aforementioned categorization, a major cause of human error that must be considered by the design engineer is design error. One must realize that during the design process, design flaws may be introduced that may lead to safety problems and accidents. During the manufacturing process, assembly errors may occur. During quality inspections, product shortcomings may not be found due to human error. During subsequent maintenance and handling by the user, errors can occur. So the designer must be aware of all these types of defects and ensure that they are considered, and hopefully eliminated, during the design process to ensure the creation of a safe design.

15.6.4 Mitigation of Human Error

Mitigating risks due to human factors should begin early in the design process. Although it is always most desirable to completely eliminate conditions that lead to human errors, it is inevitable that human errors will occur. Therefore, error containment should be considered in the design. As technology has evolved, we have become more and more interconnected, and the consequences of a failure, whether human-caused or not, can grow to enormous proportions due to this added complexity and interdependencies.

As an example, consider the August 14, 2003, failure of the power grid in the Northeastern and Midwestern United States and Ontario, Canada. The failure began when a 345-kV power line made contact with a tree in Ohio. Once the failure began, a chain reaction of events occurred due to numerous human errors and system failures. The failure propagated into the most widespread power failure in history, affecting 10 million people in Canada and 45 million people in eight US states. The power outage shut down power generation, water supplies, transportation (including rail, air, and trucking), oil refineries, industry, and communications. The failure was also blamed for at least 11 deaths.

As products and systems continue to become more complex and interwoven into our culture, designers must not only ensure that human–machine interfaces are understood and usable but must also consider the potentially far-reaching consequences that a failure might have and design products and systems in a way that will minimize these consequences.

15.6.5 Design for Error Tolerance

Error-tolerant systems and interfaces are design features worthy of consideration. Error-tolerant systems minimize the effects of human error. Error tolerance capabilities added to a system improve system reliability and hence safety as well. Human error is frequently blamed for accidents, especially high-consequence accidents. It is not uncommon to hear 60–90% of accidents attributed to human error. While we should be eliminating opportunities for human error to occur, we should also be designing products to be error tolerant. Designers need to consider the consequences of failures of their designs. Errors lead to consequences; consequences should be minimized or eliminated. This is the essence of error-tolerant design. A design that tolerates errors avoids the consequences. By providing feedback to the user on both current and future consequences, compensating for errors, and providing a system of intelligent error monitoring, a design can be made to be more error tolerant. One should note that the emphasis is on "intelligent" monitoring and feedback. The system should not just provide an indecipherable error message, such as "ERROR #404." This is the type of error message so often seen these days by the average computer user who has no idea what the problem is or how to fix it. The system should provide a more useful message that describes the problem and provides useful information to the user about the error and what should be done to resolve it.

15.7 Checklists

Checklists can be an effective tool to help a system/product designer or a system/product user consider certain criteria. These serve as reminders of important areas to think about or inspect for. These are created for a user to ensure that the product or system is operated correctly, that all procedures are followed, and that uncertainty in operation is avoided. A well-known example of a checklist is the preflight checklist, used by pilots prior to take off.

As detailed in Chapter 6, there are three main types of checklists—procedural, observational, and design. The two that apply in the human factors arena are the design checklist and the procedural checklist. The design checklist provides the designer with a checklist to follow during the design and testing of a product or system. The second, procedural checklist, provides the user with a checklist to follow when using the product or system so that nothing will be omitted.

A human factors design checklist is typically a long list of design parameters that should be considered during the product design process. These lists can be based on previous experience, lessons learned, and/or some published design guide. For example, a checklist derived from MIL-STD-1472 [8] can be used to help a design engineer ensure that a design will be usable by people for its intended purpose. A summarized list of HFE general requirements and shortened list of HFE detailed design requirements and design criteria as excerpts from MIL-STD-1742 are as follows:

General requirements

- Standardization
- Function allocation
- Human engineering design
- Fail-safe design
- Simplicity of design
- Interaction
- Safety
- Ruggedness
- Design for Nuclear, Biological, and Chemical (NBC) survivability
- Design for Electromagnetic Pulse (EMP) hardening
- Automation
- Functional use of color
- Design of aircrew systems

Examples of categories of detailed requirements and criteria

- General criteria for control display integration
- Position relationships
- Movement relationships
- Control display movement ratio
- Signal precedence
- General criteria for visual displays
- Legend lights
- Simple indicator lights
- Trans-illuminated panel assemblies
- Scale indicators
- Audio displays

- Audio warnings
- Controls for audio warning devices
- Speech transmission and reception equipment
- Operator comfort and convenience
- Speech displays
- Speech recognition
- General criteria for controls
- Discrete and continuous adjustment rotary and linear controls
- Touch-screen controls
- Workspace design

MIL-HBDK-759 [9] is another good source of HFE checklists and guidelines to supplement MIL-STD-1472. It is a companion document to MIL-STD-1472. It should be consulted for data, preferred practices, and design guidelines, including design guidelines for variations of basic hardware configurations. Much of the HFE criteria and requirements are common between both standards. There are several requirements and criteria that are included in MIL-HDBK-759, which are not covered in MIL-STD-1472. A summarized list of HFE general requirements and a short list of HFE detailed design requirements and design criteria as excerpts from MIL-HBDK-759, which are not included in MIL-STD-1742, are as follows:

- Anthropometric data and pictorials that show:
  - Dimensions for the standing body
  - Dimensions for the seated body
  - Depth and breadth dimensions
  - Dimensions for circumferences and surfaces
  - Dimensions between hands and feet
  - Dimensions between head and face
  - Range of human motion
  - Viewing distance graphs
- Direct glare from display lights
- Portability and load carrying criteria, such as weights of representative individual items that an infantryman carries in temperate hot environments

It should be realized that checklists have limitations. It is impossible for a checklist to cover all variables and combinations of conditions for all designs. Despite that shortcoming, checklists should provide guidance to the designer for the important things that need to be considered during different phases of the development process. They can also serve as tools for test engineers to verify that the product or system has been designed and produced with the user in mind. There are two checklists provided in the back of this book in Appendices A and B. While these checklists are not HFE specific, they do have human factors-related sections.

15.8 Testing to Validate Human Factors in Design

Human factors validation is just as important as the development of adequate requirements for product or system specification. It is important to test the product or system against each human factors requirement and to verify that the requirement has been met adequately. Human performance requirements should be validated in system test plans and demonstrated in usability tests, and the results addressed in test reports. The product or system should be tested by representative users to verify that it functions as planned and can be operated properly and safely by the intended user.

Acknowledgment

Substantial portions of this chapter on human factors was extracted and adapted from Design for Reliability [1].

References

[1] Raheja, D. and Gullo, L.J. (2012) *Design for Reliability*, John Wiley & Sons, Inc., Hoboken, NJ.

[2] U.S. Department of Defense (1999) *Definitions of Human Factors Terms*, MIL-STD-1908, U.S. Department of Defense, Washington, DC.

[3] *Ergonomics of Human System Interaction*, ISO 9241, International Organization for Standardization, Geneva.

[4] U.S. Army (2000) *MANPRINT in Acquisition: A Handbook*, U.S. Department of Defense, Washington, DC.

[5] Raheja, D.G. and Allocco, M. (2006) *Assurance Technologies Principles and Practices: A Product, Process, and System Safety Perspective*, 2nd ed., John Wiley & Sons, Inc., Hoboken, NJ.

[6] Booher, H.R. (2003) *Handbook of Human Systems Integration*, John Wiley & Sons, Inc., Hoboken, NJ.

[7] Dhillon, B.S. (1999) *Design Reliability: Fundamentals and Applications*, CRC Press, Boca Raton, FL.

[8] U.S. Department of Defense (2012) *Human Engineering Design Criteria*, MIL-STD-1472G, U.S. Department of Defense, Washington, DC.

[9] U.S. Department of Defense (1995) *Handbook for Human Engineering Design Guidelines*, MIL-HDBK-759C, U.S. Department of Defense, Washington, DC.

Suggestions for Additional Reading

Booher, H.R. (1990) *MANPRINT: An Approach to Systems Integration*, Van Nostrand Reinhold, New York.

Raheja, D. and Allocco, M. (2006) *Assurance Technologies Principles and Practices*, John Wiley & Sons, Inc., Hoboken, NJ.

Sanders, M.S. and McCormick, E.J. (1993) *Human Factors in Engineering and Design*, McGraw-Hill, New York.

16

Software Safety and Security

Louis J. Gullo

16.1 Introduction

Software is becoming more and more pervasive in electronic and mechanical systems as time progresses. Systems that were traditionally designed solely with hardware are being redesigned with software functions replacing many of the hardware functions, thereby transforming the systems into software-intensive systems. A software-intensive system is any system where software contributes essential influences to the design, construction, deployment, and evolution of the system as a whole [1]. Systems may be considered as software-intensive systems when greater than 50% of system's required functions are performed by software design or a combination of both hardware and software designs. Systems may be designed with hardware-only functions or software-only functions, or a combination of both hardware and software to accomplish a particular system's required function. In software-intensive systems, the safety and security features of the software become more critical than older predecessor systems where software-only functions or functions performed through a combination of hardware and software represented a small percentage of the overall system functions.

Key driving force in converting traditional hardware systems to systems composed of software functions is the need to reduce cost, to provide functional

Design for Safety, First Edition. Edited by Louis J. Gullo and Jack Dixon.
© 2018 John Wiley & Sons Ltd. Published 2018 by John Wiley & Sons Ltd.

design change flexibility, and to build trust in the system. It is widely known that it is cheaper to design systems with software functions in place of hardware functions. It is also widely accepted that software functions are more flexible and easier to change compared to their hardware counterparts. The fact that a system can build trust by adopting software functions is not as established. The sense of trust invokes high user confidence and minimal, or no, risk that the system will work when needed. If you build system trust, the system users develop confidence that the system will perform as intended whenever the system is needed. Trust evokes the sense of confidence in the system safety, system security, dependability, availability, and reliability of the system. Trust in a system is increased when system functions previously designed using hardware are redesigned using software. If we have 100% confidence in a system performing as intended without fear of harm or failure, then we have no risk in using the system. No risk means zero chance that the system will cause a mishap that injures people or damages equipment and zero chance that the system will cease to operate, causing a safety-critical or mission-critical event. Inherently, software that is tested and has no hazards nor failures should perform consistently in the manner as tested without failures or hazards no matter how many times it is tested or used, nor the number of copies made of the software.

For today's new software-intensive systems, it is necessary to analyze, test, and understand the software thoroughly to ensure a safe and secure system and to build system trust that system always works as intended without fear of disruptions or undesirable outcomes. When the software of a software-intensive system is analyzed, tested, and understood, the first decision that should be reached is whether the system software has safety-critical and security-critical functions that warrant extra rigor to analyze and test. This extra rigor involves verification that the system works as required in the system specifications and the system does not perform in a certain way. Verifying that software does not perform in a certain way could be much more costly to prove as compared with verifying that the software performs as required. For example, analysis or testing to verify that a software data buffer is working is much cheaper to perform than verifying that a software data buffer can experience a buffer overflow condition.

Hardware is physical. Software is not physical. Hardware may experience wear-out conditions. Hardware may be tested and pass its performance requirements and then fail the same test the next time it is tested due to time and stress degradation mechanisms. Software does not experience wear-out conditions, but does experience probabilistic nondeterministic failure modes. Software may be tested and pass its performance requirements and then fail the same test the next time it is tested, but for a different reason compared with hardware. Hardware may be determined to operate perfectly without failures during a test event, but if the hardware is dormant for a long period of time being exposed to environmental stresses, accumulating physical material fatigue and damage, the system may fail and experience a catastrophic hazard the next time it is operationally tested. This probabilistic physical wear-out condition is not possible for software functions in software-intensive systems. But software could experience a probabilistic mechanism related to software age and require rejuvenation. This condition of software aging is a time degradation mechanism that could be related to the time since a system is rebooted or powered off. When a system runs continuously over a long period of time, memory and data buffers fill up, and data may be lost or truncated, causing functional system problems. Software rejuvenation means that the memory or data buffers are cleared out by software system reboots. A power-on reset is one form of a system hard reboot that rejuvenates the system software. There are also soft reboots that are not as severe a form of system reboot as a hard reboot, which may accomplish the same objective. A soft reboot is a system restart without the need to interrupt system power. Windows Operating System (OS) Personal Computer (PC) users will recognize a soft reboot command to restart their PC that involves simultaneously pressing the three keys on their keyboard: CONTROL (CTRL), ALT, and DELETE keys.

In the book by Nancy Leveson titled *Safeware* [2, page 63], with the introduction of computers into the control of complex systems, a new form of complacency appears, which is the belief that software does not fail and all the coded software errors will be removed by testing. Professionals not trained in the software engineering discipline seem to believe this myth and propagate it. This myth leads to complacency and overreliance on software functions, which results in underestimating software-related risks. There is no doubt that an inherent improvement in system cost and flexibility occurs when changing a system design from mostly hardware functions to mostly software functions, but the inherent improvement in system trust, to include system safety and system security, is not so obvious.

Highly complex software-intensive systems with extensive processing power and interfaces affecting safety-critical functions require extensive planning and analysis during the Software Development Process (SDP). The same is true for security-critical functions in complex software-intensive systems. These complex software-intensive systems require special analytical tools, diverse tests in multiple use cases and environments, accurate predictive models and methods, and proven techniques to ensure the software design is safe and secure.

In some places in the world, the dividing lines between safety processes and security processes are blurred. In several European standards, safety means the same as security. For example, consider that a European transportation standard where the goal of the document is to ensure the safety and security of traveling passengers. Safe and secure both refer to the state of protection against dangers and risks. A person who is injured on a vehicle, in which he or she paid a transportation company to travel from point A to point B, could probably care less if the cause of the injury was due to an accident caused by a worn-out mechanism or was due to an act of deliberate sabotage. That person who paid to travel without injury will hold the transportation company liable for the injuries no matter the cause.

Even though many documents use these two words, safety and security, synonymously, there are other documents that refer to the subtle differences between these words. The difference is in the way the danger or risk develops or occurs. For a safety risk, the risk is accidental, unintentional, and inherent in the design of a product. This means that the risk has a probability of occurrence given the way the product is designed and used for its intended application in a normal operating condition or worst-case stress environment. For security risk, the risk is not accidental, but rather intentional or deliberate. The probability of occurrence is determined based on a threat environment while in a normal operating condition or worst-case stress environment. This probability of occurrence calculation is typically performed by reliability engineering in development of system reliability predictions using reliability metrics, such as Mean-Time-Between-Failure (MTBF) or failure rate. System safety engineers are concerned with safety risks from hazards and accidents, while reliability engineers are concerned with reliability risks associated with system failure modes and causes. The interface between software safety, system security, cybersecurity, and software reliability is very important. This interface is discussed in more detail in this chapter. The interface between system safety and system reliability is also discussed in another chapter in this book as well as discussed in the *Design for Reliability* book, companion to this book [3].

Safety and reliability engineering are inseparable. The standard Risk Assessment Code (RAC) defines the combination of the severity and likelihood to a RAC number that is assigned as a low, medium, serious, or high hazard level. The hazard level associated with the failure resulting in the hazard occurrence must be defined by the safety engineer, but reliability engineering is consulted in providing valuable data to support the safety engineer's hazard analysis. This valuable data is the probability of occurrence of a failure mode that could lead to a hazard. Based on the hazard level assignment, a determination is made whether or not additional design requirements are needed to mitigate probability of occurrence of the hazard to an acceptable hazard level so that the system or product can be used safely. Without the reliability data to determine hazard risk probability of occurrence, the risk mitigation evaluation of the hazard cannot be completed [3].

As Nancy Leveson points out [2, page 182], arguments have been made that safety is a subset of reliability or a subset of security or a subset of human engineering. Although there are admittedly some similarities in process and approach, safety has unique processes and approaches that are not included in other engineering disciplines. Safety and security engineering share a close relationship in many processes and approaches. Software safety engineering shares more processes and approaches with security engineering as compared with other engineering disciplines, like reliability and human systems integration. Software safety and cybersecurity engineering tasks are particularly similar in many respects. These similarities are discussed later in this chapter. Due to the similarities between software safety and cybersecurity, the processes and approaches of one can be leveraged by the other to the advantage of both. This advantage they both may realize is in terms of borrowing effective methodologies and tools that can be applied to both at the same time. This commonality of approaches could yield benefits in terms of manpower, productivity, and highly skilled individuals who are concerned with the design of the software as well as the safety and security features of the software.

Both safety and security are concerned with threats. Safety is concerned with threats to life and property, while security is concerned with threats to privacy, critical information, system performance, and national security. Software security, cybersecurity, and Software Assurance (SwA) are concerned with the ability to protect critical program or product information available from a system and to identify and correct design weaknesses that are system vulnerabilities that may be exploited by hackers, malicious software developers, and adversaries.

As Nancy Leveson states [2, page 183], "some of the techniques applicable to one are applicable to the other. For example, both can benefit from the use of barriers." Both safety and security benefit from the use of barriers to protect loss of life as well as loss of critical information that may be exploited with the intention to harm the system user or operator. "For security, barriers are used to prevent malicious incursions rather than accidental ones, but the technique is the same." These barriers inserted into the design are used to prevent accidental mishaps or human error, as well as intentional system sabotage or malicious intrusions into the system for unauthorized data collection or to damage the system and cause system failures. These types of system failures may be classified as mission-critical failures or safety-critical failures, or both. The techniques that warrant the need for protective design features such as barriers are virtually the same when used by safety engineers or security engineers.

There are some differences between the two disciplines, safety and security. Some of the techniques that apply to security do not apply to safety. For instance, a security technique that does not apply to safety is the "use of traps to encourage attacks against hidden defenses, or the randomization of limited defensive resources to reduce the expected success of planned attacks" [2], which affects the probability of success of anticipated planned attacks. If an accident or loss of system operation includes unauthorized system access, system modifications, or blocking authorized access to the system, then security and safety have similar goals and should be integrated in their approaches to understanding and solving the problem while realizing that the safety emphasis is understanding the causes of the accident due to unintentional reasons and security focuses on the intentional reasons.

> Security focuses on malicious actions, whereas safety is concerned with well-intended actions. The primary emphasis in security traditionally has been on preventing unauthorized access to classified information, as opposed to preventing more general malicious activities. Note, however, that if an accident or loss event is defined to include unauthorized disclosure, modification, and withholding of data, then security becomes a subset of safety. [2]

The next section attempts to show a connection between the software safety discipline and the cybersecurity and SwA disciplines. The connection may be difficult since the definitions vary depending on the document you use. This next section makes the effort to research the different documents that define these disciplines so the reader is familiar with the sources of definitions and how they compare. With this new found knowledge, the reader may decide to learn more about these engineering disciplines to make themselves more valuable to their development organizations and accept additional roles on their development teams.

16.2 Definitions of Cybersecurity and Software Assurance

Before someone knowledgeable in system software safety can accept the responsibilities of system software security, the definitions of software security, cybersecurity, and SwA must be defined and understood. The definitions for cybersecurity and SwA are not widely accepted, and there are multiple definitions in the current literature. The similarities and differences of these definitions are explored here so the reader can build an understanding and know when and where to use the definitions in the right context. Only then could the reader consider accepting the tasks of software safety with cybersecurity on a program.

Software security and cybersecurity are synonymous. Cybersecurity is also known as Information Assurance (IA). Cybersecurity as defined in DODI 8500.01, CNSS Glossary (April 2015), is the "prevention of damage to, protection of, and restoration of computers, electronic communications systems, electronic communications services, wire communication, and electronic communication, including information contained therein, to ensure its availability, integrity, authentication, confidentiality, and nonrepudiation" [4]. In the National Initiative for Cybersecurity Careers and Studies (NICCS), Cybersecurity 101 course, on the official website of the Department of Homeland Security (DHS) (NICCS/US-CERT website), cybersecurity is defined as "the activity or process, ability or capability, or state whereby information and communications systems and the information contained therein are protected from and/or defended against damage, unauthorized use or modification, or exploitation" [5]. Furthermore, the NICCS/US-CERT provides an extended definition for cybersecurity: "Strategy, policy, and standards regarding the security of and operations in cyberspace, and encompass the full range of threat reduction, vulnerability reduction, deterrence, international engagement, incident response, resiliency, and recovery policies and activities, including computer network operations, information assurance, law enforcement, diplomacy, military, and intelligence missions as they relate to the security and stability of the global information and communications infrastructure" [5]. The Committee for National Security Systems (CNSS) definition in CNSSI-4009 for cybersecurity is "the ability to protect or defend the use of cyberspace from cyberattacks with cyberspace being defined as a global domain within the information environment consisting of the interdependent network of information systems infrastructures including the Internet, telecommunications networks, computer systems, and embedded processors and controller" [6].

An important reference for system software security is NIST 800-160 [7].

NIST Special Publication 800-160 represents a comprehensive two-year interagency initiative to define systems security engineering processes that are tightly coupled to and fully integrated into well-established, international standards-based systems and software engineering processes. The project supports the federal cybersecurity strategy of "Build It Right, Continuously Monitor" and consists of a four-phase development approach that will culminate in the publication of the final systems security engineering guideline at the end of 2014. The four phases include:

- Phase 1: Development of the systems security engineering technical processes based on the technical systems and software engineering processes defined in ISO/IEC/IEEE 15288 [8].
- Phase 2: Development of the remaining supporting appendices (i.e., Information Security Risk Management (including the integration of the Risk Management Framework [RMF], security controls, and other security- and risk-related concepts into the systems security engineering processes), Use Case Scenarios, Roles and Responsibilities, System Resiliency, Security and Trustworthiness, Acquisition Considerations, and the Department of Defense Systems Engineering Process (Summer 2014).
- Phase 3: Development of the systems security engineering nontechnical processes based on the nontechnical systems and software engineering processes (i.e., Agreement, Organizational Project-Enabling, and Project) defined in ISO/IEC/IEEE 15288 [8].
- Phase 4: Alignment of the technical and nontechnical processes based on the updated systems and software engineering processes defined in ISO/IEC/IEEE 15288. [8]

This publication (NIST 800-160) addresses the engineering-driven actions necessary for developing a more defensible and survivable Information Technology (IT) infrastructure—including the component products, systems, and services that compose the infrastructure. It starts with and builds upon a set of well-established International Standards for systems and software engineering published by the International Organization for Standardization (ISO), the International Electro-technical Commission (IEC), and the Institute of Electrical and Electronic Engineers (IEEE) and infuses systems security engineering techniques, methods, and practices into those systems and software engineering processes. The ultimate objective is to address security issues from a stakeholder requirements and protection needs perspective and to use established organizational processes to ensure that such requirements and needs are addressed early in and throughout the life cycle of the system. [7]

Since there are many stakeholders involved in the systems engineering and the systems security engineering processes and those processes are relatively complex, the phased-development approach of Special Publication 800-160 will allow reviewers to focus on key aspects of the engineering processes and to provide their feedback for those sections of the publication as they are developed and released for public review. [7]

The full integration of the systems security engineering discipline into the systems and software engineering discipline involves fundamental changes in the traditional ways of doing business within organizations—breaking down institutional barriers that over time, have isolated security activities from the mainstream organizational management and technical processes including, for example, the system development life cycle, acquisition/procurement, and enterprise architecture. The integration of these interdisciplinary activities requires the strong support of senior leaders and executives and increased levels of communication among all stakeholders who have an interest in, or are affected by, the systems being developed or enhanced. [7]

To understand the difference between cybersecurity, software safety, and SwA, let us examine several established definitions for SwA. According to NIST Software Assurance Metrics and Tool Evaluation (SAMATE) project [9], SwA is

the planned and systematic set of activities that ensures that software processes and products conform to requirements, standards, and procedures to help achieve:

- Trustworthiness—No exploitable vulnerabilities exist, either of malicious or unintentional origin, and
- Predictable Execution—Justifiable confidence that software, when executed, functions as intended.

According to NASA [10], SwA is a "planned and systematic set of activities that ensures that software processes and products conform to requirements, standards, and procedures. It includes the disciplines of Quality Assurance, Quality Engineering, Verification and Validation, Nonconformance Reporting and Corrective Action, Safety Assurance, and Security Assurance and their application during a software life cycle." The NASA Software Assurance Standard also states: "The application of these disciplines during a software development life cycle is called Software Assurance."

According to the Software Assurance Forum for Excellence in Code (SAFECode) [11], SwA is "confidence that software, hardware and services are free from intentional and unintentional vulnerabilities and that the software functions as intended." From the reference, "Software Assurance: An Overview of Current Industry Best Practices," SAFECode publicly released a white paper that provides an overview of how SAFECode members approach SwA and how the use of best practices for software development helps to provide stronger controls and integrity for commercial applications.

In order to establish and assure integration of interdisciplinary activities with strong leadership support, barriers to communication effectiveness must be recognized and brought down. Some of the causes of these communication barriers are lack of established common lexicons, taxonomy, and definitions. One particular term for a common software engineering discipline used in the SDP is SwA, which means different things to different groups. One popular definition for SwA is from the Department of Defense (DoD). According to the DoD, SwA relates to "the level of confidence that software functions as intended and is free of vulnerabilities, either intentionally or unintentionally designed or inserted as part of the software." This DoD definition may seem straightforward at first glance, but, when compared with other SwA definitions from other sources, leads to some confusion.

The DHS definition of SwA is much more comprehensive. The DHS states that SwA addresses the following:

- Trustworthiness
- Predictable execution
- Conformance

According to the DHS, trustworthiness means no exploitable vulnerabilities exist either malicious or unintentionally inserted. Predictable execution means justifiable confidence that software, when executed, functions as intended. Conformance means planned and systematic set of multidisciplinary activities that ensure software processes and products conform to requirements, standards, and procedures.

Carnegie Mellon University (CMU) Software Engineering Institute (SEI) Capability Maturity Model Integration (CMMI) states the definition of SwA as "Application of technologies and processes to achieve a required level of confidence that software systems and services function in the intended manner, are free from accidental or intentional vulnerabilities, provide security capabilities appropriate to the threat environment, and recover from intrusions and failures."

The SEI is a federally Funded Research and Development Center (FFRDC) sponsored by the US DoD [12]. It is operated by the CMU. The SEI and CMU are associated with the development of the Capability Maturity Model (CMM) for software and later the CMMI for systems. The SEI works in several principal areas, to include software engineering, SwA, cybersecurity, and acquisition, and component capabilities critical to the DoD. The CMMI approach consists of models, appraisal methods, and training courses to improve SDP performance. In 2006, Version 1.2 of the CMMI Product Suite included the release of CMMI for Development, which was the first of three constellations defined in CMMI Version 1.2. The other two constellations include CMMI for Acquisition and CMMI for Services. The CMMI for Services constellation was released in February 2009. Another management practice developed by the SEI is the CERT Resilience Management Model (CERT-RMM). The CERT-RMM is a capability model for operational resilience management. Version 1.0 of the Resilience Management Model was released in May 2010.

SwA as defined in CNSS Instruction 4009 [6] is "the level of confidence that software is free from vulnerabilities, either intentionally designed into the software or accidentally inserted at any time during its lifecycle, and that the software functions in the intended manner."

A paper was written by the Consortium for IT Software Quality (CISQ) titled "How to Deliver Resilient, Secure, Efficient, and Easily Changed IT Systems in Line with CISQ Recommendations" [13]. The CISQ is the global standard for software quality. This paper, which is cited as a reference by the Object Management Group (OMG), explains what steps are needed to deliver secure, efficient, reliable, and easy-to-change complex IT systems, with coding and architecture that comply with CISQ recommendations and emerging standards. Based on 20 years of research in software engineering and business IT, CISQ requirements for reliability, performance efficiency, security, and maintainability are the core of the CISQ standards and recommendations. CISQ highlights in the paper the lack of correlation between good coding practices at the unit level and the value for the business. The paper illustrates the technical reasons for how a software application made from a myriad of high-quality constituents can turn into a fragile, unpredictable, and dangerous system, potentially disrupting a vital system or business process. It includes technical points supporting the need for a system-level architectural analysis of source code and applications' inner structure to deliver high-quality business applications. Finally, CISQ provides an introduction to the software measurement and analysis solutions available on the market.

In the CISQ ARiSE presentation titled "Advances in Measuring and Preventing Software Security Weaknesses" [14], Robert Martin from the MITRE Corporation describes how software design weaknesses may be prevented or eliminated. If the software design weakness case exists that proves it will not be possible to prevent or eliminate the weakness, then a method is developed to predict and measure its probability of occurrence minimizing the risk of security vulnerabilities and resultant software safety hazards. Employing methods that detect and assess software design weaknesses to prevent or predict system vulnerabilities, ensuring that the system is not exploited, attacked, or susceptible to safety-critical hazards, then the user's confidence in the software-enabled system performance increases.

CISQ performed a comparison of the definitions of SwA and published their results for public review [14]. An excerpt from the CISQ comparison of SwA definitions from ARiSE, which is Advanced Research in Software Engineering, is shown in Figure 16.1.

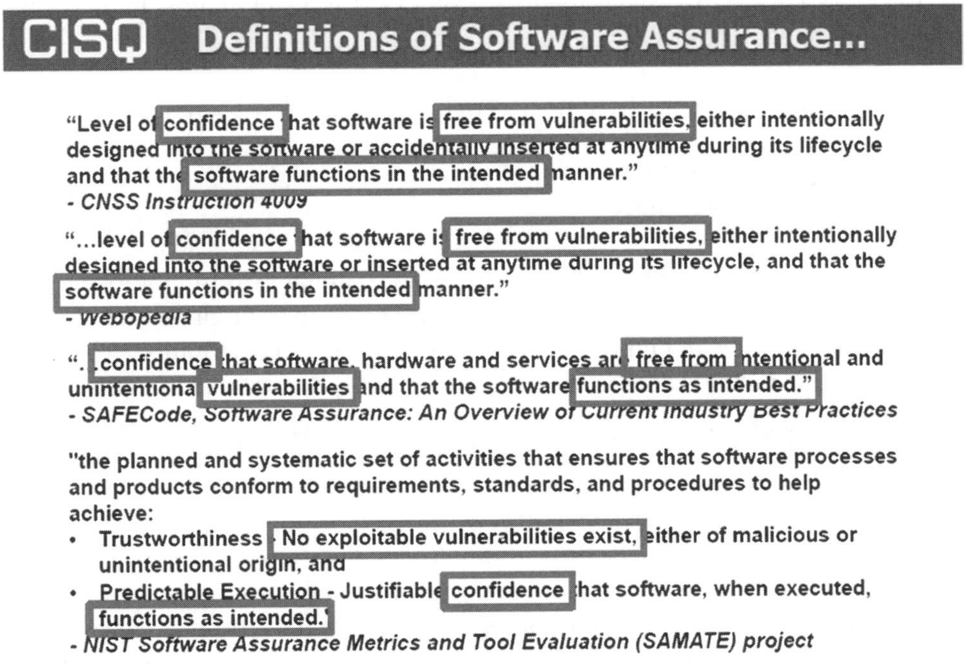

CISQ Definitions of Software Assurance...

"Level of confidence that software is free from vulnerabilities, either intentionally designed into the software or accidentally inserted at anytime during its lifecycle and that the software functions in the intended manner."
- *CNSS Instruction 4009*

"...level of confidence that software is free from vulnerabilities, either intentionally designed into the software or inserted at anytime during its lifecycle, and that the software functions in the intended manner."
- *Webopedia*

"...confidence that software, hardware and services are free from intentional and unintentional vulnerabilities and that the software functions as intended."
- *SAFECode, Software Assurance: An Overview of Current Industry Best Practices*

"the planned and systematic set of activities that ensures that software processes and products conform to requirements, standards, and procedures to help achieve:
• Trustworthiness - No exploitable vulnerabilities exist, either of malicious or unintentional origin, and
• Predictable Execution - Justifiable confidence that software, when executed, functions as intended.'
- *NIST Software Assurance Metrics and Tool Evaluation (SAMATE) project*

Figure 16.1   CISQ comparison of SwA definitions from ARiSE

As you can see from the CISQ comparison, the definition of SwA is not consistent from one reference definition to another. The definitions are similar, but there are some important differences. One of these differences is that the SwA discipline is strictly applied to cybersecurity and does not impact software reliability and dependability. This means only those functions impacted by cyber threats are a concern in the overall software design. The author believes this is a stovepiped view and poor interpretation of the meaning of SwA, which applies to all software functions in a system, no matter if a design weakness vulnerability exists, which may be exploited by a cyber threat. The author prefers that SwA should be the engineering discipline that ensures that all software functions in the intended manner, regardless of inherent design weaknesses that may be deemed as cybersecurity issues exploitable by external means or normal failure modes that may occur by normal operator or system usage.

Another term, which is used in recent literature similar to the term, SwA, is integrity or integrity level. Integrity could be defined in a similar way as SwA. As a result, engineers may refer to SwA and software integrity as synonyms or use these terms in very similar ways applied interchangeably in context. To clear up some confusion on the difference in these two terms, the following definition of integrity level from IEEE/IEC 15026 [15] is provided.

"Integrity level" [15]—What the integrity level fulfills or claims, namely, that the system or element meets:

- A certain target for a property such as risk, reliability, or occurrences of dangerous failures
- Within specified uncertainty limitations
- Under specified conditions

This definition of integrity level seems to be more similar to the definition of "reliability" as stated by the Institute of Electrical and Electronics Engineers (IEEE), and then it is similar to the definition of SwA. If we go back to the definition of cybersecurity, as defined in DODI 8500.01, CNSS Glossary [4], we see that integrity is part of cybersecurity, along with availability, authentication, confidentiality, and nonrepudiation. Integrity is not unique to SwA nor cybersecurity. It relates to any form of risk assessment that is possible and feasible in today's engineering field.

The OMG is an open membership, not-for-profit consortium dedicated to producing and maintaining specifications for interoperable enterprise applications. The OMG membership roster includes many of the most successful and innovative companies in the computer industry, as well as those at the forefront of using technology to gain a competitive edge in their business. All have made the commitment to actively participate in shaping the future of enterprise, Internet, real-time, and embedded systems [16].

According to OMG, SwA is "justifiable trustworthiness in meeting established business and security objectives." OMG's SwA Special Interest Group (SIG) [16] works with Platform and Domain Task Forces and other software industry entities and groups external to the OMG to coordinate the establishment of a common framework for analysis and exchange of information related to software trustworthiness by facilitating the development of a specification for a SwA Framework that will:

- Establish a common framework of software properties that can be used to represent any/all classes of software so software suppliers and acquirers can represent their claims and arguments (respectively), along with the corresponding evidence, employing automated tools (to address scale)
- Verify that products have sufficiently satisfied these characteristics in advance of product acquisition so that systems engineers/integrators can use these products to build (compose) larger assured systems with them
- Enable industry to improve visibility into the current status of SwA during development of its software
- Enable industry to develop automated tools that support the common framework

OMG contributed to the creation of the DHS Software Assurance Common Body of Knowledge (SwA CBK) [17]. While the term "software assurance" potentially refers to the assurance of any property or functionality of software, the emphases currently encompass safety and security and integrate practices from multiple disciplines while recognizing that software must, of course, be satisfactory in other aspects as well, such as usability and mission support.

Next, let us examine the safety and security engineering from OMG DHS Reference 3.5. In recent years, the software safety community has more examples of successful experience with producing high-confidence software than the software security community. The safety community's experience provides lessons for software security practitioners, but the engineering safety problem differs from the security one in a critical way—it presumes nonexistence of maliciousness. Today, security is a concern for most systems as software has become central to the functioning of organizations, and much of it is directly or indirectly exposed to the Internet or to insider attack as well as to subversion during development, deployment, and updating. While safety-oriented systems so exposed now must also face the security problem, this subsection speaks of traditional safety engineering that does not address maliciousness [17].

In the OMG DHS Reference 3.5.2 [17], OMG discusses combining safety and security engineering.

When both are required, a number of areas are candidates for partially combining safety and security engineering concerns including:

- Goals
- Solutions
- Activities
- Assurance case:
  - Goals/claims
  - Assurance arguments
  - Evidence
- Evaluations.

Software assurance for the software-intensive system will verify system operation in an acceptably secure manner. Software assurance should present definitive evidence, supported by process, procedure, and analysis, that a system and its software will be acceptably secure throughout its life cycle, including termination. Software assurance should demonstrate that within the totality of the environment in which the software will operate, any problems created by the software itself failing to operate as required, have been identified and assessed and that any necessary amelioration has been made. Subsequent to contract award, the contractor shall modify the case as development progresses, and knowledge of the system and its software increases. All the lifecycle activities, resources, and products, which contribute to or are affected by the security of the software, also need to be covered. [17]

Now, we focus on bridging the gap between "What are software safety, software assurance, and security engineering" and "How do we make software safety, software assurance, and security better in our designs." There are two methodologies to design for cybersecurity and software assurance. These methodologies are cyber hardening and cyber resilience. In its simplest sense, cyber hardening is the means to prevent cyberattacks to your system from external entities by reducing attack surfaces, by strengthening interfaces using intrusion detection and encryption as examples, by requiring strict password protection and user authentication protocols, and by employing good security awareness and safeguards during system design and system production manufacturing and testing. Cyber hardening allows the user to anticipate system compromise, defend against a cyberattack, provide attack countermeasures, and achieve mission success in a cyber-contested environment.

Cyber resiliency is defined using a common reference from the MITRE Corporation called the Cyber Resiliency Engineering Framework (CREF) [18]: "Cyber resiliency engineering is a part of mission assurance engineering, and is informed by a variety of disciplines, including information system security engineering, resilience engineering, survivability, dependability, fault tolerance, and business continuity and contingency planning" [18]. It also encompasses features of system and software reliability [3]. Reliability is the ability of a design to survive internal system electrical and mechanical stress conditions and external environmental stress conditions over a period of time and keep on working. Resilience is the ability of a design to survive intentional attacks, intrusions, interruptions, and stresses from persons over a period of time and keep on working. The difference between reliability and resilience is that reliability focuses on all the causes of system software failures that are inherent in the design of the system software or unintentional customer-induced overstress conditions, while resilience focuses on the causes of system software failures that are induced by external infiltration and malicious attempts to gain access to the system and interfere with the system performance.

The DoD is very serious about their intent to develop new systems that are cyber resilient. The Engineered Resilient Systems (ERS) is a project funded by the US DoD. In his presentation on ERS Power of Advanced Modeling and Analytics in Support of Acquisition [19], Jeff Holland gave his ideas about how ERS lowers DoD system acquisition risks. ERS defines architecture and workflow products. The ERS workflow includes requirements and system modeling, trade-space creation and analysis, and alternative analysis. The ERS products include SysML model builder, ERSTAT, CREATE, Conceptual Model Builder (CMB), Environmental Simulator (EnvSim), Machine Assisted Design, Decision Dashboard, FACT-X, Statistical Analysis Tools, Big Data Analysis and Visualization, Analysis of Alternatives (AoA), and Mission Context Analysis [19]. By use of ERS capabilities, the multiple types of design weaknesses will be eliminated or largely reduced by detecting, predicting, and preventing inherent design failure modes and causes of security vulnerabilities.

Many design weaknesses impact software reliability, software safety, SwA, and cybersecurity equally. A design weakness such as a data buffer overflow could impact the reliability and safety of the system, as well as introduce a vulnerability that may be exploited by a bad actor and cause a cybersecurity and SwA issue. The MITRE Corporation did a great job at preparing a list of common software design weaknesses that may be cybersecurity vulnerabilities as well as failure causes for unsafe systems and poor system reliability. This list is called the Common Weakness Enumeration (CWE) [20].

CWE is a community-developed dictionary of software weakness types. CWE is a SwA strategic initiative that provides a unified, measurable set of software weaknesses that is enabling more effective discussion, description, selection, and use of software security tools and services that can find these weaknesses in source code and operational systems as well as better understanding and management of software weaknesses related to architecture and design. There are hundreds of software design weaknesses and vulnerabilities catalogued and described in the CWE. A CWE-related effort is the Common Attack Pattern Enumeration and Classification (CAPEC) [21]. CAPEC™ is a comprehensive dictionary and classification taxonomy of known attacks that can be used by analysts, developers, testers, and educators to advance community understanding and enhance defenses. CAPEC artifacts may be accessed from the MITRE website [20].

16.3 Software Safety and Cybersecurity Development Tasks

The SDP includes a variety of tasks performed by software development engineers that are applicable to the software safety process. For instance, software configuration control is important to be performed during the SDP to benefit the overall software design as well as benefit the software safety process. There may be other tasks that the software safety engineer performs, which is typically not included in the SDP. This difference means that the software safety engineer must integrate these software safety tasks into the existing SDP without causing a disruption to the SDP. The best way to integrate safety tasks into the SDP without causing disruption to the software developers is to leverage the existing SDP and make improvements that benefit both the software developer and the safety engineer. As Nancy Leveson recommends in her book titled *Safeware* [2, chapter 12, page 251], there are several general safety processes that may be applied to the SDP:

> The general safety process applied to software development is similar to that applied to any component, especially control components. The basic software system safety tasks include the following [2]:

> - Trace identified system hazards to the software-hardware interface. Translate the identified software-related hazards into the requirements and constraints on software behavior.
> - Show the consistency of the software system safety constraints with the software requirements specification. Demonstrate completeness of the software requirements with respect to system safety properties.
> - Develop system-specific software design criteria and requirements, testing requirements, and computer-human interface requirements based on the identified software system safety constraints.
> - Trace safety requirements and constraints to the code.
> - Identify the parts of the software that control safety-critical operations and concentrate safety analysis and test efforts on those functions and on the safety-critical path that leads to their execution.

- Identify safety-critical components and variables to code developers, including critical inputs and outputs (the interface).
- Develop a tracking system within the software and system configuration control structure to ensure traceability of safety requirements and their flow through documentation.
- Develop safety-related software test plans, test descriptions, test procedures, and test case requirements and additional analysis requirements.
- Perform any special safety analyses such as computer-human interface analysis, software fault tree analysis, or analysis of the interface between critical and noncritical software components.
- Review test results for safety issues. Trace identified safety-related software problems back to the system level.
- Assemble safety-related information (such as caution and warning notes) for inclusion in design documentation, user manuals, and other documentation. [2]

There are various software safety development tasks that may be useful to cybersecurity engineering. One software safety task output is the safety-critical component list. Safety-critical components may be derived from mission-critical component lists. Any failure of a mission-critical component that may lead to a safety hazard, severity 1 or severity 2 effect, will be entered on a safety-critical component list. Mission-critical component lists are normally prepared by reliability engineering. Severity is defined as system effect that results from the hazard or failure. The severity drives the priority of the problem to get resolved. The priority refers to the necessity for a design change with respect to software functionality. The priority is enumerated as (1) critical, (2) major, (3) minor, (4) annoyance, and (5) other. Table 16.1 shows the definitions of the five priority classes.

A critical component may be a hardware component or a software component. A software component may be any code that can be deployed independently. Software components may be called Computer Software Configuration Items (CSCIs), or Software Configuration Items (SCIs), or Software Components (SCs), or units, or packages, or modules, or processes, or functions, or routines, or methods, or programs. Usually, software components are composed of multiple functions or processes, but a single SC may contain a single process or function if that process or function is complex and composed of many interfaces and thousands of lines of code (KLOCs) or thousands of source lines of code (KSLOCs). A software component, such as a software function, whose failure mode leads to a safety-critical system failure or hazard may be considered as critical as a hardware component, such as a microprocessor, whose failure mode also produces a safety-critical system hazard or failure effect.

Table 16.1 Priority classifications

Priority	Description
Critical (priority 1)	A problem that: • Is safety critical with catastrophic effects involving loss of life • Prevents the accomplishment of an operational or mission essential capability • Jeopardizes security or other requirement designated critical
Major (priority 2)	A problem that: • Is safety critical with damage to equipment without loss of life or personal injury • Adversely affects the accomplishment of an operation or mission essential capability and no work-around solution is known • Adversely affects technical, cost, or schedule risks to the project or the life cycle support of the system, and no work-around solution is known
Minor (priority 3)	A problem that: • Adversely affects the accomplishment of an operational or mission essential capability, but a work-around solution is known • Adversely affects technical, cost, or schedule risks to the project or the life cycle support of the system, but a work-around solution is known
Annoyance (priority 4)	A problem that: • Results in user/operator inconvenience or annoyance and does not affect a required operational or mission essential capability • Results in inconvenience or annoyance for development or support personnel, but does not prevent the accomplishment of those responsibilities
Other (priority 5)	Any other effect, such as cosmetic or enhancement change

All eleven of the software safety development tasks cited from *Safeware* [2] may be applied to cybersecurity engineering. Any person trained to perform these tasks for a software safety analysis may also perform these tasks for a cybersecurity engineering analysis. A person translates these eleven tasks from a software safety focus to a cybersecurity vulnerability focus. This translation is shown as follows:

1. Trace identified system design weaknesses that are potential cybersecurity vulnerabilities that may be exploited by a cyber threat to the software–hardware interface. Translate the identified potential cybersecurity vulnerabilities into cybersecurity design enhancements and protection mechanisms written into the hardware and software requirements and constraints on software behavior.

2. Show the consistency of the cybersecurity design constraints with the software requirements specification. Demonstrate completeness of the software requirements with respect to cybersecurity design properties.

3. Develop system-specific software design criteria and requirements, testing requirements, and computer–human interface requirements based on the identified cybersecurity design constraints.

4. Trace cybersecurity design requirements and constraints to the code.

5. Identify the parts of the software that control safety-critical functions and operations, or mission-critical functions, or software functions that may open back doors for unauthorized access and may allow intrusion. Concentrate cybersecurity design analysis and test efforts on those functions and on the critical design paths and information flows that lead to proper system execution that may also allow for exploitation, intrusion, or disruption by external cyber threats.

6. Identify cybersecurity-critical software functions, components, and variables to code developers, including critical inputs and outputs (the interface).

7. Develop a tracking system within the software and system configuration control structure to ensure traceability of cybersecurity requirements and their flow through documentation.

8. Develop cybersecurity-related software test plans, test descriptions, test procedures, and test case requirements and additional analysis requirements.

9. Perform any special cybersecurity analyses such as computer–human interface analysis, software fault tree analysis, or analysis of the interface between critical and noncritical software components to determine the presence of high-risk cybersecurity design weaknesses.

10. Review test results for cybersecurity issues. Trace identified cybersecurity-related software problems back to the system level.

11. Assemble cybersecurity-related information (such as caution and warning notes) for inclusion in design documentation, user manuals, and other documentation.

In 1990, the Electronic Industries Association (EIA) G-48 System Safety Committee published the Safety Engineering Bulletin No. 6B, titled "System Safety Engineering in Software Development" and the *Software System Safety Handbook* [22]. The G-48 System Safety Committee created the procedures, methodology, and criteria for the application of system safety engineering to systems, subsystems, and equipment. The purpose of the document is "…to provide guidelines on how a system safety analysis and evaluation program should be conducted for systems, which include computer-controlled or monitored functions. It addresses the problems and concerns associated with such a program, the processes to be followed, the tasks which must be performed, and some methods, which can be used to effectively perform those tasks" [22].

This *Software System Safety Handbook* references IEEE 1228 [23]. The IEEE published IEEE STD 1228-1994, "IEEE Standard for Software Safety Plans," for the purpose of describing the minimum acceptable requirements for the content of a software safety plan. This standard contains four clauses. Clause 1 discusses the application of the standard. Clause 2 lists references to other standards. Clause 3 provides a set of definitions and acronyms used in the standard. Clause 4 contains the required content of a software safety plan. An informative annex is included and discusses software safety analyses. IEEE STD 1228-1994 is intended to be "wholly voluntary" and was written for those who are responsible for defining, planning, implementing, or supporting software safety plans. This standard closely follows the methodology of MIL-STD-882B, Change Notice 1. The current military standard version of 882 at the time of this printing is MIL-STD-882E. This tells you how many revisions old the IEEE STD 1228 is. We could sarcastically conclude that it took the military standard over 20 years before it adopted the software safety practices that were first published in the IEEE STD 1228.

As stated in Ref. [22] *Software System Safety Handbook* from G-48 reference,

A software design flaw or run-time error within safety-critical functions of a system introduces the potential of a hazardous condition that could result in death, personal injury, loss of the system, or environmental damage. Appendix F provides abstracts of numerous examples of software-influenced accidents and failures. The incident examples in Appendix F include the following:

- F.1—Therac Radiation Therapy Machine Fatalities
- F.2—Missile Launch Timing Error Causes Hang-Fire
- F.3—Reused Software Causes Flight Controls Shut Down
- F.4—Flight Controls Fail at Supersonic Transition
- F.5—Incorrect Missile Firing Due to Invalid Setup Sequence
- F.6—Operator Choice of Weapon Release Over-Ridden by Software Control.

To avoid or eliminate the risks of these types of incidences, a software safety engineer for their particular program should learn the various types of software hazard analysis tools that exist and that could be brought to bear on preventing issues for their program's benefit and the benefit of their customers.

Table 16.2 contains a list of software hazard analysis tools that are used by various experts in accomplishing system software analysis tasks as documented in the *Software System Safety Handbook*. You will notice the number of tools and techniques that seem to be similar and may appear redundant or ambiguous. Some tools seem silly and probably don't belong on the list, such as the last entry on the bottom of the right column. This table demonstrates the lack of a standardized approach for accomplishing software safety hazard analysis tasks. The wise engineer understands what these tools and techniques are in terms of their similarities

Table 16.2 Reprinted from *Software System Safety Handbook* from G-48

Software hazard analysis tools

No.	Tool/technique	No.	Tool/technique
8	Fault tree analysis	1	Hierarchy tool
4	Software prelim hazard analysis	1	Compare and certification tool
3	Traceability analysis	1	System cross-check matrices
3	Failure modes and effects analysis	1	Top-down review of code
2	Requirements modeling/analysis	1	Software matrices
2	Source code analysis	1	Thread analysis
2	Test coverage analysis	1	Petri-net analysis
2	Cross reference tools	1	Software hazard list
2	Code/module walk-through	1	BIT/FIT plan
2	Sneak circuit analysis	1	Nuclear safety cross-check anal
2	Emulation	1	Mathematical proof
2	Subsystem hazard analysis	1	Software fault hazard analysis
1	Failure mode analysis	1	MIL-STD 8S2B, series 300 tasks
1	Prototyping	1	Topological network trees
1	Design and code inspections	1	Critical function flows
1	Checklist of common SW errors	1	Black magic
1	Data flow techniques		

No., cumulative total from those responding to the 1988 survey.

and differences and uses the right tool for the right task. Don't use a screwdriver to hammer a nail into a wooden board.

16.4 Software FMECA

One example of a useful tool and technique that is included in the above list (fourth on the list) is the Failure Modes and Effects Analysis (FMEA). A derivative and enhancement methodology of the FMEA is the Failure Modes, Effects, and Criticality Analysis (FMECA). The Software FMECA is performed on software designs to analyze the effect of software hazards or failure modes that could lead to hazards, which are identified from historical data from prior software development programs, empirical evidence from software testing or software simulations, actual software caused mishaps or failures from customer use applications, or any of various types of software design analyses. Experienced software engineers in the design process must be included in the Software FMECA team. This team will explore and postulate what the effect would be on the product/system output(s), given that a software element is experiencing an error, anomaly, failure, timing issue, process hang, or data corruption, for example. This multi-perspective

Software FMECA team is required to enable a comprehensive software design review of the potential effect of the software hazard or failure mode. The Software FMECA team composition must include sufficient perspective to understand the complete product or system operation. The Software FMECA team must include participants who understand the operation of the software as well as the software's role in the overall product or system operation. The scope of the Software FMECA significantly influences what the hazard or failure effect will be and substantiates the potential design changes to mitigate the effects and the subsequent severity rating for each effect before and after a design change is incorporated into the software design.

Table 16.3 is a modified version of an example SW FMEA from the original table 3, chapter 7, DfR book [3]. It provides examples of the potential effects for some software failure modes for a notional vehicle navigation software-intensive system.

The Software FMECA is to identify the potential root or underlying cause(s) of each previously identified failure mode. The nature of the potential root or underlying cause(s) varies greatly by the nature of the failure mode and the level of abstraction of the Software FMECA. Refer to Table 16.4 (Re: the cause column is modified from the original table 4, chapter 7, DfR book, the failure mode column is copied from the table 16.3). Table 16.4 depicts some examples of software root causes for each of the example software failure modes shown in Table 16.3.

Sometimes, software safety hazard and failure causes defined in an SW FMEA, such as described in Table 16.4, may be resolved or eliminated by a well-written software requirement. The following section provides examples of software safety requirements that resolve failure causes discovered in SW FMEA analysis reports.

16.5 Examples of Requirements for Software Safety

Software safety is achieved when identified software hazards and critical failure modes are eliminated or risk of occurrences is mitigated. This elimination or risk mitigation is guaranteed when software design requirements are written and executed in the code. The following are examples of software requirements that could be designed into a system to prevent future system hazards or safety-critical failures:

- Require the software design to define safe states for each operation, function, or process within the software code, and require the use of software comments embedded into the code where safe states are designated and safety-critical functions are processed.

Table 16.3 Potential software failure mode effects for software elements of a navigation system

Example software failure modes	Example effects of the corresponding software failure mode
Use case	
User of a navigational sensor is unable to update coordinates for airborne vehicle position	Sensor reports the wrong position and causes vehicle to travel in the wrong direction
Software architecture physical view component	
The processor handling all sensor data and calculations experiences a failure, which deprives the vehicle navigation processor of real-time sensor data updates	Without real-time sensor data updates, the navigation processor software may cause incorrect course adjustments that may lead to dangerous aerodynamic maneuvers such as steep stall angles during sudden climbs that lead to an engine stall, potentially producing a vehicle crash
Software system call tree	
A system call to the mission planning subsystem experiences response time lag due to a mission planning subsystem priority interrupt and time to respond for exception handling	The mission planning subsystem does not conduct sensor data update to react to rapidly changing environments, which could cause a vehicle disaster
McCabe sub-tree	
A unique logical McCabe sub-tree execution within a software call tree temporarily freezes, throwing an exception, and does not update sensor readings at specified time intervals, causing a delay to the next sensor reading	The vehicle sensor reading is delayed, but will only affect the vehicle operation if a sudden environmental condition occurs, which requires a sudden course adjustment and correction. If environmental conditions are stable, there is no impact to navigation performance
Interaction between two software objects	
Two objects exist as dependent functions in the navigation subsystem: (1) event handler and (2) sensor communicator. If the sensor communicator sends a medium priority message to the event handler on their functional interface, and the event handler ignores the message until the end of an ongoing, lengthy, low priority task, the message may require a time-dependent operation that gets more critical to the vehicle system safety as time progresses	The effect of the delayed message processing by the event handler may be serious or minor, depending on the real-time constraints on the software system. In an urgent situation such as collision avoidance, the effect of the delayed message processing could be catastrophic
Pointers	
A pointer is misdirected and points to an incorrect memory address, which then provides a false value needed for navigation	The false value could cause the vehicle to go off course and miss its objective, or the route to its destination could have a disastrous result

Table 16.4 Example causes of potential failure modes

Failure modes	Potential root or underlying cause(s)
Use case	
User of a navigational sensor is unable to update coordinates for vehicle position	Requirements ambiguity exists about when a user is able to change the coordinates for the position
Software architecture physical view component	
The processor handling all sensor data and calculations experiences a failure, which deprives the vehicle navigation processor of real-time sensor data updates	The sensor processor experiences an intermittent failure due to power surge in the hardware electronics
Software system call tree	
A software system call to the mission planning subsystem experiences a lack of response to the call due to a mission planning subsystem priority interrupt and exception handling	The mission planning subsystem fails to respond to the system call because it is programmed to react in minutes rather than seconds, for example, two different scales of time and priority are programmed in the system
McCabe sub-tree	
One of the unique logical McCabe sub-tree executions within a software call tree temporarily freezes, throwing an exception, and does not update sensor readings at specified time intervals, causing a delay to the next sensor reading	Due to inexact timing margin, a race condition surfaced within the McCabe sub-tree, which causes a delay in the sensor reading update to the main memory and sent to the main navigation processor during the next read cycle
Interaction between two software objects	
Two objects exist as dependent functions in the navigation subsystem (1) event handler and (2) sensor communicator. If the sensor communicator sends a medium priority message to the event handler on their functional interface, and the event handler ignores the message until the end of an ongoing, lengthy, low priority task, the message may require a time-dependent operation that gets more critical to the vehicle system safety as time progresses	The programming logic in the event handler mistakenly assigns a very low priority to the message received from the sensor communicator
Pointers	
A pointer is misdirected and points to an incorrect memory address, which then provides a false value needed for navigation	The code containing the pointer has faulty conditional logic that resets the pointer to an invalid area of memory

- Require that the system software shall power up system initialization functions with a predefined power-up sequence ending in a predetermined safe state with safety-critical functions enabled, and indicate a "System Ready" status.
- Require timing constraints and limits in all safety design features. For instance, a safety requirement with a timing constraint is "safety interlock circuit shall engage 10 seconds (+/− 1 sec) after safety interlock switch is actuated."
- Require the software design to notify the operator when an operational transition is about to occur between safe and unsafe states or transition between safe and unsafe modes
- Safety-critical software shall revert to a safe state when a safety-critical function is terminated either manually by an operator or automatically by the system.
- The safety-critical software shall be designed such that the operator may cancel current processing with a single action and the system shall revert to a predesignated safe state.
- The software shall detect failures in external safety-critical hardware input/output hardware devices and interfaces and shall revert to a safe state upon their occurrence.
- A failure of the software shall not cause system damage nor personal injury during performance of a certain function at a specific period of time under certain stresses or conditions.
- Normal operation of the software shall not cause system damage nor personal injury during performance of a certain function at a specific period of time under certain stresses or conditions.
- Accuracy of data flowing into and out of safety-critical software processes (e.g., SCIs, components, units) shall be checked by other software processes using a master–master or master–slave configuration within a fault-tolerant architecture.
- There shall be periodic status messages (at least once per 10 messages) between software components (SCI) within an allowable message traffic error rate to verify the integrity of network communications.
- Integer values specified as floating point or fixed point. It is very likely that integers specified to be fixed point will result in an error that could be avoided if specified as floating point. A detailed example of this situation is described in the following paragraph.

16.6 Example of Numerical Accuracy Where 2 + 2 = 5

When does a mathematical error in a simple arithmetic calculation, such as $2 + 2 = 5$, actually be correct? If a software process misses a step in a procedure to prepare numbers as inputs to an equation, such as truncating decimals or to round up or round down to the whole number or a computation performed with invalid numbers. If 2 is a whole number and 2 is set as the value of input variable A and input variable B, then $A + B = 4$, where 2 is added to 2. However, what if the input value for A was taken from a measurement where two decimal places are allowed? If the input is transmitted to A and B as 2.42, so that $2.42 + 2.42 = 4.84$. Then that result is 5 if rounding up is a separate process step in the procedure to execute the calculation and generate a result. The same may apply where $4 + 4 = 9$. However, there is another calculation step that could result in $4 + 4 = 10$. The step in processing the inputs to A and B might truncate any value to the right of the decimal point, leading to a mandatory round-down result. What if the input transmitted to A and B was 4.8, so that $4.8 + 4.8 = 9.6$? If rounding up is performed on the inputs and the output result, then the answer is 10.

16.7 Conclusions

There are many opportunities to leverage the standards, frameworks, and methodologies through the many concepts and references cited in this chapter to synergize engineering disciplines of software safety and software security. One would hope that the reader recognizes the benefits to synergizing multidisciplined engineering functions to reduce cost in system and product development while ensuring elimination or minimization of the risks of software-related hazards and cybersecurity vulnerabilities.

Acknowledgments

The author wishes to acknowledge the contributions from Robert Stoddard in providing much of the Software FMECA source material for this chapter, which was originally published in our book, *Design for Reliability*, Wiley 2012 [3].

References

[1] Institute of Electrical and Electronics Engineers, *IEEE Recommended Practice for Architectural Description of Software-Intensive Systems*, IEEE-Std-1471-2000, IEEE, New York, 2000.

[2] Leveson, N. G., *Safeware: System Safety and Computers. A Guide to Preventing Accidents and Losses Caused by Technology*, Pearson (Addison-Wesley), Reading, MA, 1995.

[3] Raheja, D. and Gullo, L., *Design for Reliability*, John Wiley & Sons, Inc., Hoboken, NJ, 2012.

[4] Committee on National Security Systems (CNSS) Glossary, CNSSI No. 4009, National Security Agency, Ft Meade, MD, April 6, 2015.

[5] National Initiative for Cybersecurity Careers and Studies (NICCS), Cybersecurity 101 Course, on the Official Website of the Department of Homeland Security, https://niccs. us-cert.gov/awareness/cybersecurity-101; http://www.standardscoordination.org/ sccinitiativeskeyconcepts/cybersecurity (Accessed August 10, 2017).

[6] National Information Assurance (IA) Glossary, CNSS Instruction No. 4009, Committee for National Security Systems, National Security Agency, Ft Meade, MD, April 26, 2010.

[7] Ross, R., Oren, J. C., and McEvilley, M., *Systems Security Engineering: An Integrated Approach to Building Trustworthy Resilient Systems*, NIST Special Publication 800-160, National Institute of Standard and Technology, Gaithersburg, MD, 2014.

[8] ISO/IEC/IEEE 15288 (2015) *Systems and Software Engineering—System Life Cycle Processes*, Institute of Electrical and Electronic Engineers (IEEE), New York.

[9] NIST, Software Assurance Metrics and Tool Evaluation (SAMATE) project; Main Page: SAMATE project, Retrieved May 8, 2013 from https://Samate.nist.gov (Accessed on August 10, 2017).

[10] United States National Aeronautics and Space Administration, *Software Assurance Standard*, NASA-STD-2201-93, NASA, Washington, DC, 1992.

[11] SAFECode, Software Assurance: An Overview of Current Industry Best Practices, February 2008, Retrieved May 8, 2013 from, http://www.safecode.org/publication/ SAFECode_BestPractices0208.pdf (Accessed on August 10, 2017).

[12] CMU SEI CMMI, http://www.sei.cmu.edu/ (Accessed on August 10, 2017).

[13] OMG, Reference: How to Deliver Resilient, Secure, Efficient, and Easily Changed IT Systems in Line with CISQ Recommendations, http://www.omg.org/CISQ_compliant_ IT_Systemsv.4-3.pdf (Accessed on August 10, 2017).

[14] Martin, R. A., CISQ ARiSE Presentation, Advances in Measuring and Preventing Software Security Weaknesses, MITRE Corp, SEI, CMU, OMG, http://www.omg. org/news/meetings/tc/tx-14/special-events/cisq-presentations/CISQ-Seminar-2014-6-17-ROBERT-MARTIN-Advances-in-Measuring-and-Preventing-Software-Security-Weaknesses%20.pdf (Accessed on August 10, 2017).

[15] IEEE-Std-15026-1-2010, *Systems and Software Engineering: Systems and Software Assurance—Part 1: Concepts and Vocabulary*, Institute of Electrical and Electronic Engineers (IEEE), New York, 2011. (Adoption of ISO/IEC TR 15026-1 Part 1:2010.)

[16] OMG Software Assurance (SwA) reference links: http://adm.omg.org/Software Assurance.pdf; http://swa.omg.org/docs/softwareassurance.v3.pdf; Omg Swa Sig, Retrieved May 8, 2013 from, *Swa.omg.org* February 26, 2010.

[17] Redwine, S. T., Ed. *Software Assurance: A Guide to the Common Body of Knowledge to Produce, Acquire, and Sustain Secure Software*, Department of Homeland Security (DHS), Washington, DC, 2006.

[18] Cyber Resiliency Engineering Framework (CREF) from MITRE, https://www.mitre. org/publications/technical-papers/cyber-resiliency-engineering-framework (Accessed on August 10, 2017).

[19] Engineered Resilient Systems, paper by Jeff Holland at NDIA conference, March 2015, http://www.defenseinnovationmarketplace.mil/resources/ERS_COI_March_2015. pdf (Accessed on August 10, 2017).

[20] Common Weakness Enumeration (CWE), sponsored by the National Cyber Security Division of the U.S. Department of Homeland Security, Copyright 2011, The MITRE Corporation, https://cwe.mitre.org/ (Accessed on August 10, 2017).

[21] Common Attack Pattern Enumeration and Classification (CAPEC), The MITRE Corporation, https://capec.mitre.org/ (Accessed on August 10, 2017).

[22] DoD Joint Software System Safety Committee of the Joint Services System Safety Panel and the Electronic Industries Association (EIA), G-48 Committee, Software System Safety Handbook, December 1999, http://www.system-safety.org/Documents/Software_System_Safety_Handbook.pdf (Accessed on August 10, 2017).

[23] IEEE Computer Society. Software Engineering Standards Committee.; Institute of Electrical and Electronics Engineers; IEEE Standards Board, *IEEE Standard for Software Safety Plans*, IEEE STD 1228-1994, Institute of Electrical and Electronic Engineers (IEEE), New York, 1994.

17

Lessons Learned

Jack Dixon, Louis J. Gullo, and Dev Raheja

17.1 Introduction

To quote the Spanish philosopher, essayist, poet, and novelist George Santayana, "Those who cannot remember the past are condemned to repeat it." Santayana lived between 1863 and 1952. These words are still relevant today. In the safety arena, we want to remember the past because we don't want to condemn the world to repeat our accidents in the future. No one wants to see or experience another Bhopal, Chernobyl, 3-Mile Island, Challenger, train derailment, or airplane crash. It is bad enough to experience any accident for the first time, but is much worse to live through an accident that happens for a second time.

So, how do we ensure we do not repeat our traumatic history? To answer this, we simply study our lessons learned. What do we mean by lessons learned? A lesson learned is "…knowledge or understanding gained by experience. The experience may be positive, as in a successful test or mission, or negative, as in a mishap or failure…A lesson must be significant in that it has a real or assumed impact on operations; valid in that is factually and technically correct; and applicable in that it identifies a specific design, process, or decision that reduces or eliminates the potential for failures and mishaps, or reinforces a positive result" [1]. A lesson learned is the study of similar types of data and knowledge discovered from past events to prevent future traumatic recurrences or enable great successes.

Many in the safety community recognize the importance of lessons learned, and create archives that capture accident and incident histories. The idea is to preserve historical records of safety problems and their resolution from past programs. These archives of past safety problems should be made widely available so that others can benefit by researching them and can learn from them so that the same problems are not repeated in future products or systems. An archive of lessons learned safety cases is another source of data to aid in conducting system safety analysis to supplement the various types of system safety analyses previously discussed in this book. Even though it is one of many sources for system safety analyses, it is a very important source to be used as part of a complete, successful system safety effort.

 Paradigm 10: If You Stop Using Wrong Practices, You Are Likely to Discover the Right Practices

17.2 Capturing Lessons Learned Is Important

We have found that engineering development practices were used in the past that caused problems that were found later in a system or product life cycle many years after development ended. The correction of these practices was not realized until the offending practices were stopped. It can take many years of constant repeat problem occurrences before this realization is made. This problem of repeat occurrences happens because data related to the events leading up to each problem occurrence were not being captured and understood to determine patterns and cause and effect relationships. Data may have been available, but were not analyzed and considered for its value. Obviously, it is important to learn from our mistakes, but to do this effectively, and to help others learn from the mistakes made, lessons learned must be captured, analyzed, saved, and made readily available for others to use. Let the data and subsequent lessons learned determine the right practices to use. As Trevor Kletz said,

Publication is not, however, the complete answer. What is published is soon read, filed, and forgotten, and the accidents happen again, even in the same company and the same facility as they occurred before. Organizations have no memory; only people have memories and after a few years they move on, taking their memories with them. I wish I could say that the situation is improving but, if anything, as a result of downsizing and earlier retirement, it is getting worse. It may therefore be worth summarizing some of the actions that could be taken to improve corporate memories:

- Include in every instruction, code, and standard a note on the reasons for it and accounts of accidents which would not have occurred if the instruction, etc. had been followed.
- Never remove equipment before you know why it was installed. Never abandon a procedure before you know why it was adopted.
- Describe old accidents as well as recent ones in safety bulletins and discuss them at safety meetings. 'Giving the message once is not enough.'
- Follow up at regular intervals to see that the recommendations made after accidents are being followed, in design as well as operations.
- Remember that the first step down the road to an accident occurs when someone turns a blind eye.
- Include important accidents of the past in the training of undergraduates and company employees.
- Devise better retrieval systems so that we can find, more easily than at present, details of past accidents, in our own and other companies, and the recommendations made afterwards. We need a system that will use the information already entered for other purposes to remind operators, designers, and people preparing permits-to-work of hazards they may be overlooked. The computer will be active and the human will be passive. [2]

17.3 Analyzing Failure

It is important that failures be properly analyzed so that we may learn from them. After a failure occurrence, we must endeavor to find the root cause of the failure, not just the superficial reasons like "failed to follow procedure." Digging deeper will ensure that not only are the real causes discovered, but also the right lesson will hopefully be learned and the right solution will be put in place to prevent the same failure from occurring in the future.

Why is failure analysis often short changed? Because examining our failures in depth is emotionally unpleasant and can chip away at our self-esteem. Left to our own devices, most of us will speed through or avoid failure analysis altogether. Another reason is that analyzing organizational failures requires inquiry and openness, patience, and a tolerance for causal ambiguity. Yet managers typically admire and are rewarded for decisiveness, efficiency, and action—not thoughtful reflection. That is why the right culture is so important. Things take time to cook. The culture should allow for the time to cook the ideas, reflect on past lessons learned, and develop the right solutions.

The challenge is more than emotional; it's cognitive, too. Even without meaning to, we all favor evidence that supports our existing beliefs rather than alternative explanations. We also tend to downplay our responsibility and place undue blame on external or situational factors when we fail. [3]

Finding the reasons for failure in complex situations can be difficult. Research by Edmondson "…has shown that failure analysis is often limited and ineffective—even in complex organizations like hospitals, where human lives are at stake. Few hospitals systematically analyze medical errors or process flaws in order to capture failure's lessons. Recent research in North Carolina hospitals, published in November 2010 in the New England Journal of Medicine, found that despite a dozen years of heightened awareness that medical errors result in thousands of deaths each year, hospitals have not become safer" [3].

Fortunately, there are shining exceptions to this pattern, which continue to provide hope that organizational learning is possible. At Intermountain Healthcare, a system of 23 hospitals that serves Utah and south-eastern Idaho, physicians' deviations from medical protocols are routinely analyzed for opportunities to improve the protocols. Allowing deviations and sharing the data on whether they actually produce a better outcome encourages physicians to buy into this program. [3]

As mentioned earlier, digging deeper is critical to understanding our failures. The best way to accomplish this is to use a team. The team should be diversified, comprised of people drawn from different disciplines and viewpoints. A good example of digging deeper using the team approach can be found after the Space Shuttle Columbia accident, which occurred on February 1, 2003. All seven crew members were killed when Columbia disintegrated over Texas and Louisiana as it reentered Earth's atmosphere: "A team of leading physicists, engineers, aviation experts, naval leaders, and even astronauts devoted months to an analysis of the Columbia disaster. They conclusively established not only the first-order cause, which was a piece of foam had hit the shuttle's leading edge during launch, causing fractures in the ceramic tile outer structure, but also second-order causes: A rigid hierarchy and schedule-obsessed culture at NASA, which made it especially difficult for engineers to speak up about anything, but the most rock-solid concerns" [3].

 Paradigm 2: Be Courageous and "Just Say No"

Again, this emphasized the importance of the proper organizational culture in not only investigating failures but also preventing them in the first place. Edmondson in researching errors and other failures in hospitals "…discovered substantial differences across patient-care units in nurses' willingness to speak up about them. It turned out that the behavior of midlevel managers—how they responded to failures and whether they encouraged open discussion of them, welcomed questions, and displayed humility and curiosity—was the cause. [*She has*] seen the same pattern in a wide range of organizations" [3].

A prime example of the problem of culture was illustrated by the Columbia accident mentioned earlier.

NASA managers spent some two weeks downplaying the seriousness of a piece of foam's having broken off the left side of the shuttle at launch. They rejected engineers' requests to resolve the ambiguity (which could have been done by having a satellite photograph the shuttle or asking the astronauts to conduct a space walk to inspect the area in question), and the major failure went largely undetected until its fatal consequences 16 days later. Ironically, a shared, but unsubstantiated belief among program managers, that there was little they could do, contributed to their inability to detect the failure. Post-event analyses suggested that they might indeed have taken fruitful action. But clearly leaders hadn't established the necessary culture, systems, and procedures. [3]

 **Paradigm 9: Taking No Action Is Usually Not an Acceptable Option**

17.4 Learn from Success and from Failure

Traditionally, system safety efforts focus on identifying what can go wrong and then designing ways to prevent or mitigate the identified risk to an acceptable level. System safety has been successful in many ventures, including military systems, aerospace systems, and nuclear power plants. In system safety, we are used to dealing with and predicting failures. As a result, we typically associate lessons learned with failures, but we can also learn from our successes. If we have a very successful project, product, or system, we should not only celebrate its success, but we should learn from it. What was done and why was it so successful?

A lesson is knowledge or understanding gained by experience. The experience may be positive (a best practice), as in a successful test, mission, exercise, or workshop, or negative, as in a mishap or failure. Successes and failures are both considered sources of lessons. [4]

NASA in their publication *One Hundred Rules for NASA Project Managers* [5] succinctly emphasizes the lessons to be learned from successes:

Rule 93: Things that fail are lessons learned for the future. Occasionally things go right: these are also lessons learned. Try to duplicate that which works.

There should be a balance between the good and the bad. Lessons learned should not only focus on the mistakes that have been made, but they also need to account for all the good things that happened along the way to success. Otherwise, all of the activities performed, the processes conducted, and decisions made that enabled the project or product to succeed would be lost. Losing what worked might be worse than ignoring the things that went wrong or the mistakes that were made. So, instead of just focusing on the failures, it is important to identify what worked and ensure that those things are repeated in future efforts.

 Paradigm 10: If You Stop Using Wrong Practices, You Are Likely to Discover the Right Practices

If It Isn't Broken, Fix It Anyway

Gino and Pisano in an article in *Harvard Business Review* state that many people ask: "'…if it isn't broken, why fix it?' Consequently, when we succeed, we just focus on applying what we already know to solving problems. We don't revise our theories or expand our knowledge of how our business works. Does success mean 'it isn't broken'? Not necessarily. The reality is that while a success (or a string of successes) may mean you're on the right track, you can't assume this to be true without further testing, experimentation, and reflection. You should use success to breed more success by understanding it" [6].

They go on to observe that "There is nothing wrong with toasting your success. But if you stop with the clinking of the champagne glasses, you have missed a huge opportunity. When a win is achieved, the organization needs to investigate what led to it with the same rigor and scrutiny it might apply to understanding the causes of failure. Recognize that this may be an uncomfortable process. You may learn, for instance, that success was achieved only by happenstance" [6]. You might also find that you were lucky and the success was achieved due to circumstances outside your control.

The best way to accomplish this analysis of all the things that led to the success, or failure, is by conducting in-depth reviews of the process or product success and/or failure. For example,

The military holds "After-Action Reviews" (AARs) of each combat encounter and combat-training exercise, irrespective of the outcome. AARs are debriefs that, when used properly, generate specific recommendations that can be put to use immediately. As in business, the reasons for success or failure in combat often are not clear. Companies can employ the same process, which is relatively straightforward. Like sports coaches and players who convene right after a game to review a team's performance, AAR participants meet after an important event or activity to discuss four key questions: What did we set out to do? What actually happened? Why did it happen? What are we going to do next time?

The challenge, of course, is to apply the same degree of rigor whether things are going well or badly. When things go well, our biggest concerns are how to capture what we did and how to make sure we can repeat the success. Replication is important. We need to spread good practices throughout our organizations. But if the chief lesson from a successful project is a list of things to do the same way the next time, consider the exercise a failure. You must constantly strive for continuous improvement. Don't be satisfied with status quo.

Tools like Six Sigma and Total Quality Management have taught us to dig into root causes of problems. Why not use the same approach to understand the root causes of success? Institute a phase in the process where each factor that contributed to success is classified as "something we can directly control" or "something that is affected by external factors." Factors under your control can remain part of your winning formula. But you need to understand how external factors interact with them. [6]

17.5 Near Misses

It is the one that almost happened. Near misses are sometimes called close calls, near hits, narrow escapes, or precursors. They provide another type of lessons learned. Unfortunately, near misses are a largely untapped, but critically important, resource for preventing accidents.

According to the definition used by the National Safety Council, "A Near Miss is an unplanned event that did not result in injury, illness, or damage – but had the potential to do so" [7].

Because nothing bad happened (i.e., no injury, no illness, no damage, or no loss), near misses are often ignored. Because accidents are usually preceded by near-miss events, ignoring these close calls means lost opportunities to prevent accidents. Since there is no loss of life or property, a near miss provides a no-cost learning opportunity. Why not use it as a way to prevent catastrophes? We suggest you collect the data on near misses, just as you would on actual mishaps and accidents. A pattern may emerge that can be used for predictions of accidents and prognostic reasoning for accidents.

While near misses can play a part in many business-related activities, they are extremely important from a safety standpoint. As reported in the *Harvard Business Review*, "…near misses are relevant to managers at all levels in their day-to-day work, as they can also presage lesser, but still consequential problems. Research on workplace safety, for example, estimates that for every 1,000 near misses, one accident results in a serious injury or fatality, at least 10 smaller accidents cause minor injuries, and 30 cause property damage but no injury. Identifying near misses and addressing the latent errors that give rise to them can head off the even the more mundane problems that distract organizations and sap their resources" [8].

Every organization should be collecting and analyzing near misses. They should have a near-miss reporting system that encourages blame-free reporting. Don't shoot the messenger! While not assigning blame is probably the most important aspect of achieving successful reporting, there are other barriers that often prevent reporting of near misses. Bridges [9] reports that

The barriers to getting near misses reported…are:

- Fear of disciplinary action.
- Fear of teasing by peers (embarrassment).
- Lack of understanding of what constitutes a near miss versus a non-incident.
- Lack of management commitment and lack of follow-through once a near miss is reported.
- An apparently high level of effort is required to report and to investigate near misses compared to low return on this investment.
- There is No Way to investigate the thousands of near misses per month or year!
- Disincentives for reporting near misses (e.g., reporting near misses hurts the department's safety performance).
- Not knowing which accident investigation system to use (or confusing reporting system).
- Company discourages near-miss reporting due to fear of legal liability if these are misused by outsiders.

Once near misses are reported, they must be analyzed to determine their root cause. This root cause analysis is needed to identify the system defects that resulted in the close call as well as determining any factors that may helped prevent a more catastrophic loss.

17.5.1 Examples of Near Misses That Ended in Disaster

Near misses are the precursors of most failures and should not be disregarded. "The space shuttle Columbia's fatal re-entry, British Petroleum's (BP's) Gulf of Mexico Deep Water Horizon oil rig disaster, Toyota's automobile stuck accelerators …all were preceded by near-miss events that should have tipped off managers to impending crises. The problem is that near misses are often overlooked—or, perversely, viewed as a sign that systems are resilient and working well" [8].

Near Misses with BP Oil Well

Consider the BP Gulf oil rig disaster. As a case study in the anatomy of near misses and the consequences of misreading them, it's close to perfect. In April 2010, a gas blowout occurred during the cementing of the Deepwater Horizon well. The blowout ignited, killing 11 people, sinking the rig, and triggering a massive underwater spill that would take months to contain. Numerous poor decisions and dangerous conditions contributed to the disaster: Drillers had used too few centralizers to position the pipe, the lubricating 'drilling mud' was removed too early, managers had misinterpreted vital test results that would have confirmed that hydrocarbons were seeping from the well. In addition, BP relied on an older version of a complex fail-safe device called a blowout preventer that had a notoriously spotty track record.

Why did Transocean (the rig's owner), BP executives, rig managers, and the drilling crew overlook the warning signs, even though the well had been plagued by technical problems all along (crew members called it 'the well from hell')? We believe that the stakeholders were lulled into complacency by a catalog of previous near misses in the industry—successful outcomes in which luck played a key role in averting disaster. Increasing numbers of ultra-deep wells were being drilled, but significant oil spills or fatalities were extremely rare. And many Gulf of Mexico wells had suffered minor blowouts during cementing (dozens of them in the past two decades); however, in each case chance factors—favorable wind direction, no one welding near the leak at the time, for instance—helped prevent an explosion. Each near miss, rather than raise alarms and prompt investigations, was taken as an indication that existing methods and safety procedures worked. [8]

Toyota Speed Warning

On August 28, 2009, California Highway Patrol officer Mark Saylor and three family members died in a fiery crash after the gas pedal of the Lexus sedan they were driving in stuck, accelerating the car to more than 120 miles per hour. A 911 call from the speeding car captured the horrifying moments before the crash and was replayed widely in the news and social media.

Up to this point, Toyota, which makes Lexus, had downplayed the more than 2,000 complaints of unintended acceleration among its cars it had received since 2001. The Saylor tragedy forced the company to seriously investigate the problem. Ultimately, Toyota recalled more than six million vehicles in late 2009 and early 2010 and temporarily halted production and sales of eight models, sustaining an estimated $2 billion loss in North American sales alone and immeasurable harm to its reputation.

Complaints about vehicle acceleration and speed control are common for all automakers, and in most cases, according to the National Highway Traffic

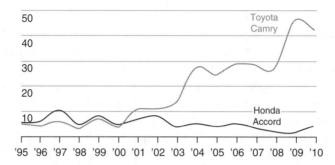

Figure 17.1 Foreshadowing of Toyota problems. Percentage of customer complaints having to do with speed control. Source: National Highway Traffic Safety Administration.

Safety Administration, the problems are caused by driver error, not a vehicle defect. However, beginning in 2001, about the time that Toyota introduced a new accelerator design complaints of acceleration problems in Toyotas increased sharply, whereas such complaints remained relatively constant for other automakers (see Figure 17.1). Toyota could have averted the crisis if it had noted this deviation and acknowledged the thousands of complaints for what they were—near misses. [8]

These studies in near misses illustrate that to avoid the larger failures (i.e., accidents), attention must be paid to fixing the smaller failures. Most accidents are the result of a series of close calls that went unnoticed until their stars aligned in just the wrong way. The questions you should ask yourself are:

Why did these close calls go unnoticed?
What can I do differently to be alert to the data that forewarns of the accidents?
What should I focus on to recognize the factors that lead up to near misses?
Am I using the wrong practices to be able to collect and analyze the right data to predict and prevent near misses and close calls?

 Paradigm 10: If You Stop Using Wrong Practices, You Are Likely to Discover the Right Practices

Strong workplaces rely on strong people that attract, focus, and keep the most talented employees. Talented employees need to be constantly challenged and energized to feel worthy in an organization. Your workplaces processes may be working well for your daily tasks, performance objectives, and team activities, but they can always be improved upon. Strong managers and teams know they must constantly seek improvements in their processes and people to pursue perfection. Talented employees must be challenged to uncover weaknesses in processes and products and do things better. They should want to push back on the status quo and look for opportunities to continuously improve and innovate. Continuous Improvement (CI) comes by measuring what is important in your lives, in your organizations, in your processes, and in your systems and products and then using those measurements to determine what needs to be changed. What gets measured gets improved. Measurements are used to focus on what is important. Measurements with statistical data analysis result in data-driven decisions toward discovery of the right practices.

In the book titled *First, Break All the Rules* [10], the authors describe how you can measure human capital in the section of the book titled "The Measuring Stick." In this section, the authors recite a story about Isaac Newton to reveal how to pry apart a strong workplace to understand the core elements.

In 1666, Isaac Newton closed the blinds of his house in Cambridge and sat in a darkened room. Outside, the sun shone brightly. Inside, Isaac cut a small hole in one of the blinds and placed a glass prism at the entrance. As the sun streamed through the hole, it hit the prism and a beautiful rainbow fanned out on the wall in front of him. Watching the perfect spectrum of colors playing on his wall, Isaac realized that the prism had pried apart the white light, refracting the colors to different degrees. He discovered that white light was, in fact, a mixture of all the colors in the visible spectrum, from dark red to deepest purple; and that the only way to create white light was to draw all of these different colors together into a single beam. [10]

In *First, Break All the Rules*, the authors used statistical analysis to perform the same type of experiment as conducted by Sir Isaac Newton for a different application. The authors collected large amounts of workplace measurements and data. They used statistical analysis techniques to understand variabilities and search for patterns in the data. The authors drilled down and filtered their data in measuring the strength of a workplace to discover 12 simple questions to capture the most important information about the qualities of the workplace to make it strong. These questions are [10]:

1. Do I know what is expected of me at work?
2. Do I have the materials and equipment I need to do my work right?
3. At work, do I have the opportunity to do what I do best every day?
4. In the last seven days, have I received recognition or praise for doing good work?
5. Does my supervisor, or someone at work, seem to care about me as a person?

6. Is there someone at work who encourages my development?
7. At work, do my opinions seem to count?
8. Does the mission/purpose of my company make me feel my job is important?
9. Are my coworkers committed to doing quality work?
10. Do I have a best friend at work?
11. In the last six months, has someone at work talked to me about my progress?
12. This last year, have I had opportunities at work to learn and grow?

The authors wanted to find questions that would differentiate the most productive departments. Only the strongest workplaces in the most productive departments can handle paradigm shifts in stopping their use of practices they have grown accustomed to using and searching for better practices to address causes of near misses and close calls. These 12 questions serve as a measuring stick that allows you to determine the strength of your workplace and link individual and team results to business outcomes. It is the strength of your workplace that results in improvements in your workplace. These improvements can be enacted by individuals at a grassroots level or by organizations as top-down process improvement initiatives. For widespread adoption of systemic improvements, organizations need to establish functional disciplines, similar to system safety engineering, whose job it is to ensure the right organizational common process changes take place to eliminate the possibility of catastrophic near misses and accidental close calls.

17.6 Continuous Improvement

As with system safety, there is another related discipline that strives to improve products and processes. It is called Continuous Improvement. While CI is most often applied to manufacturing processes and Quality Management Systems (QMS), its precepts are also applicable to design and development processes (e.g., Design for Six Sigma (DFSS) and ISO 9001), the topic of lessons learned, and system safety processes.

The Institute of Quality Assurance defined CI as a gradual never-ending change that is:

… focussed on increasing the effectiveness and/or efficiency of an organisation to fulfil its policy and objectives. It is not limited to quality initiatives. Improvement in business strategy, business results, customer, employee and supplier relationships can be subject tocontinual improvement. Put simply, it means 'getting better all the time. [11]

W. Edwards Deming was a pioneer of the field of quality and implemented statistical control and CI ideas in Japan in the 1950s. "Many in Japan credit Deming as one of the inspirations for what has become known as the Japanese post-war economic miracle of 1950 to 1960, when Japan rose from the ashes of war to start Japan on the road to becoming the second largest economy in the world through processes partially influenced by the ideas Deming taught" [12].

"Deming saw [*continuous improvement*] as part of the 'system' whereby feedback from the process and customer were evaluated against organisational goals" [13]. That is just a variation on the theme of lessons learned where we seek feedback from users, designers, and so on with the goal of improving the system, or product, and preventing accidents.

The Fourteen Points

When teaching the Japanese, and later in the United States, Deming developed 14 points for transformation of management. The points are also credited with launching the Total Quality Management movement although Deming did not use the term. The 14 points are adapted from Deming's book, *Out of the Crisis* [14]:

1. Create constancy of purpose toward improvement of product and service, with the aim to become competitive and to stay in business, and to provide jobs.
2. Adopt the new philosophy. We are in a new economic age. Western manage-ment must awaken to the challenge, must learn their responsibilities, and take on leadership for change.
3. Cease dependence on inspection to achieve quality. Eliminate the need for inspection on a mass basis by building quality into the product in the first place.
4. End the practice of awarding business on the basis of price tag. Instead, min-imize total cost. Move toward a single supplier for any one item, on a long-term relationship of loyalty and trust.
5. Improve constantly and forever the system of production and service, to improve quality and productivity, and thus constantly decrease costs.
6. Institute training on the job.
7. Institute leadership. The aim of supervision should be to help people and machines and gadgets to do a better job. Supervision of management is in need of overhaul, as well as supervision of production workers.
8. Drive out fear, so that everyone may work effectively for the company.
9. Break down barriers between departments. People in research, design, sales, and production must work as a team, to foresee problems of production and in use that may be encountered with the product or service.

10. Eliminate slogans, exhortations, and targets for the work force asking for zero defects and new levels of productivity. Such exhortations only create adversarial relationships, as the bulk of the causes of low quality and low productivity belong to the system and thus lie beyond the power of the work force.
    • Eliminate work standards (quotas) on the factory floor. Substitute leadership.
    • Eliminate management by objective. Eliminate management by numbers, numerical goals. Substitute leadership.
11. Remove barriers that rob the hourly worker of his right to pride of workmanship. The responsibility of supervisors must be changed from sheer numbers to quality.
12. Remove barriers that rob people in management and in engineering of their right to pride of workmanship. This means, inter alia, abolishment of the annual or merit rating and of management by objective.
13. Institute a vigorous program of education and self-improvement.
14. Put everybody in the company to work to accomplish the transformation. The transformation is everybody's job.

Seven Deadly Diseases

While the 14 points express Deming's philosophy of transformational management, he also developed the seven deadly diseases that describe the barriers that management faces in improving effectiveness and in implementing CI:

1. Lack of constancy of purpose to plan product and service that will have a market and keep the company in business, and provide jobs.
2. Emphasis on short-term profits: short-term thinking (just the opposite from constancy of purpose to stay in business), fed by fear of unfriendly takeover, and by push from bankers and owners for dividends.
3. Evaluation of performance, merit rating, or annual review.
4. Mobility of management; job hopping.
5. Management by use only of visible figures, with little or no consideration of figures that are unknown or unknowable.
6. Excessive medical costs.
7. Excessive costs of liability, swelled by lawyers that work on contingency fees. [15]

While these points are largely focused on improving management and manufacturing, many have the same themes discussed throughout this chapter, and indeed, this entire book. Many apply directly to the lessons learned process and to the system safety process, as well.

17.7 **Lessons Learned Process**

CI can only be effectively and efficiently achieved within any organization when that organization develops a lessons learned process and rewards those that use this process to achieve value-added benefits. In order to capture lessons learned and to actually learn from them, an organization must have a Lessons Learned Program (LLP) or a lessons learned project to develop its lessons learned process. The following provides a brief summary of the components necessary to have a viable LLP.

Capture

Once there has been an accident or failure, the first thing to do is capture the information to start the lessons learned process. The collection of pertinent information can be accomplished during the investigation of the accident. Additional information can be gathered through interviews, questionnaires, surveys, document review, or data mining. It is always best to have a team of experts with diverse backgrounds involved to provide varied perspectives. As discussed earlier in this chapter, lessons learned don't necessarily come only from failures. The same process of capturing information to support an LLP also applies to collecting information on successes or near misses. Near misses should be reported and data collected and entered into the lessons learned library or archive. After a successful project, product, or system development, an end-of-project retrospective should be conducted to collect lessons learned from the particular successes.

Analyze

Once essential information regarding the failure is collected, analysis of the situation to determine the root cause(s) is the next step. The analysis should answer the "who, what, when, where, and why" questions to identify the root causes of the failure or the success. In addition to ferreting out the root cause, the analysis should result in the corrective actions that need to be taken to prevent future occurrences as well as lesson(s) to be learned from the incident. This is the most difficult and time-consuming part of the lessons learned process but should ultimately lead to the improvement of the system/product when the lesson(s) learned is implemented.

Disseminate

One of the most important aspects of a successful LLP is the dissemination and sharing of the information. This sharing of the knowledge gained must include the participants in the system/product development, manufacturing, and users.

In addition, providing the lessons learned to the widest audience possible so that everyone may learn from the mistakes made. Share! "The next best thing to learning from your own lessons is to learn from other's lessons – Gains without much pain!" [16] Sharing can be accomplished through many means including reports, briefings, tools, databases, and websites.

Archive

Maintaining the lessons learned for future use always involves some form of archiving of the information. This can simply be a library of written reports or a more sophisticated tool or method using a digital repository or database. Examples of digital repositories are object databases or relational databases that can be customized for each individual applications and needs. Regardless of what approach is used, the information must be easily retrievable in order to be valuable to future users. The data going into the database must be complete, accurate, high quality, and peer reviewed to prevent "Garbage In – Garbage Out." The archived and stored data must be valuable to many types of users. There must also be a long-term commitment to maintaining the archive; otherwise it will become irrelevant and unused.

17.8 Lessons Learned Examples

This section provides a small sampling of different types of failures and the lessons learned from them. We are not trying to provide an exhaustive list of failures and their lessons, but we have attempted to provide a variety of examples that illustrate different types of lessons learned.

17.8.1 Automobile Industry Lessons Learned from the Takata Airbag Recall

Chapter 4 briefly discussed the Takata airbag failures that resulted in 16 deaths and the largest recall in history. The airbag inflators were failing during airbag deployment, causing explosive ruptures of the inflator's steel housing, producing shrapnel that caused injuries, and sometimes death, to the car's occupants.

The following excerpts from the *Expert Report of Harold R. Blomquist*, created for National Highway Traffic Safety Administration (NHTSA), provide a quick summary of the reasons for the airbag failures [17]:

> …the inflator design permits moist air to slowly enter the inflator, where the moisture-sensitive propellant slowly degrades physically due to temperature cycling. During subsequent air bag deployment in a crash, the damaged propellant burns more rapidly than intended, and overpressurizes the inflator's steel housing causing fragmentation.

The affected Takata inflator which does

(...not contain a desiccant) deploys by igniting the booster propellant, ...which in turn ignites the main propellant... These propellants are comprised of solid powders that are blended together and pressed to form the propellant part, akin to pharmaceutical tablets. ...The main propellant contains [*Phase Stabilized Ammonium Nitrate*] PSAN, a moisture-sensitive oxidizer that...degrades the propellant part if exposed to moisture and temperature cycling.

Assessment...indicates that defective...inflators are generally free of known manufacturing defects. In other words, it appears the inflators were manufactured as intended and passed all lot acceptance tests and vehicle manufacturer specifications.

The report concludes,

...that all non-desiccated Takata frontal driver and passenger inflators contain a propellant that degrades over time. The degradation is principally the result of long term daily temperature cycling of moist propellant.

The final conclusions of the report are

...that the inflator ruptures occur through the following factors and sequence of events:

a. Affected inflators are inadequately sealed for protection of the moisture sensitive PSAN-based main propellant;
b. Allowing moist air to enter the inflator air space; and
c. Causing damage to the physical structure of the main propellant consisting of the formation of pores/channels.
d. Over the course of years, the extent of damage progresses by a slow process driven by daily temperature fluctuations.
e. Then, during combustion, the extremely hot gas enters the pores/channels;
f. Which causes a transition from layer-by-layer burning to burning en masse that over-pressurizes the steel shell to cause catastrophic failure (rupture) with fragmentation hazard to vehicle occupants.

Lesson Learned

The lesson learned here is that even though the product was manufactured as intended, met the specifications, and passed all lot testing, the choice of a moisture-sensitive propellant for use in a product that would be exposed to various environments for years was the wrong choice. Secondarily, using moisture-sensitive propellant and having no desiccant and inadequate seals in the design also is a lesson to be learned.

Paradigm 3: Spend Significant Effort on Systems Requirements Analysis

Continuing with failures in the automobile industry, we have the General Motors (GM) recall for faulty ignition switches.

17.8.2 Automobile Industry Lessons Learned from the 2014 GM Recall

One common safety issue is related to automobile product recalls due to safety of a critical component used in the operation of an automobile. All the issues related to automobile safety are too numerous to cover in great detail in one book, let alone one chapter of this book. One particular issue that was studied for this book was the GM recall involving ignition switches that mainstream media publicized in 2014. We studied the GM ignition switch recall from its inception in 2014 and continued to track the GM recall progress up to current events just prior to publication of this book.

17.8.2.1 Summary of the 2014 GM Recall

On May 16, 2014, the US government fined GM Corporation $35 million for failing to disclose defects in automobile ignition switches. This amount was a record fine. GM delayed over 10 years to disclose the presence of the defect to the government and to the public. The defect originally only affected a small list of several types of small car designs, but that list of affected car designs gradually grew to many other GM makes and models. The defect was linked to at least 13 deaths as stated by GM representatives.

GM recalled more than 2.5 million vehicles in 2014 for problems with its automotive ignition after multiple driving accidents caused more than 40 deaths according to one source. And the death toll continued to rise. GM established a compensation program for its customers. As of January 31, 2015, GM received more than 2200 claims for injuries and deaths as a result of the ignition system failure. Even though GM ultimately accepted responsibility for the personal injuries and deaths, the company was slow to react when the crisis began. GM originally knew of the potential for this defect at least 10 years earlier. It really bothers people when a company the size of GM, who made $500M in 1Q2014, does not react fast enough to customer data on safety-critical failures and not practicing due diligence to conduct a product recall for faulty ignition switches to prevent deaths and injuries. The company obviously calculated the risks of doing an early recall. The company probably factored the costs of the recall and compared these costs with the costs of unreliability and doing nothing. And they decided to wait. The cost to be considered is not just the recall expenses and legal expenses, including fines and fees, but the loss of life and loss of customer trust. Imagine the effect to GM by this type of betrayal of public trust and the loss in revenue from future sales.

"The mistakes that led to the ignition switch recall should never have happened. We have apologized and we do so again today," said GM CEO Mary Barra. "We have faced our issues with a clear determination to do the right thing both for the short term and the long term. I believe that our response has been unprecedented in terms of candor, cooperation, transparency and compassion." GM Chairman Theodore M. Solso said, "GM's Board of Directors took swift action to investigate the ignition switch issue and we have fully supported management's efforts to regain the trust and confidence of customers and regulators, and to resolve the Justice Department's investigation. GM's Board and leadership recognize that safety is a foundational commitment, and the changes the company has made in the last 15 months have made it much stronger" [18].

17.8.2.2 Failure Cause and Effects

The ignition switch, also known as an actuator, was found to have a manufacturing defect. The defect is related to an ignition lock actuator housing manufactured with a critical design dimension parameter and an outer mechanical assembly diameter that exceeds the actuator mechanical dimension specifications. The defective ignition switches were designed and manufactured with lower than required torque resistance. In certain situations, the switch could move easily out of the "Run" position into "Accessory" or "Off" positions. When the switch moved out of "Run," it could disable the car's frontal airbag, increasing the risk of death and serious injury in certain types of crashes in which airbags were otherwise designed to deploy. In the primary recall, the oversized ignition lock actuator housing causes the ignition key to get stuck in the "Start" position. If the vehicle is driven with the key stuck in the "Start" position and experiences a "significant jarring event," the ignition lock cylinder could move into the "Accessory" position, affecting engine power, steering, and braking. Airbag sensing may be affected depending on the sequence of events, related to the timing of the key movement into the "Accessory" position relative to a vehicle accident or crash. There is a finite probability that airbag sensing at the time of a front-end collision or crash could result in the airbags not deploying.

The switches were assembled into small model cars, such as the Chevrolet Cobalt and Saturn Ion. The effects of the ignition switch failures lead to driver and passenger fatalities and injuries caused by disabled airbags during crashes and the driver's inability to control the car while driving due to the following:

1. Disables the power-assisted steering and brakes
2. Cars slip out of the "run" position
3. Cars' engines shut down

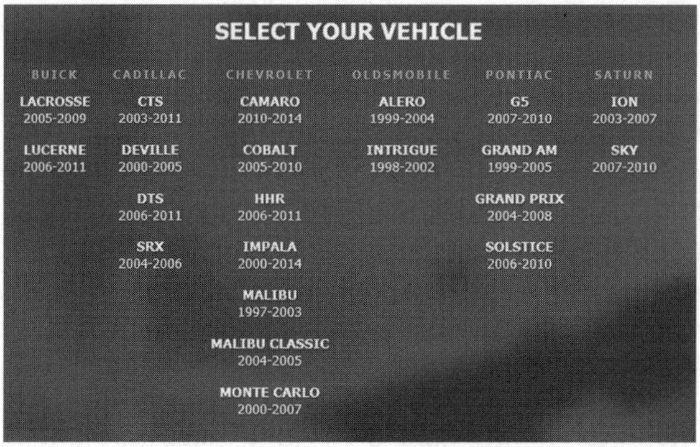

Figure 17.2 GM models and years effected by the ignition switch recall

Besides Chevrolet Cobalt and Saturn Ion, GM discovered other models with the defect. The GM recall covers certain Chevrolet Silverado light-duty and heavy-duty pickups, as well as Avalanche, Tahoe, and Suburban; GMC Sierra light-duty and heavy-duty pickups, as well as Yukon and Yukon XL; and Cadillac Escalade, Escalade ESV, and Escalade EXT. It concerns models from 2011 and 2012 and 2007–2014 vehicles that have been repaired with defective parts. GM discovered these other models in an internal review following warranty party returns.

GM set up a website specific for customers of GM vehicles with potential for a recall due to the defective ignition switch. The website is called GM Ignition Recall Safety Information [19]. The website shows the models and years of the cars affected by the recall. Figure 17.2 shows the listing of these cars and years.

One would think this should change the way CEOs and industry think about problem avoidance and management.

17.8.2.3 Corrective Actions Taken by GM

As stated earlier, GM reached a monumental agreement with the US Attorney's Office on May 16, 2014, regarding the company's handling of the ignition switch defect. GM reached a settlement with the US Attorney's Office for the Southern District of New York in the form of a deferred prosecution agreement. Under the agreement, the US Attorney's Office agrees to defer prosecution of charges against GM related to the ignition switch defect and recall for three years. If GM satisfies the terms of the agreement, federal prosecutors will then seek dismissal of the charges with prejudice. The agreement includes a requirement that GM cooperate with the federal government and establish an independent monitor to review and assess the company's policies and procedures in certain discrete areas relating to safety issues and recalls. GM will also pay a $900 million financial penalty associated with this agreement and will record a charge for this amount in the third quarter. The agreement states that the government's decision to defer prosecution was based on the actions GM has taken to "demonstrate acceptance and acknowledgement of responsibility for its conduct," including [18]:

- Conducting a swift and robust internal investigation
- Furnishing investigators with information and a continuous flow of unvarnished facts
- Providing timely and meaningful cooperation more generally in the government's investigation
- Terminating wrongdoers
- Establishing a full and independent victim compensation program that is expected to pay out more than $600 million in awards

"Reaching an agreement with the Justice Department does not mean we are putting the issue behind us," Barra said. "Our mission has been to take the difficult lessons from this experience and use them to improve our company. We've come a long way and we will continue to build on our progress." With GM's response following the ignition switch recalls in February and March 2014, GM pledged its full cooperation to authorities investigating the matter. GM's Board also retained former US Attorney Anton Valukas to conduct an independent investigation. In June 2014, the Valukas Report was provided to the NHTSA, the US Department of Justice, and members of both the US House of Representatives and Senate. The results of the investigation were later made public. Barra also discussed the Valukas Report with GM employees in a global town hall meeting, during which she said, "We aren't simply going to fix this and move on. We are going to fix the failures in our system… and we are going to do the right thing for the affected parties" [18].

After the recall, GM began making far-reaching changes to its vehicle quality and safety organizations [18]. Some of the changes and improvements are as follows:

- GM created a new position, Vice President, Global Vehicle Safety, with global responsibility for the development of GM vehicle safety systems, confirmation, and validation of safety performance, as well as post-sale safety activities, including recalls. Decisions about vehicle safety and recalls are now elevated to some of the highest levels of the company.
- Approximately 200 employees joined the Global Safety organization in 2014, including more than 30 new safety investigators in North America whose roles are to help identify and quickly resolve potential safety issues.

- A data analytics team was created to search for emerging issues using internal and external sources, including those reported to NHTSA.
- The company created a "Speak up for Safety" program, designed to give employees and dealers an easy and consistent way to report potential vehicle or workplace safety issues or suggest safety-related improvements.
- GM's Global Vehicle Engineering organization was reorganized to improve cross-system integration, deliver more consistent performance across vehicle programs, and address functional safety and compliance in vehicle development.
- GM has implemented new policies and procedures to expedite the repair of recalled vehicles and to improve its certified pre-owned vehicle program.
- The company also established the GM Ignition Compensation Claims Resolution Facility, which is independently administered by Attorney Kenneth Feinberg. The facility was designed to settle claims brought by people who suffered physical injuries or lost family members in accidents that may have been related to the ignition switch.
- It has awarded settlements even to those whose claims involved contributory negligence, as well as claims that would have been barred by bankruptcy court rulings [18].

In February 2014, GM conducted a recall of approximately 700,000 vehicles affected by the defective switch. In March 2014, the recall population had grown to more than two million vehicles. Since February 2014 and the inception of the federal criminal investigation, GM has taken exemplary actions to demonstrate acceptance and acknowledgement of responsibility for its conduct. GM, among other things, conducted a swift and robust internal investigation, furnished the government with a continuous flow of unvarnished facts gathered during the course of that internal investigation, voluntarily provided, without prompting, certain documents and information otherwise protected by the attorney–client privilege, provided timely and meaningful cooperation more generally in the federal criminal investigation, terminated wrongdoers, and established a full and independent victim compensation program that has to date paid out hundreds of millions of dollars in awards [20].

17.8.2.4 Court Case

Attorney General Loretta E. Lynch, Secretary Anthony Foxx of the Department of Transportation, US Attorney Preet Bharara of the Southern District of New York, Administrator Mark R. Rosekind of the NHTSA, Inspector General Calvin L. Scovel III of the US Department of Transportation (DOT-OIG), Special Inspector General Christy Goldsmith Romero of the Special Inspector General for the

Troubled Asset Relief Program (SIGTARP), and Assistant Director in charge Diego Rodriguez of the FBI's New York Field Office announced the filing of criminal charges against General Motors Company (GM or the company), an automotive company headquartered in Detroit, that has designed, manufactured, assembled, and sold Chevrolet, Pontiac, and Saturn brand vehicles, among others. GM is charged with concealing a potentially deadly safety defect from its US regulator, the NHTSA, from the spring of 2012 through February 2014 and, in the process, misleading consumers concerning the safety of certain of GM's cars. The models equipped with the defective switch were the 2005, 2006, and 2007 Chevrolet Cobalt; the 2005, 2006, and 2007 Pontiac G5; the 2003, 2004, 2005, 2006, and 2007 Saturn Ion; the 2006 and 2007 Chevrolet HHR; the 2007 Saturn Sky; and the 2006 and 2007 Pontiac Solstice [20].

US Attorney Bharara announced a deferred prosecution agreement with GM under which the company admits that it failed to disclose a safety defect to NHTSA and misled US consumers about that same defect. The admissions are contained in a detailed statement of facts attached to the agreement. The agreement imposes on GM an independent monitor to oversee GM's reporting of safety issues and public statements, including the review and assessment of policies, practices, and procedures related to GM's safety-related public statements, and sharing of engineering data and recall processes. The agreement also requires GM to transfer $900 million to the United States by no later than September 24, 2015, and agree to the forfeiture of those funds pursuant to a parallel civil action also filed today in the Southern District of New York [20].

"For nearly two years, GM failed to disclose a deadly safety defect to the public and its regulator," said US Attorney Bharara. "By doing so, GM put its customers and the driving public at serious risk. Justice requires the filing of criminal charges, detailed admissions, a significant financial penalty, and the appointment of a federal monitor. These measures are designed to make sure that this never happens again." "Today's action strengthens NHTSA's efforts to protect the driving public," said Administrator Rosekind. "It sends a message not only to GM, but to the entire auto industry, that when it comes to safety, telling the full truth is the only option." "GM concealed a safety defect from consumers and regulators, which put drivers at risk," said Assistant Director in Charge Rodriguez. "The resolution of this case shows that safety should never take a backseat to expediency" [20].

"General Motors not only failed to disclose this deadly defect, but as the Department of Justice investigation shows, it actively concealed the truth from NHTSA and the public," said Transportation Secretary Foxx. "Today's announcement sends a message to manufacturers: deception and delay are unacceptable, and the price for engaging in such behavior is high." The criminal charges are contained in an information (the information) alleging one count of engaging in a scheme to conceal material facts from NHTSA and one count of wire fraud. If GM abides by all of the terms of the agreement, the government will defer prosecution on the information for three years and then seek to dismiss the charges [20].

In 2004 and 2005, as GM employees, media representatives, and GM customers began to experience sudden stalls and engine shutoffs caused by the defective switch, GM considered fixing the problem. However, having decided that the switch did not pose a safety concern, and citing cost and other factors, engineers responsible for decision-making on the issue opted to leave the defective switch as it was and simply promulgate an advisory to dealerships with tips on how to minimize the risk of unexpected movement out of the Run position. GM even rejected a simple improvement to the head of the key that would have significantly reduced unexpected shutoffs at a price of less than a dollar a car. At the same time, in June 2005, GM made public statements that, while acknowledging the existence of the defective switch, gave assurance that the defect did not pose a safety concern [20].

The following is a short timeline of significant events involving the GM Recall beginning in 2005 and ending in 2014, as researched and published by NPR [21]:

- **March 2005:** GM rejects a proposal to fix the problem because it would be too costly and take too long.
- **May 2005:** A GM engineer advises the company to redesign its key head, but the proposal is ultimately rejected.
- **July 29, 2005:** Maryland resident Amber Marie Rose, 16, dies when her 2005 Chevrolet Cobalt crashes into a tree after the ignition switch shuts down the car's electrical system and the airbags fail to deploy.
- **December 2005:** GM issues a service bulletin announcing the problem, but does not issue a recall.
- **March 2007:** Safety regulators inform GM of the issues involved in Amber Rose's death; neither GM nor the safety regulators open a formal investigation.
- **April 2007:** An investigation links the fatal crash of a 2005 Chevrolet Cobalt in Wisconsin to the ignition defect, but regulators do not conduct an investigation.
- **September 2007:** A NHTSA official emails the agency's Office of Defects Investigation (ODI) recommending a probe looking into the failure of airbags to deploy in crashes involving Chevrolet Cobalts and Saturn Ions, prompted by 29 complaints, four fatal crashes, and 14 field reports.
- **October 26, 2010:** *Consumer Reports* says GM is considered "reliable" based on scores from road tests and performance on crash tests.
- **2012:** GM identifies four crashes and four corresponding fatalities (all involving 2004 Saturn Ions) along with six other injuries from four other crashes attributable to the defect.
- **June 2013:** A deposition by a Cobalt program engineer says the company made a "business decision not to fix this problem," raising questions of whether GM consciously decided to launch the Cobalt despite knowing of a defect.

- **End of 2013:** GM determines that the faulty ignition switch is to blame for at least 31 crashes and 13 deaths.
- **Febraury 7, 2014:** GM notifies NHTSA "that it determined that a defect, which relates to motor vehicle safety, exists in 619,122 cars."
- **Febraury 13, 2014:** GM officially recalls 2005–2007 Chevrolet Cobalts and 2007 Pontiac G5s.
- **March 17, 2014:** GM recalls 1.55 million vans, sedans, and sport utility vehicles.
- **March 17, 2014:** Barra states in a video apology that "something went very wrong" in GM's mishandling of the crisis. She says the company expected about $300 million in expenses in the current quarter to cover the cost of repairing three million vehicles.
- **March 18, 2014:** GM appoints a new safety chief.
- **March 20, 2014:** The House Energy and Commerce Committee's Subcommittee on Oversight and Investigations schedules a hearing for April 1, titled "The GM Ignition Switch Recall: Why Did It Take So Long?"
- **March 28, 2014:** GM recalls an additional 824,000 vehicles (including all model years of the Chevrolet Cobalt and HHR, the Pontiac G5 and Solstice, and the Saturn Ion and Sky), stating ignition switches could be faulty; the new total number of recalled vehicles in the United States is 2,191,146.
- **April 1–2, 2014:** Barra and NHTSA Acting Administrator David Friedman testify at House and Senate hearings on the handling of the recall. Barra apologizes to family members whose loved ones have died from the defect.
- **April 10, 2014**: GM starts a Speak Up for Safety campaign, aimed at encouraging employees to say something when they see a potential safety issue for customers.
- **April 10, 2014**: GM adds ignition lock cylinders to its safety recall of 2.2 million older model cars in the United States.
- **May 16, 2014**: The government announces GM will pay a record $35 million civil penalty after NHTSA determined the automaker delayed reporting the ignition switch defect.
- **June 5, 2014:** An internal inquiry by Anton Valukas, a former US attorney, into the ignition switch recall finds an 11-year "history of failures" and "a pattern of incompetence and neglect," Barra says.

Lessons Learned

We discussed the GM recall of 2014 related to a defective component and how GM admits to failing to disclose deadly safety defects in its cars to consumers and US regulators. Several lessons to learn here include doing a better design, watching tolerances, and doing hazard analysis to find and eliminate hazards. However, the main lesson learned here is one that many politicians belatedly learn—the cover-up is worse than the crime!

(Note: For more details about the GM ignition safety switch recalls, we suggest you refer to the GM recall website, cited in Ref. [19], or contact General Motors customer service at 800-222-1020. If you prefer, you can also contact the NHTSA at 888-327-4236.)

Next we enter the scary world of medical safety looking at the things that can go wrong with medical devices and what could happen to you during a hospital visit.

17.8.3 Medical Safety

The medical device recalls for unsafe performance have been steadily going up for the last 10 years. Good safety design practices must be used to reverse the trend. There is a similar scenario with the hospitals. Many hospitals are years behind in keeping patients safe from medical mistakes and using new technologies leveraged from other safety-critical industries such as defense, aerospace, and the nuclear industry. According to the National Patient Safety Foundation, about 400,000 patients die each year from medical mistakes [22].

The following topics will be discussed in this section:

Medical device concerns
- Insufficient use of risk analysis tools, such as those from the international standard, ISO 14971
- Examples of unsafe devices
- Insufficient knowledge to design for safety
- Insufficient verification and validation
- Regulations that are guidelines, not prescriptive
- New technologies that are constantly challenging
- Recommendations

Hospital safety concerns
- Inadequate measures on safety of patients
- System that is too complex
- Examples of fatal hospital mistakes
- New technologies that are not understood
- Insufficient regulatory requirements for safety
- Insufficient incentives to improve safety
- Need to change culture of CI to continuous innovation
- Recommendations

17.8.3.1 Medical Device Concerns Details

Insufficient Use of Risk Analysis Tools Covered in the ISO 14971

ISO 14971 is an international standard that describes the application of risk management to medical devices. The use of ISO 14971 is not mandatory, but the FDA acknowledges it as a good standard. It expects device designers to use the standard.

The result is that most companies use only one tool out of five. Very few use more than one tool based on the author's knowledge of the industry for over 20 years.

The tool most companies use is Failure Modes and Effects Analysis (FMEA). Unfortunately many companies do not use it for preventing design failures by creating good requirements during specification writing. Instead, they use it after the specifications are approved. They miss the chance of finding defects in the specifications.

Examples of Unsafe Medical Devices

Many devices are unsafe because of production defects, maintenance mistakes, user mistakes, and software malfunctions. There is an average of three recalls of medical devices each day. Examples include the following:

- An FDA recall was made on Magnetic Resonance Imaging (MRI) machines. Someone brought a metal container in the MRI room; the machine malfunctioned because of powerful magnets in the machine.
- There were recalls of cardiac pacemakers because a user died in front of a cashier in a shopping mall. The shop's cash register created electromagnetic field, resulting in very high breathing rate.

Insufficient Knowledge to Design for Safety

A patient is never harmed from just one cause. There are always at least two causes [23]. This knowledge is not widely known nor mentioned in regulations. So industries keep making design mistakes and production mistakes not knowing that design mistakes can be tolerated if they could make the design fail-safe by preventing the second cause responsible for the harm. We talked about the recall of MRI because of a metal container in the area. They could have prevented the harm by either disallowing anything metallic in the area or by designing to tolerate the metals in the area.

Insufficient Verification and Validation

One major problem in verification and validation of safety is the timing of the risk analysis. Most designers conduct the risk analysis after the specification and the conceptual design are approved, which is too late. The author's experience with new devices over 25 years shows as many as 200 requirements are missing or are vague in a product like a radiology system or an automated surgical system. If that many requirements are ignored, designers can only do verification and validation on partial requirements. No wonder the recalls are increasing as the complexity increases.

Regulations Are Guidelines, Not Prescriptive

FDA regulations are not supposed to be prescriptive because they can become counterproductive to innovation, but there are sufficient guidelines to develop a good design. They emphasize sound design through risk analysis. Unfortunately guidelines are often misinterpreted to only do what is necessary instead of challenging the design. There is often more effort spent on documentation required by the FDA audits, resulting in less time for robust designs. Another problem with regulations is that no independent verification and validation is required such as in Department of Defense standards. This results in biased verification and validation.

New Technologies Are Constantly Challenging

Medical technology is growing rapidly. Wearable technologies, use of too much interoperability among devices, surgical robots, 3D printing of organs of the body, and billing from electronic health records are some examples. In addition, cyber threats and hackers are a constant threat. They change patient records, harm patients by changing settings in the monitoring devices, and hold hospitals hostages after stealing medical records and patient identities. Risk analysis must include broader wider-reaching topics that can affect patient care in many ways that often includes unknown risks.

Recommendations

Organizations need to avoid costly medical device litigations and having to defend themselves in lawsuits by doing the right things to design safe devices and leverage design processes and products from other industries to improve patient care. They must work to prevent medical device recalls costing millions of dollars and lost company reputations. Here are some recommendations:

- Before working hard, find out what are the right things to do. Working hard won't help if you are working on wrong things.
- Always conduct thorough risk analysis using Preliminary Hazard Analysis (PHA) and FMEA before approving specifications.
- If the entire team is not challenging the specification and the conceptual design, learn to say "no" to the "yes" people.
- Look for independent verification and validation to be the stopgap measures in discovering defects during the design and production processes and ensuring freedom of defects in the medical product designs when delivered to customers.
- The only standard of performance should be zero defects. Learn to do it right the first time. (See Philip Crosby's book *Quality is Free*, cited in suggestions for additional reading at the end of this chapter.)

17.8.3.2 Hospital Safety Concerns Details

Inadequate Measures on Safety of Patients

A government agency, Agency for Healthcare Research and Quality (AHRQ), has created a survey of patient experience. Hospitals are measured on the results of these surveys and are provided incentive payments for survey completions. The problem is the survey measures experience based on how was the food quality, facilities restful sleep, how well the nurses satisfy their physical and emotional needs, and amenities available. There is very little on how safe was the patient and the results of their care, such as receiving right diagnosis, wrong surgery, wrong medications, response on emergency situations, and not responding to alarms [24]. Also the submission of surveys is voluntary. If the survey results are substandard, hospitals don't have to submit. The big problem is that dead patients don't respond [25]. Their families also often don't respond because they are depressed and don't have time for filling out surveys.

System Is Too Complex

Each patient is different. Each doctor may have their own opinion based on the quality of communication with the patient and the accuracy of medical records. Patients often don't tell everything to the nurse who is documenting the patient's information. Each doctor may follow his/her intuition in care. Each process can have too much variation with multiple caregivers. Nurses hear many alarms a day resulting in alarm fatigue, and they do not always respond to the urgent needs of the patients. Each surgeon may have their own method and protocol for surgery. 700,000 patients get infections from the doctors, caregivers, and poorly sanitized beds and furniture. There are often multiple devices talking to each other, and they result in wrong alerts, false alarms, false conclusions, and so on. Hospitals are about 20 years behind on using system safety science compared with the aerospace industry. Hospitals are complex organizations and need to apply system safety practices to improve patient safety.

Examples of Fatal Hospital Mistakes

Patients have died because a nurse was unable to respond because of too many alarms and distractions. Patients have died because of improper settings of patient monitoring systems. Some examples of fatal hospital mistakes include:

- A young mother was in an ICU with her son monitoring his treatment. She could not sleep because too many alarms were sounding. She requested that the nurse reduce the number of alarms. The nurse inadvertently silenced ALL the alarms. The son was dead when the mother woke up.

- A hospital housekeeper unplugged two patients' devices to plug in power to her vacuum cleaner.
- Some hospital housekeepers don't carry the right type of sanitizers for preventing a type of infection called *Clostridium difficile* that may be residing in drapes and on patient's bed rails.
- Many patients die from wrong radiation settings.
- About 10% patients get wrong surgery. About 40 times a week, wrong surgeries are performed.
- Babies often get the wrong type of blood thinner, resulting in deaths.

As mentioned at the beginning, over 400,000 patients die from such mistakes each year. This amount of needless deaths must be addressed and steps taken by the medical industry to reduce this number. The following are examples of what steps may be taken to turn the trend around.

New Technologies Are Not Understood

If the medical device organizations are inefficient at risk analysis, hospitals are even worse. There is very little analysis done on new technologies. They trust suppliers to do the risk analysis. Many do not even question the suppliers. For example, a radiology technician forgot to turn down the radiation level; this author recommended that the hospital call the supplier and ask them to incorporate an automatic reset to safe level after each use. Their response was "Asking a supplier to change the software is not our job." Hospitals should leverage the supply chain management organizations from other industries to take control of their supplier base. Hospitals must hold themselves accountable to know the risks of the products they purchase from their suppliers and to require design changes to improve the designs of these products.

Insufficient Regulatory Requirements for Safety

The Joint Commission requires hospitals to conduct the FMEA only once a year on a critical process. Most hospitals just do that much. Why not conduct the risk analysis on all the processes? Almost all the processes are life threatening. Hospitals should use the standard ISO 14971 for all their medical devices. 100% of this standard can be used by hospitals to manage the risks of their products.

Insufficient Incentives to Improve Safety

According to a *New York Times* article, hospitals that made more mistakes also made more money [26]. This report was based on the research article from *The Journal of the American Medical Association*. The authors were from Boston Consulting Group, Harvard School of Medicine and Public Health, and Texas Health Association (a nonprofit group). The reason for poor performance is because insurers pay hospitals for longer stays and extra care, the article says. It suggests that changing the system to stop rewarding poor care may help; and until this happens, hospitals have little incentive to improve.

Need to Instill a Culture of Continuous Improvement and Continuous Innovation

Health care needs to prevent medical device defects used for patient care quickly and permanently. Minor defects can become life threatening. The medical community will do well if they would leverage the benefits from other industries for performance of process control steps for CI as well as continuous innovation. Johns Hopkins Hospital had a central line catheter infection rate of about 14% for years until they adopted an idea from the airline industry. It requires a small checklist by a coworker (copilot in the airline) who verifies that every step in the checklist has been done correctly. The hospital assigned nurses to verify the checklist, while the surgeons insert the catheter into the patients' chest for feeding tubes. The infection rate dropped down to 0% within three months. The only cost was the nurse observing the surgeon for about 2 minutes. Reference [23] contains this case history and contains nine simple ways to innovate.

Recommendations for the Future

- Hospitals must learn system safety principles from the existing literature such as Ref. [27] or from the military standard—MIL-STD-882 [28].
- To minimize the risks due to complexity, use systems engineering practices recommended by the White House Technical Advisory Council [29].
- Hospitals must learn to do risk analysis on every process even though there can be over 1000 processes.
- An insurance system should be in place to reward only good performers on patient safety.
- Senior management needs to eliminate the medical mistakes as much as possible as their number one priority.

Now we will move on to a series of short lessons learned that illustrate additional problems.

17.8.4 Hoist Systems

NASA conducts many studies, not all necessarily the result of a particular failure. One such study summarizes a number of specific and generic lessons that have been "learned" as a result of the fact-finding activities of the Aerospace Safety Advisory Panel [30]. The Panel was established by congressional statute. The personnel (with exception of staff) are non-NASA. The Panel reports to both the NASA Administrator and the Congress. The Panel members are truly independent, and they have "no axe to grind." One of the items considered in this particular study was hoists and presents a good example of developing lessons learned even when no failure has occurred—reviewing the process and looking for improvements that can be made.

NASA uses cranes and hoists at several locations to move heavy items. The following is a short excerpt from the report of the Aerospace Safety Advisory Panel.

Findings:

Single failure points exist on many of the major hoist systems at both KSC and VLS. These hoists are used to handle and move sophisticated and sometimes potentially dangerous hardware such as the Orbiter, the Solid Rocket Booster segments, and so on. In most cases waivers have been requested and granted, and in fact these have become deviations (not temporary as a waiver but permanent) to the specifications. JSC-07700, Volume X, paragraph 3.5.1.2.1.1 requires all ground support equipment to be designed to fail-safe.

Lessons Learned:

Equipment is designed and manufactured without due regard of the environment in which the hoists will be working nor with the critical type of hardware involved. Much of this cannot be helped because equipment is not manufactured specifically for NASA's use, but must be modified to meet NASA's requirements. Waiving or accepting single failure points requires greater attention from management and the quality assurance/safety personnel.

Recommendations:

With known requirements it should be incumbent upon the KSC and VLS users to eliminate as many single failure points through very early knowledge of requirements and procurement actions. As with any other equipment, FMEA's and resultant critical items should be defined, hazards noted and a final risk assessment made. The risk assessment must include and be dependent upon impacts to the critical hardware that will interface with the hoist systems.

Lesson Learned

Here we can see that single point failures are lurking in the system. Requirements were ignored. Waivers, meant to be temporary, are becoming permanent, and management's attention is lacking. In addition, hazard analysis and risk assessment are not being properly conducted.

 Paradigm 3: Spend Significant Effort on Systems Requirements Analysis

17.8.5 Internet of Things

This lesson learned relates to the value of software safety and cyber security as discussed in Chapter 16. According to the Oxford English Dictionary, the Internet of Things (IoT) is

The interconnection via the Internet of computing devices embedded in everyday objects, enabling them to send and receive data. [31]

Today there are many devices that fall into this category. When people think about the IoT, they think about all the household gadgets that can be interconnected and communicated with by external means. These can include all sorts of networkable smart devices such as lights, thermostats, coffee makers, security systems, front door locks, vacuum cleaners, smoke detectors, refrigerators, baby monitors, coffee pots, and so on, all connected together by the Internet and controlled by your cell phone.

On October 21, 1916, the *New York Times* reported that

Major websites were inaccessible to people across wide swaths of the United States on Friday after a company that manages crucial parts of the internet's infrastructure said it was under attack. Users reported sporadic problems reaching several websites, including Twitter, Netflix, Spotify, Airbnb, Reddit, Etsy, SoundCloud and The New York Times.

The company, Dyn, whose servers monitor and reroute internet traffic, said it began experiencing what security experts called a distributed denial-of-service attack just after 7 a.m. Denial of Service (DoS) is one of many cyberattack measures that face us today. Reports that many sites were inaccessible started on the East Coast, but spread westward in three waves as the day wore on and into the evening. And in a troubling development, the attack appears to have relied on hundreds of thousands of internet-connected devices like cameras, baby monitors and home routers that have been infected—without their owners' knowledge—with software that allows hackers to command them to flood a target with overwhelming traffic. [32]

Later reporting by Reuters indicated, "Up to 10,000 webcams will be recalled in the aftermath of a cyber attack that blocked access last week to some of the world's biggest websites, Chinese manufacturer Hangzhou Xiongmai Technology Co told Reuters on Tuesday" [33].

While we typically think of the IoT as being consumer-based products, there are many industrial and commercial IoT products that are or can be applied for other purposes, such as:

A typical corporate building is made up of systems: security systems, fire/smoke/water-alarm systems, heating/cooling systems, lighting systems. These systems and devices can be made to communicate intelligently, both with each other and with building managers. They can provide a huge boost in convenience, savings, safety, and environmental payoff.

At this very moment, gas companies are installing sensors on remote pumping stations in Alaska, so that their engineers can monitor the machinery's health with an app instead of driving out there for inspection. Tire companies are embedding sensors into their tires, and sharing the collected data to trucking companies to save fuel and money. Sensors in municipal water utilities can predict when machines will fail, so they can be fixed before disaster strikes. [34]

Lesson Learned and Recommendations

Investigations have exposed some of the flaws in the system and steps have been taken by various organizations involved with IoT technology to understand what can go wrong as IoT technology is proliferated. The following are adapted from Ref. [35]:

Many IoT devices require their users to do the hard work of securing them. When security problems are found, vendors release firmware patches that can fix some problems, but do most end users know how to patch firmware on an IoT device? Vendors should provide automatic-updating for security fixes.

Many devices are shipped with the same passwords and many are commonly used, like "admin" or "password". Manufacturers should ship every product with a unique password.

Consumers should have a simple way of identifying secure/safe IoT devices when shopping. The example of Underwriters Laboratories' (UL) safety labels often comes up, followed by a request that UL do just that for cybersecurity. UL and other organizations are investigating this possibility, but it will take sometime in the next few years before this happens.

There are no product-safety regulations for IoT devices like there are for food, air travel, etc. The government will have to step in and fix the problem. The Federal Trade Commission (FTC) has been studying this problem and has reported on it in a 2015 report, Internet of Things: Privacy & Security in a Connected World. In the meantime, the FTC can pursue companies for deceptive practices for misleading customers about their devices' security.

So far, most of these attacks involving the IoT have been security or privacy related, but it is only a matter of time before they become safety related. As we discussed in Chapter 16, there is a fine line between security issues and safety.

Think about hackers controlling the air in truck tires as that semitruck passes you at 70 mph as it rolls down the highway. Or, what could they do with the gas sensors monitoring natural gas pumping stations? Or, what could they do to our utility services, such as our power grid or water supply?

17.8.6 Explosion in Florida

The following accident summary and analysis is based on information in Refs. [36] and [37].

On December 19, 2007, a powerful explosion and subsequent chemical fire killed four employees and destroyed T2 Laboratories, Inc. (T2), a chemical manufacturer in Jacksonville, Florida. The explosion also injured 32, including four employees and 28 members of the nearby public. Debris from the exploded reactor was found a mile away, and the explosion damaged buildings within one quarter mile of the facility. The US Chemical Safety and Hazard Investigation Board (CSB) investigated the accident and found that the explosion was due to a runaway exothermic reaction that occurred in a chemical batch reactor during the production of methylcyclopentadienyl manganese tricarbonyl (MCMT), a compound used as an octane-increasing gasoline additive.

The process required cooling, and the cooling system intermittently injected water into a cooling jacket surrounding the reactor based on the rate of reaction temperature increase. The CSB determined that the runaway reaction was the result of a loss of cooling that led to uncontrolled temperature and pressure increases that ultimately resulted in the explosion. Although the exact cause of the cooling loss could not be determined, the CSB noted that the cooling water system lacked redundancy, making it susceptible to single point failures. The CSB also stated that no emergency source of cooling existed. In addition, no precautionary procedures existed for loss of cooling as a system failure mode or hazard, so when the operators noticed the increasing reactor temperature, they did not know what to do. The CSB stated that no hazard analysis had been conducted on this process. Because they did not perform the analysis, T2 did not recognize the runaway reaction hazard associated with the MCMT. Such an analysis would likely have determined the need for additional hazard controls to prevent a runaway reaction. A pressure relief system existed in the system, but that relief system was sized for normal operations, and not for the worst-case stress conditions seen in this accident.

Near Misses

As an interesting aside, it should be noted that the CSB investigation revealed earlier problems with the T2 process…near misses?

The runaway reaction on December 19, 2007, was not the first unexpected exothermic reaction that T2 experienced. Three of the first 10 MCMT batches resulted in unexpected exothermic reactions. In each instance the batch recipe was slightly different, and T2 did not repeat the recipes in an effort to isolate the problem, instead changing recipes in each of the first 10 batches. T2 announced successful commercial operation to its stakeholders after Batch 11.

In 2005, at Batch 42, T2 increased the batch size by one third. There are no records of additional chemical or process analysis conducted as part of this recipe change, which may have introduced significant new risks. A greater volume of reactants increased the energy that the reaction could produce and likely altered cooling and pressure relief requirements.

When the MCMT process yielded unexpected results in early batches, T2 did not halt production, investigate causes, and redesign the process. Instead, T2 attempted to control unexpected reaction results online through operator controls or minor alterations to continue running the process as it was already constructed. As demand grew, T2 increased batch size and frequency with no additional documented hazard analysis.

Lessons Learned

Bad design
- The cooling system employed by T2 was susceptible to single point failures due to a lack of design redundancy.
- T2 sized the reactor relief devices based on anticipated normal operations, without considering potential emergency conditions. The MCMT reactor relief system was incapable of relieving the pressure from a runaway reaction.
- T2 changed batch recipes and failed to investigate deviations thus ignoring near-miss opportunities.
- T2 increased the batch sizes and failed to perform any hazard analysis on the changes.

No process hazard analysis
- Companies developing chemical processes for production, like T2, must fully research the hazards involved. Hazards should be identified in each phase of commercial development and actions must be taken to assess risk and mitigate or eliminate potential consequences. Process Hazard Analysis (PrHA) must be conducted during the development phase. The analysis identifies hazards and helps establish operating limits and identify operating strategies to prevent runaway reactions. One of T2's design consultants identified the need to perform a Hazard and Operability Study (HAZOP) during scale-up. A comprehensive HAZOP likely would have identified the need for testing to determine the thermodynamic and kinetic nature of the reaction, as well as the limitations of the cooling and pressure relief systems. The CSB found no evidence that T2 ever performed the HAZOP.

17.8.7 ARCO Channelview Explosion

The following accident summary and analysis is based on information in Ref. [38].

A wastewater tank at the ARCO chemical plant in Channelview, TS, exploded during the restart of a compressor on July 5, 1990. The explosion killed 17 people, and damages were estimated to be $100 million. During normal operations a 900,000-gallon wastewater tank contained process wastewater from propylene oxide and styrene processes. Upstream of the tank, peroxides and caustic could mix in the piping. A nitrogen purge kept the vapor space in the tank inert, and a compressor drew the hydrocarbon vapors off before the waste was disposed of in a deep well.

On the day of the accident, the nitrogen purge had been significantly reduced due to maintenance, and a temporary oxygen analyzer failed to detect the buildup of a flammable atmosphere in the tank. When the compressor was restarted, flammable vapors were sucked into the compressor and ignited. The flashback of the flame into the headspace of the tank caused the explosion.

Causes. The wastewater tank was not considered part of the operating plant. Hence, the management and workers did not understand that a chemical reaction was taking place in the tank, generating oxygen. The lack of understanding enabled a series of poor decisions, such as discontinuing the nitrogen purge, poor design and location of the temporary oxygen probe, no management of change review of these decisions, and no pre-start-up safety review.

Lessons Learned

- The systems approach is necessary to ensure that ALL components of the system (including the wastewater tank in this case) are considered in the design and development of safe procedures.
- Chemicals that enter any wastewater tank are still prone to reaction.
- Ensure that proper management of change procedures are followed before any maintenance work is performed. In this incident, the workers did not know that a chemical reaction that could produce an oxygen buildup was taking place in the tank. Therefore, they did not comprehend the importance of continuing an effective nitrogen purge.
- Conduct pre-start-up safety reviews to identify hazards.
- Ensure workers know and understand the operation and the procedures necessary to keep it safe.

Paradigm 8: Develop a Comprehensive Safety Training Program to Include Handling of Systems by Operators and Maintainers

17.8.8 Terra Industries Ammonium Nitrate Explosion

The following accident summary and analysis is based on information in Ref. [38].

On December 13, 1994, a massive explosion occurred in the Ammonium Nitrate (AN) portion of Terra Industries' fertilizer plant in Port Neal, IA, killing four people and injuring 18. Serious damage to other parts of the plant caused the release of nitric acid into the ground and anhydrous ammonia into the air. The plant produced nitric acid, ammonia, ammonium nitrate, urea, and urea-ammonium nitrate.

The explosion occurred after the process had been shutdown and the ammonium nitrate solution was left in several vessels. Multiple factors contributed to the explosion, including strongly acidic conditions in the neutralizer, application of 200-psig steam to the neutralizer vessel, and lack of monitoring of the plant when the process was shut down with materials in the process vessels.

The Environmental Protection Agency (EPA) investigation concluded that the conditions that led to the explosion occurred due to the lack of safe operating procedures. There were no procedures for putting the vessels into a safe state at shutdown or for monitoring the process vessels during shutdown. The EPA also noted that no hazard analysis had been done on the AN plant and that personnel interviewed "indicated they were not aware of many of the hazards of ammonium nitrate."

Lessons Learned

- Operating procedures need to cover all phases of operation. In this case, the lack of procedures for shutdown and for monitoring the equipment during shutdown led operators to perform actions that sensitized the ammonium nitrate solution and provided energy to initiate the decomposition reaction.
- Because there had been no hazard identification study, personnel did not know about the conditions that sensitize ammonium nitrate to decomposition.

17.9 Summary

This chapter has emphasized the critical need to learn from our mistakes so that they are not repeated. Every organization should develop and utilize a LLP, should collect and pay attention to near misses, and should always strive to CI.

Trevor Kletz, the world-renowned expert in process safety, is often quoted as saying "Organizations don't have memory—only people do." This chapter has provided examples of lessons learned and some philosophy of concerning them in order to underscore the importance of using lessons learned to avoid future failures.

This common worry should be replaced by a new paradigm—one that recognizes the inevitability of failure in today's complex work organizations. Those that catch, correct, and learn from failure before others do will succeed. Those that wallow in the blame game will not. [3]

References

[1] Secchi, P. (Ed.) (1999) *Proceedings of Alerts and Lessons Learned: An Effective Way to Prevent Failures and Problems* (Technical Report WPP-167). European Space Agency, Noordwijk, the Netherlands.

[2] Kletz, T. (1999) The Origins and History of Loss Prevention, *Process Safety and Environmental Protection (Transactions of Institution of Chemical Engineers)*, Vol. 77, No. Part B, 109–116.

[3] Edmondson, A. C. (2011) Strategies for Learning from Failure, *Harvard Business Review*, Vol. 89, No. 4, 48–55.

[4] Center for Army Lessons Learned (2011) *Establishing a Lessons Learned Program*, Handbook 11-33, U.S. Army Combined Arms Center, Center for Army Lessons Learned, Ft. Leavenworth, KS.

[5] Madden J. and Stewart, R. (2014) *One Hundred Rules for NASA Project Managers*, NASA Goddard Space Flight Center, Greenbelt, MD.

[6] Gino, F. and Pisano, G. P., Why Leaders Don't Learn from Success, Harvard Business Review, April 2011.

[7] National Safety Council, Near Miss Reporting Systems, National Safety Council, Itasca, IL, 2003.

[8] Tinsley, C. H., Dillon, R. L., and Madsen, P. M. (2011) How to Avoid Catastrophe, *Harvard Business Review*, Vol. 89, No. 4, 90–99.

[9] Bridges, W. G. (2012) Gains from Getting Near Misses Reported, Presentation at the 8th Global Congress on Process Safety, Houston, TX, April 1–4, 2012.

[10] Buckingham, M. and Coffman, C. (1999) *First, Break All the Rules*, Simon and Schuster, New York, NY.

[11] Fryer, K. J., Antony, J., and Douglas, A. (2007) Critical Success Factors of Continuous Improvement in the Public Sector, *The TQM Magazine*, Vol. 19, No. 5, 497–517.

[12] W. Edwards Deming, https://en.wikipedia.org/wiki/W._Edwards_Deming (Accessed on August 10, 2017).

[13] Continual Improvement Process, https://en.wikipedia.org/wiki/Continual_improvement_process (Accessed on August 10, 2017).

[14] Deming, W. E. (1982) *Out of the Crisis*, Massachusetts Inst Technology, Center for Advanced Engineering Study, Cambridge, MA.

[15] The W. Edwards Deming Institute, Seven Deadly Diseases of Management, https://deming.org/management-system/deadlydiseases (Accessed on August 10, 2017).

[16] Midha, A. (2005) How to Incorporate "Lessons Learned" for Sustained Process Improvements, Presentation at the NDIA CMMI Technology Conference, Denver, CO, November 13–17, 2005.

[17] National Highway Traffic Safety Administration (NHTSA) (May 2016) *Expert Report of Harold R. Blomquist*, EA15-001, Air Bag Inflator Rupture, Washington, DC.

[18] GM Ignition switch recall issue press release, http://www.slideshare.net/sagar1122/gm-ignition-switch-recall-issue-press-release (Accessed on August 10, 2017).

[19] GM Recall Website, http://www.gmignitionupdate.com/product/public/us/en/GMIgnitionUpdate/index.html (Accessed on August 10, 2017).

[20] U.S. Department of Justice, U.S. Attorney of the Southern District of New York Announces Criminal Charges against General Motors and Deferred Prosecution Agreement with $900 Million Forfeiture, September 17, 2015 https://www.justice.gov/opa/pr/us-attorney-southern-district-new-york-announces-criminal-charges-against-general-motors-and (Accessed on August 10, 2017).

[21] National Public Radio (NPR), Timeline: A History Of GM's Ignition Switch Defect, March 31, 2014 (Sources: *General Motors, National Highway Traffic Safety Administration, House Energy and Commerce Committee*, The New York Times, Automotive News, *Bloomberg, NPR research*) http://www.npr.org/2014/03/31/297158876/timeline-a-history-of-gms-ignition-switch-defect (Accessed on August 10, 2017).

[22] Daniels, E., 440,000 Deaths Annually from Preventable Hospital Mistakes, The National Trial Lawyers web site, January 21, 2015, http://www.thenationaltriallawyers.org/2015/01/hospital-deaths/ (Accessed on August 10, 2017).

[23] Raheja, D. (2011) *Safer Hospital Care*, CRC Press/Taylor & Francis, New York, NY.

[24] Robbins, A., The Problems with Satisfied Patients, The Atlantic.com web site, April 17, 2015 http://www.theatlantic.com/health/archive/2015/04/the-problem-with-satisfied-patients/390684/ (Accessed on August 10, 2017).

[25] Firger, J., 12 Million Americans Misdiagnosed Each Year, CBS News, April 17, 2014 http://www.cbsnews.com/news/12-million-americans-misdiagnosed-each-year-study-says/ (Accessed on August 10, 2017).

[26] Grady, D., Hospitals Profit from Surgical Errors, New York Times, April 6, 2013, http://www.nytimes.com/2013/04/17/health/hospitals-profit-from-surgical-errors-study-finds.html (Accessed on August 10, 2017).

[27] Raheja, D. and Allocco, M. (2006) *Assurance Technologies Principles and Practices*, Wiley-Interscience, Hoboken, NJ.

[28] MIL-STD-882E, System Safety, 2012.

[29] Cassel, C., Penhoet, E., and Savitz, M., New PCAST Report Says "Systems Engineering" Can Improve Health Care, White House Home Blog, May 29, 2014, https://www.whitehouse.gov/blog/2014/05/29/new-pcast-report-says-systems-engineering-can-improve-health-care (Accessed on August 10, 2017).

[30] Roth, G. L. (1986) *Lessons Learned: An Experience Data Base for Space Design, Test and Flight Operations*, National Aeronautics and Space Admin (NASA), Washington, DC.

[31] Internet of Things, https://en.oxforddictionaries.com/definition/internet_of_things (Accessed on August 10, 2017).

[32] Perlroth, N., Hackers Used New Weapons to Disrupt Major Websites Across U.S., New York Times, October 21, 2016.

[33] Jiang, S. and Finkle, J., China's Xiongmai to Recall up to 10,000 Webcams After Hack, Reuters, October 26, 2016.

[34] Pogue, D., Here's the Real Money-Maker for the Internet of Things, http://finance. yahoo.com/news/david-pogue-on-the-internet-of-things-120124623.html (Accessed on August 10, 2017).

[35] Pegoraro, R., Hackers Are Taking over Your Smart Devices, Here's How We Can Stop Them, http://finance.yahoo.com/news/hackers-are-taking-over-your-smart-devices-heres-how-we-can-stop-them-193329567.html;_ylt=AwrBT4FonmJYNREA.4pXNyoA;_ylu=X3oDMT EyZjM2YmNiBGNvbG8DYmYxBHBvcwMxBHZ0aWQDQjI5MTNfMQRzZWMDc3I-(Accessed on August 10, 2017).

[36] Hardy, T., Resilience: Case Studies and Lessons Learned, GCA Paper No. 2014-001, January 6, 2014.

[37] U.S. Chemical Safety and Hazard Investigation Board, Investigation Report, T2 Laboratories, Inc. Runaway Reaction, Report No. 2008-3-I-FL, September 2009.

[38] Ness, A. (2015) *Lessons Learned from Recent Process Safety Incidents*, Center for Chemical Process Safety, American Institute of Chemical Engineers (AIChE), New York, NY.

Suggestions for Additional Reading

Crosby, P. (1979) *Quality is Free*, McGraw-Hill, New York, NY.

Deming, W. E. (1982) *Quality Productivity and Competitive Position*, Massachusetts Inst Technology, Center for Advanced Engineering, Cambridge, MA.

Imai, M. (1986) *Kaizen: The Key to Japan's Competitive Success*, Random House, New York, NY.

Lawson, D. (2005) *Engineering Disasters: Lessons to be Learned*, ASME Press, New York, NY.

Raheja, D. (2011) *Safer Hospital Care: Strategies for Continuous Innovation*, Productivity Press, New York, NY.

Walton, M. (1986) *The Deming Management Method*, Dodd, Mead & Company, New York, NY.

18

Special Topics on System Safety

Louis J. Gullo and Jack Dixon

18.1 Introduction

Safety is the number 1 priority for many industries and marketplaces, such as automotive, commercial aviation, aerospace, and medical systems, to name a few. Safety is a subject that gets the attention of customers and consumers alike across a wide swath of our culture when a popular and expensive product is prone to safety issues.

This final chapter delves into several special topics and applications to consider the future of system safety. There is no better industry marketplace to address the future of system safety features than the commercial aviation and automobile industries. We examine the historical and current safety data of both industries to see what it tells us about the historical trends and the future probabilities of fatal accidents. The data from commercial air travel supports the need to change motor vehicle design direction in favor of implementing new design for safety enhancements. We see how commercial air travel data shows the trends improving for safer flights in the future, while motor vehicle statistics demonstrate fatality trends getting worse. We explore the safety design benefits from commercial air travel that could be leveraged by automobile manufacturers and developers of new ground transportation systems. It is safe to predict that automobile systems

Design for Safety, First Edition. Edited by Louis J. Gullo and Jack Dixon.
© 2018 John Wiley & Sons Ltd. Published 2018 by John Wiley & Sons Ltd.

will become safer due to adoption of design features found in commercial air travel. These commercial air travel features introduce autonomous system behaviors that prevent in-flight catastrophes by avoiding head-on collisions or near misses. This chapter will also consider future improvements in commercial aerospace travel.

18.1.1 Why Are Many Commercial Air Transport Systems Safe?

Today's commercial air transport systems are safe because of robust autopilot designs, Flight Control Computers (FCCs), Flight Management Systems (FMSs), rapid telecommunications and telemetry, advanced guidance and navigation systems including Inertial Reference Units (IRUs) and Global Positioning Satellite (GPS) systems, fault-tolerant architecture, and multiple levels of system redundancies for critical system functions. Much of these designs were created specifically for flight safety, which are integrated with other systems, such as Prognostics and Health Management (PHM) capabilities, Traffic Collision Avoidance System (TCAS), Wind Shear Computer (WSC), and assorted other autonomous system sensors and functions that make the vehicles not only safer but more reliable and efficient. These flight systems and other similar types of systems are the reasons why today's commercial aircraft systems are so safe.

Every design decision in commercial air transportation was made after careful consideration of its impact on safety. With nearly one billion flight hours accumulated since commercial air travel began, a steady stream of information into big data clouds and data lakes (e.g., Hadoops) provides the commercial air industry with invaluable current knowledge. This knowledge is used to understand and react to the stresses and normal in-service use conditions for each aircraft that is used for airline business revenue generation to constantly improve the design of airplanes.

The traditional mechanical controls for many contemporary jetliners have been replaced by electronic controls. These electronic controls for guiding airplanes are called Fly-by-Wire (FBW) control systems. FBW systems are standard components of large aircraft including the latest versions of the Boeing 777, the Boeing 787, the Airbus A330, the Airbus A340, and the Airbus A380. "As planes transition from machine to computer, the day of 'the brawny guy' pulling on the yoke is over, says Missy Cummings, an associate professor of aeronautics and astronautics at the Massachusetts Institute of Technology and a former US Navy fighter pilot. 'We don't need Chuck Yeager anymore.' The modern pilot is a manager of information, and technology plays the muscular role on the flight deck" [1].

"Many planes today can operate in a geographic window so exact that their horizontal position remains within 'a wingspan, with vertical deviation less than the height of the tail,' says Ken Shapero, director of marketing for GE Aviation.

The linking of onboard and on-the-ground systems creates highways in the sky where nobody veers out of their lanes. 'Automation determines the trajectory of the airplanes, and for the most part, air traffic controllers let the airplanes fly,' notes Steve Fulton, a former airline pilot who founded the navigation company Naverus, acquired by GE Aviation in 2009. Challenging terrain, low visibility, bad weather—the kinds of hazards that can close airports and divert airplanes—will no longer cause chaos. 'It's a whole different world,' Fulton says" [1].

18.1.2 How Many Aircraft In-Flight Accidents and Fatalities Occur in Recent Times and over History?

"On Thursday, May 19, EgyptAir flight MS804, traveling from Paris to Cairo, crashed into the Mediterranean Sea. All 66 passengers and crew members aboard were killed. Terrorism is suspected" [2].

The EgyptAir A320-200 aircraft, Flight MS804, was on a scheduled international flight from Paris, France, to Cairo, Egypt, when air traffic control in the area noticed radar contact with MS804 was lost. The aircraft was cruising at 37,000 feet and over the eastern Mediterranean Sea north of the Egyptian coast and crashed into the sea. There were 56 passengers and 10 crew members on board. No one survived. The EgyptAir crash was the fifth major airline crash worldwide since the beginning of 2016 [2].

According to the data collected by aviation-safety.net, in the year 2016, there were 18 accidents worldwide, with 320 fatalities. Considering the same worldwide data using averages over a 10 year period, there were 28 accidents per year and 662 fatalities per year. This summary data illustrates an approximate 50% decline in accidents per year and fatalities per year. The data shows fatality rates trending down over the last 20 years. A detailed explanation of the raw data is explained on the aviation-safety.net website [3].

The most recent US aircraft accident occurred on October 28, 2016, when FedEx Express DC-10 aircraft, tail number N370FE, Flight FX910, was on a US domestic flight from Memphis, TN, to Fort Lauderdale, FL. At some point during the landing, the left main landing gear collapsed, and the aircraft came to rest near the landing runway. A fire broke out, seriously damaging the left side of the aircraft. The two crew members were not injured. The aircraft was not carrying any passengers [4].

Also on October 28, 2016, American Airlines 767-300 aircraft, tail number N345AN, flight number AAL383, was on a scheduled US domestic flight from Chicago, IL, to Miami, FL. During the takeoff, the right engine experienced an uncontained failure, and the flight crew aborted the takeoff. The aircraft caught fire in the area of the right engine, and the aircraft occupants evacuated the aircraft. There were eight minor injuries among the 161 passengers and nine crew members [4].

The risks of airline travel accidents around the world are very small, but the risks are even smaller and look really good if you board a scheduled passenger airliner in the United States. Since September 11, 2001, there have been four commercial aircraft accidents in the United States that have resulted in fatalities. That means that in nine of the past 13 years, the odds of being involved in an accident that included a fatality have been zero [5].

The United States hasn't seen a crash with major loss of life since 2001, when a regional plane, American Airlines Flight 587, slammed into the Queens neighborhood of Belle Harbor, New York, killing 265. More recently, 49 died in 2006 in the takeoff crash of a regional jet, Comair Flight 5191, in Lexington, Kentucky. The last fatal airline accident in the United States was a crash landing of Asiana Airlines Flight 214. Asiana Airlines Flight 214, a Boeing 777, crashed in San Francisco on July 6, 2013. It resulted in three deaths out of 307 passengers and crew on board. Prior to that, the last accident was the February 12, 2009, crash of a regional aircraft, Colgan Air Flight 3407, near Buffalo, New York, which resulted in the deaths of all 50 people on board [5, 6].

Given the data trends, these facts may still cause fear or concern for some people to think that flying is not safe. This cause for concern is based on the accidents that occur without examining all the data. Some of this data that should be understood in order to alleviate any fears or concerns include the data trends (as you have seen earlier), the accident comparisons to other forms of travel, and the amount of aircraft and aircraft industry design for safety improvements that have been made over the history of commercial air travel. Besides the technological improvements over the years, the training of the pilots and flight crews has improved immensely. One case to point out is the recent motion picture titled "Sully," which is a real-life story about a pilot whose skilled flying and quick thinking saved the lives of his passengers after his aircraft collided with a flock of birds on takeoff.

'Technology is no substitute for experience, skill and judgment,' explains Chesley 'Sully' Sullenberger, who sat at the controls of a highly automated Airbus A320 on the day that he and first officer Jeff Skiles put US Airways Flight 1549 down in New York's Hudson River. One hundred and fifty-five people survived the flight, known as the Miracle on the Hudson—a feat that Sullenberger attributes to a lifetime of flying, as well as preparation, anticipation and focus. [1]

The level of security and safety in the commercial airline industry is mainly judged by examining specific types of fatal incidents and compliance with existing regulations. A recent report published by the airline safety and product rating review website Airline Ratings identifies the top 20 safest commercial airliners using criteria such as safety and security certifications, being blacklisted by the Federal Aviation Administration (FAA) or other foreign transportation agencies and the number (or absence) of fatal accidents in the past 10 years. It's important to note, however, that according to the International Air Transport Association, only six percent of airline accidents in 2015 included fatalities. This fact seriously skews the measurement of risks. Risk measurement should also, in my view, take into account close calls and incidents in which passengers are hurt, even if they aren't killed. [2]

"During the 1950s and 1960s, fatal accidents occurred about once every 200,000 flights," says Julie O'Donnell, a spokeswoman for Boeing. "Today, the worldwide safety record is more than 10 times better, with fatal accidents occurring less than once in every two million flights" [1].

Unless you've avoided television and the Internet entirely over the past few years, it has been almost impossible not to be aware of the recent dramatic and tragic airplane crashes. From the still-mysterious disappearance of Malaysia Airlines Flight 17 to the deliberate crash of Germanwings Flight 9525 by a pilot, the details of each crash are unsettling and full of sorrow and heartbreak. It's enough to make even seasoned travelers wonder if flying is getting steadily less safe. Although the past few years have featured a few high-profile crashes, if you take the long view it becomes clear that the airline industry actually has a very good safety record—and it's getting better, not worse. [7]

Driving in an automobile on the ground is the obvious first comparison to airline travel. The National Safety Council notes that the odds of losing a life in a motor vehicle crash are 1 in 112, 0.9%, while the odds of losing a life in a plane crash are much lower, at 1 in 96,566, 0.001% [7]. This means there is almost three orders of magnitude improvement in survival odds by flying in a commercial aircraft over a certain distance compared with driving the same distance in a four-wheeled automobile. The odds are even better in favor of commercial air flight when compared with ground transportation in a two- or three-wheeled vehicle, such as a motorcycle.

The National Safety Council offers these odds on other methods of dying, all of which are significantly more likely to happen than being killed in a plane crash [7]:

- Being assaulted by a firearm: 1 in 358
- Being electrocuted: 1 in 12,200
- Walking down the street: 1 in 704
- Falling: 1 in 144
- Overdosing on a prescription painkiller: 1 in 234

Airline safety experts believe that the commercial air travel industry is likely to maintain and even improve these statistics as technology improves, considering replacement of older aircraft with newer models, and advanced air travel processes developing in regions such as parts of Asia strive to match the air safety record of Europe and the United States [7].

'Commercial airliners crashed at a rate of one for every 3.1 million flights worldwide in 2015—making last year one of the safest on record', industry officials said Monday. The four fatal crashes counted in the statistics all involved turboprop aircraft and taken together killed 136 people, according to the report by the International Air Transport Association. None of the crashes occurred in the U.S., where the last fatal crash of a U.S. passenger airliner was in February 2009. Any commercial flight is relatively safe, according to Arnold Barnett, a professor of statistics at Massachusetts Institute of Technology, who crunched numbers from 2008 through mid-2014 to calculate the risks of flying. The risk is higher on flights that start from airports in developing countries—one death in a million flights—versus flights from industrialized nations—at one death in 23.9 million flights, he said. Still, he said, a kid in an airport is more likely to become president or win an Olympic gold medal than to perish on the next flight. Last year's rate of losing 0.32 planes for every million flights to major accidents was higher than the 0.27 rate in 2014. But the fatality rate improved on previous years, when airlines averaged 17.6 fatal accidents per year and 504 deaths for the period of 2010 through 2014, according to IATA. [7]

"According to the International Air Transport Association (IATA), in 2015 there was one commercial jet accident per 4.5 million flights. This was in line with 2014, when the number was one accident per 4.4 million flights, and better than 2013 (one accident per 2.4 million flights). In 2015, there were just four fatal accidents with a total of 136 fatalities. (Note that the Germanwings crash and a suspected terrorist attack on Metrojet Flight 9268 are not included in these stats because they were judged not to be accidents; these add an additional 374 deaths.) Compare that to the period from 2010 through 2014, which had an average of 17.6 fatal accidents and 504 fatalities per year. In 2015 the 510 total fatalities were out of more than 3.5 billion journeys. For perspective, ABCNews.com reports that on average you would need to fly every day for 55,000 years in order to be involved in a fatal crash" [7].

An article in the Wall Street Journal lays out the trend toward tremendously improved safety very clearly (a paywall applies). In fact, older travelers may be taken aback a bit by the risk they survived in the past; one of the most powerful stats noted in the WSJ article states that if we had the same accident rate today as in 1973, there would be a fatal crash every other day. [7]

Some other critical stats to help you get over your fear of being in a plane crash [7]:

- The crash of Germanwings Flight 9525 was the first major crash in Western Europe since 2008.
- The Germanwings aircraft was an Airbus 320; with 3,673 of these aircraft in operation worldwide, assuming each plane flies once daily (many do more than one flight), there should be a 320 taking off or landing every 11–12 seconds. At that kind of frequency, one crash is a very low number.
- Know that almost 96 percent of passengers involved in plane crashes make it out alive, according to the ABC News story linked above, and some fatalities could be prevented if passengers knew what to do. [7]

According to the Aviation Safety Network, there were 990 fatalities from 21 accidents in 2014. 2013 produced 265 fatalities in 29 accidents, and 2012 had 475 fatalities worldwide from 23 accidents. Even though 2014 was a bad year in terms of total fatalities, the last several years have produced the fewest fatal accidents in modern history. Then when you consider that 537 of these fatalities were from the two Malaysia tragedies, one of which was an act of war and the other may or may not turn out to meet the definition of an accident, the number of fatalities caused by unsafe air travel gets even lower compared with other periods of history [5].

The point is that (last year aside) the number of fatalities attributed to air travel has been declining rapidly over the past 30 years. Considering that the number of flights and passenger miles has increased nearly as quickly, the actual *rate* of fatalities continues to fall dramatically, by any measure. To simplify the data, there are now approximately three times as many passenger miles flown as there were 30 years ago, but about one half the number of fatalities, on average. [5]

Out of the 30 million commercial flights in 2014, there were 21 fatal accidents, which means that you had a 0.000007 chance of being onboard any one of those flights, or roughly 1 in 1.43 million. If you are a very frequent flier and you boarded 100 flights last year, your chances were about one in 14,300 that one of those flights would experience a fatality, although not necessarily that you would be one of those fatalities. For a more typical traveler who takes 10 flights a year, the risk declines to about one in 142,000. [5]

The world has seen the safest overall period in aviation history, according to the aviation safety industry. As shown by the above statistics, the numbers of yearly aviation deaths and major plane crashes worldwide and the United States and

Canada have been dropping for decades. "Last year, 265 people were killed in flight incidents—the safest year in aviation since 1945. This year, the worldwide number of aviation deaths has more than doubled, but it's still relatively low. There have been 761 deaths in 12 commercial aviation accidents in 2014, according to the Aviation Safety Network, one of several organizations that tracks these statistics. Its data—spanning 1946 to the present—include hijackings, sabotage and shoot-downs. With the exception of the 9/11 attacks, it's hard to know whether the loss of three airliners in seven days is unprecedented, said Rudy Quevedo, global program director of Washington's Flight Safety Foundation. Crunching those numbers would 'take us some doing and would be very labor-intensive,' he said. 'It is a rare event.' " [6]

Clearly, the bigger the airplane, the more lives can be lost. In 1972, the worst year on record, there were 55 crashes. The Russian crash of Aeroflot Flight 217 killed 174 people, and 155 were killed in the Spanish crash of a Convair 990 Coronado. In recent years, the data show that we're having fewer fatal accidents overall, said Quevedo. "It's a perfectly safe system. So while it's an unfortunate tragedy to have these recent three crashes together, it doesn't shock me, because you could go two or three years in a row without having one. It all equals out in the end" [6].

Measuring the number of crashes or deaths alone doesn't offer an accurate safety snapshot, Quevedo said. You also have to factor in the overall amount of aviation traffic [6]. The aviation industry monitors the total worldwide airport departures by all commercial aircraft. This data can be used to normalize the risks of accidents by factoring the accident rates based on usage of the systems. To accomplish this normalization, the aviation industry collects the quantity of accidents and quantity of departures over a specified period of time for each air transport carrier. From this data, the industry calculates the aviation accident rate, which results from the number of accidents divided by the number of departures over a specified period of time.

Last year, the rate was 0.24 out of one million departures. That means less than one accident for every one million flights. "That number proves that the chances of being in a fatal aircraft accident are extremely rare," Quevedo said. Andrew Charlton, managing director of Aviation Advocacy, a Swiss strategic consulting firm, put it differently: "The single most dangerous part about flying is driving to the airport" [6].

Using the departure data, Boeing was able to demonstrate trends decreasing when measuring the annual fatal accident rate per million departures per year. In 1959 the US and Canadian aircraft accident rate was approximately 40 per million departures. Over the last 20 years, that same accident rate has substantially decreased, hovering in the range of 0.0–0.5 per million departures [8].

18.2 Airworthiness and Flight Safety

Airworthiness is the measure of an aircraft's suitability for safe flight. Certification of airworthiness is initially conferred by a certificate of airworthiness from a national aviation authority and is maintained by performing the required maintenance actions [9].

The application of airworthiness defines the condition of an aircraft and supplies the basis for judgment of the suitability for flight of that aircraft, in that it has been designed with engineering rigor, constructed, maintained, and is expected to be operated to approved standards and limitations, by competent and approved individuals, who are acting as members of an approved organization and whose work is both certified as correct and accepted on behalf of the State [9].

An example of an aircraft that was not legally airworthy is Larry Walters' "Lawn chair flight" in 1982. In the United States, Title 14, Code of Federal Regulations, Subchapter F, Part 91.7 states: "a) No person may operate an aircraft unless it is in an airworthy condition. b) The pilot in command of a civil aircraft is responsible for determining whether that aircraft is in condition for safe flight. The pilot in command shall discontinue the flight when unairworthy mechanical, electrical, or structural conditions occur" [10].

The true definition of the word "airworthy" was never included in the Code of Federal Regulations until the 14 CFR Part 3, General Requirements, was established. The definition was included in the guidance, such as Advisory Circulars and Orders, but never in the Rule. Part 3 defines the definition of airworthy as the aircraft conforms to its type design and is in a condition for safe flight [10].

A more generic and non-process-oriented definition is required. Airworthiness is defined in JSP553 Military Airworthiness Regulations (2006) Edition 1 Change 5 [9] as

The ability of an aircraft or other airborne equipment or system to operate without significant hazard to aircrew, ground crew, passengers (where relevant) or to the general public over which such airborne systems are flown.

An example of a method used to delineate "significant hazard" is a risk reduction technique used by the military and used widely throughout engineering known as As Low As Reasonably Practicable (ALARP). This is defined as

The principle, used in the application of the Health and Safety at Work Act, that safety should be improved beyond the baseline criteria so far as is reasonably practicable. A risk is ALARP when it has been demonstrated that the cost of any further Risk reduction, where cost includes the loss of capability as well as financial or other resource costs, is grossly disproportionate to the benefit obtained from that risk reduction. [10]

18.3 Statistical Data Comparison Between Commercial Air Travel and Motor Vehicle Travel

Now let's consider safety statistics for automotive travel and compare these statistics to commercial aviation statistics. This section and the following sections of this chapter discuss data on automobile ground travel accidents in the United States compared with commercial air travel accidents in the United States, how commercial aviation made those improvements, and how the automobile industry could leverage safe air travel design capabilities that might translate into automobile design for safety improvements.

18.3.1 How Many Motor Vehicle Accidents Occurred Recently and in the Past?

Total US fatal motor vehicle crash totals were 32,166 in 2015, in which 35,092 deaths occurred. This resulted in an average across the United States of 10.9 deaths per 100,000 people. The fatality rate per 100,000 people ranged from 3.4 deaths per 100,000 people in the Washington, DC area to 24.7 deaths per 100,000 people in Wyoming. Some of the states with fatality rates above the average were Alabama (17.5), Arkansas (17.8), Kentucky (17.2), Mississippi (22.6), Montana (21.7), North Dakota (17.3), Oklahoma (16.4), and South Carolina (20.0) [11].

The National Highway Traffic Safety Administration (NHTSA) Office of Vehicle Safety Research's mission is to strategize, plan, and implement research programs to continually further the Agency's goals in reduction of crashes, fatalities, and injuries. Our research is prioritized based on potential for crash/fatality/injury reductions and is aligned with congressional mandates and Department of Transportation (DOT) and NHTSA goals. The National Center for Statistics and Analysis (NCSA), an office of the NHTSA, is responsible for providing a wide range of analytical and statistical support to NHTSA and the highway safety community at large [12]. The following data is provided by the NHTSA on their website.

Automotive fatality statistics for the United States show a declined from 1975 with 44,525 fatalities to 2015 with 35,092 fatalities. The fatality rate also declined over the same time period, from 3.35 to 1.12 fatalities per 100 million Vehicle Miles Traveled (VMT). The 2015 crash statistics for 2015 not only show 35,092 US fatality total but also includes a total of 32,166 fatal crashes, 1,715,000 injury-related crashes, 2,443,000 people injured, and approximately 13 million crashes. The economic cost is estimated at $242 billion [13].

The percent change in fatalities by same quarter from year to year between 2006 and 2015 is shown in Figure 18.1 [13]. From this graph, one may correctly conclude that motor vehicle accidents that result in deaths are on the rise, especially

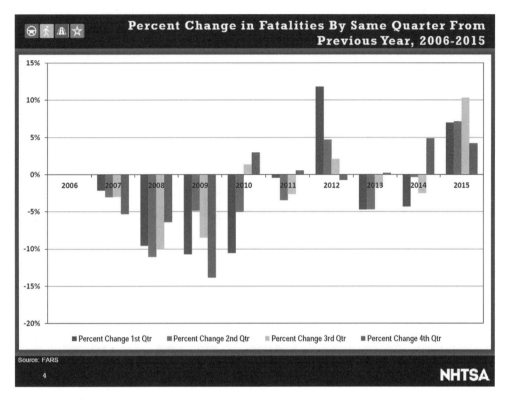

Figure 18.1   Percent change in motor vehicle fatalities from 2006 to 2015

looking at the data over the last four years. The overall number of fatalities increased 7.2% between 2014 and 2015. Figure 18.2 shows a chart of the annual percentage change data in terms of VMT between 1975 and 2015. This chart shows a sudden data spike up in terms of miles traveled in the last 10 years. VMT is an estimator of the number of opportunities for accidents. Predictions for future accidents correlate with the number of VMT. This spike can serve as a prediction estimator for the future probabilities of continuing fatality rate increases over the foreseeable future.

18.3.2 When Do Systems Improve Safety?

Automobile manufacturers usually improve their vehicles' system safety features when a large number of accidents occur that could have been prevented with a system design modification and/or when regulatory authorities intervene. This is proven in the case of seat belts and airbags. The same is true for local government department of transportation improvements to the safety features on their roads that they responsible to maintain. As an example, at residential crossroads, traffic

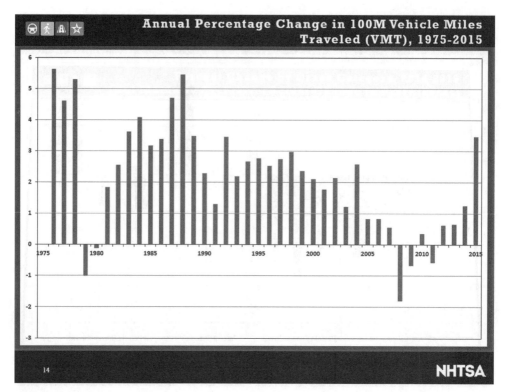

Figure 18.2 Annual percentage change in Vehicle Miles Traveled (VMT) between 1975 and 2015

intersections usually don't start with traffic lights. It is usually following a major traffic accident when people are killed that a traffic signal will be installed.

The safety design benefits from commercial air travel could be leveraged by commercial automobile manufacturers and developers of new ground transportation systems, like smart highways, vehicular infrastructure integration, Integrated Vehicular Health Monitoring (IVHM), and Intelligent Transportation Systems (ITS), to enable safe operation of autonomous cars on all forms of roadways with various degrees of human intervention. With very little hypothetical conjecture, it is safe to say that more and more systems will become safer due to adoption of design features found in commercial air travel. These features introduce autonomous system behaviors that include prevention of in-flight catastrophes by avoiding head-on collisions or near misses using an aircraft avoidance system called TCAS and severe weather avoidance like wind shear using the WSC. The outputs from the TCAS and WCS are part of the suite of sensors used for system-wide autonomous behavior. This autonomous behavior is prevalent in most modern large commercial passenger aircraft designed with specific types of aircraft flight control systems, including FCCs, FBW control systems, and autopilot systems with operating mechanisms to control aircraft direction, speed, altitude, pitch, roll, and yaw during a flight.

Autonomous Vehicle (AV) controls will inherently make automobiles safer. Insurance industries believe this and encourage automobile companies to invest in safer designs through artificial intelligence, sensors, intelligent highways, and smart cars.

18.4 Safer Ground Transportation Through Autonomous Vehicles

AVs will make us safer in the future, without a doubt. While AVs significantly improve transportation safety, they should offer immense social, economic, and environmental benefits. One of the ways AVs will provide socioeconomic and vehicle owner cost benefits is in reduction of the cost of insurance premiums and vehicle collision expenses. These economic cost benefits are possible through the use of embedded health monitoring and early warning sensors and the use of online real-time sensor data processing and algorithms for predictive analytical models to control cars to avoid or prevent automobile accidents.

Referencing the report titled "Safety and Liability of Autonomous Vehicle Technologies," Richard Ni and Jason Leung discuss their research ways to develop safe operations for AVs. Their research explores system safety requirements and performance standards that are needed to ensure safe operations of AVs. Furthermore, they study and make recommendations how a new car manufacturer liability scheme should be created and recommend how individual owner insurance policies and the legal system should be modified in light of the different levels of AV technology [11].

> As driving functions become increasingly automated, not only do technical specifications and safety regulations become increasingly outdated, but there is also a shift in responsibility from the human driver to the vehicle itself. This motivates a new look at both the safety and liability regimes so that maximum benefit may be derived from AVs. [11]

The goal of Richard Ni and Jason Leung as described in their report titled "Safety and Liability of Autonomous Vehicle Technologies" is to provide recommendations for California's Department of Motor Vehicles (DMV) to assist AV integration in the California legal system, which may become the groundwork for safety and liability regulations of AVs at the federal level. If they are successful with new groundbreaking State of California laws and regulations, this should impact many other states and enable widespread changes in the transportation safety posture of the United States as well as many other countries.

As stated by Richard Ni and Jason Leung in their report, "In the safety subsection, our case studies on airbags and autonomous aerial technology provide a framework that we use to develop our recommendations from possible alternative recommendations. In the liability subsection, we distinguish tort liability from manufacturer liability and use a case study on the 2001 Firestone and Ford tire controversy and interviews with representatives of car companies and technology experts to justify our recommendations for liability regulations" [11].

Autonomous Vehicle (AV) technologies have the potential to revolutionize transportation. As these technologies increasingly become competent in taking over driving functions and transition from the testing to commercial phase, safety regulations become increasingly important. Furthermore, the anticipated shift in responsibility for driving from the driver to the vehicle itself motivates an analysis of the liability concerns. The continuum of this technology can be best summarized by the National Highway Traffic and Safety Association (NHTSA) [14] in five levels of automation. We will use this Five Level Hierarchy throughout this paper" [15]:

- **Level 0:** The human driver is in complete control of all functions of the car.
- **Level 1:** One function is automated.
- **Level 2:** More than one function is automated at the same time (e.g., steering and acceleration), but the driver must remain constantly attentive.
- **Level 3:** The driving functions are sufficiently automated that the driver can safely engage in other activities.
- **Level 4:** The car can drive itself without a human driver.

Policymakers have become progressively more aware of the huge economic, social and environmental changes that AVs may present. They face many policy questions, many of which will determine the adoption and impact of AV technology. This paper will focus on suggesting appropriate policy principles to guide policymakers to decide when this technology should be permitted on the roads and an appropriate liability regime. [15]

Safety and liability regulations of level 2 (in which some driving functions are automated), level 3 (in which most driving functions are automated), and level 4 (in which all driving functions are automated), with respect to developing new AV technologies, need to be well understood before prevalent design for safety measures can be incorporated into automobiles on a wide scale.

On March 13th, 2004, the first Defense Advanced Research Projects Agency (DARPA) challenge was held in the Mojave Desert, with a prize of $1 million for any team whose self-driving car could complete the 150 mile course. The first place teamed lasted only 7.32 miles, but this was cause for excitement. In the following year's challenge, five teams successfully completed the course, and in 2007, DARPA organizers revamped the challenge to simulate an urban environment. As of April 2014, Google has boasted that its self-driving cars have completed around 700,000 autonomous miles, including many in urban or suburban environments. Major car companies such as GM, Mercedes Benz, Audi, and many others have followed with their own research on autonomous vehicles, incorporating elements of autonomy such as parking assistance and lane keeping. Other major tech companies are becoming involved as well Tesla, for example, recently announced that their model S sedans are taking the next leap forward in automated technology with a limited autopilot mode and exploration of other autonomous features. Some car companies believe that the technology for full autonomy could exist by 2025. [15]

In 2013, the NHTSA found that 34,080 Americans lost their lives in traffic accidents. They've asserted that the vast majority of accidents occurred as a result of human error, with estimates at around 93%. A plurality (40.1%) of accidents occur as a result of recognition errors, such as inadequate surveillance, and decision errors (37.0%) are not far behind. Level 1 and 2 technologies, such as forward collision warnings and lane departure warnings, can help reduce the accidents, and fully automated vehicles have the potential to do far more to help save tens of thousands of lives annually in the US alone. Other forms of transportation like bicycles, trains and airplanes have significantly lower death rates per mile travelled, but cars are still the most popular method of transportation for most people. Thus, it's vitally important to develop technology to reduce the rate of automobile-related accidents. [15]

Overall, the number of automobile crashes in the United States has been gradually declining, but the astounding number of fatalities and injuries still poses a major public health problem. One can imagine an ideal world where technology has improved to the point where our vehicles are automated and accidents do not occur. Over time, integration of autonomous vehicle technology into our society could save millions of lives and billions of dollars in property damage and public health costs. One research paper estimates that even "at 10 percent market penetration, the technology has the potential to save over 1,000 lives per year and offer tens of billions of dollars in economic gains." Ultimately, lives saved are the biggest potential benefit of greater integration of AVs into our society. [15]

As discussed in Chapter 16 of this book, poor cybersecurity can lead to poor design for safety and the introduction of hazards that may potentially cause accidents. As more automation and connectivity is added to automobiles, hackers may gain access to a car's control system and cause fatal accidents while a driver is driving their car. A good example of this situation is discussed in an article published in *Wired* magazine in 2015 [16]. In this article, two hackers demonstrate how they can wirelessly carjack their friend's vehicle while their friend is test driving on an interstate highway. In the demonstration, the hackers take control of their friend's car and cause the car to malfunction and stop on a busy highway, with no recourse for the driver, other than to call his hacker friends on his phone and beg them to stop the cyber hack demonstration. The two hackers have spooked the automotive industry and plan to inspire legislation from the US lawmakers, Senators Ed Markey and Richard Blumenthal, in their plan to introduce an automotive security bill and gain approval in 2017, or thereabouts. It is hoped that the new legislation will call on the NHTSA and the Federal Trade Commission to set new security standards and create a privacy and security rating system for consumers. "Controlled demonstrations show how frightening it would be to have a hacker take over controls of a car," Markey wrote in a statement to *Wired*. "Drivers shouldn't have to choose between being connected and being protected. We need clear rules of the road that protect cars from hackers and American families from data trackers" [16].

To mitigate the risks of cyberattackers by hackers attempting to gain access to automobile control systems, a software safety/security engineer could conduct an analysis similar to a Fault Tree Analysis (FTA), as discussed in an earlier chapter in this book. This type of FTA analysis is called an attack tree analysis. An attack tree analysis would be created in a similar manner as an FTA using combinational logic gates to arrange functional and timing dependences on an event diagram with cut sets to show resultant effects of an attack vector. This is just one example of an analysis methodology that could be used to prevent motor vehicle accidents initiated by wireless intruders and cyber threats to our safe transportation on busy roadways.

Another issue to explore is the plans for commercial space travel and how the Federal Aviation Administration (FAA) is preparing to take on new regulations for this new responsibility.

18.5 The Future of Commercial Space Travel

As spacecraft become more complex and autonomous, and space travel is handed off to the DOT to facilitate commercial space travel, the need for safety-critical functions and reliable fault protection will become more and more of a high priority. When coupled with the additional requirement of limiting cost, the task of implementing fault protection on spacecraft becomes extremely challenging. The domain knowledge about spacecraft fault protection should be captured and stored in a reusable, component-based spacecraft architecture. The spacecraft-level fault protection strategy can then be created by composing generic component specifications; each with component-level fault protection included. The resulting design can be validated by formal analysis and simulation before any costly implementation begins. As spacecraft technology improves, and the technology is applied in new marketplaces, new methodologies for fault protection data storage, data processing, predictive models, Prognostics and Health Monitoring (PHM), and data telemetry may be added [17].

The United States is developing plans working with the FAA and the DOT for commercial space travel in the near future. The FAA is preparing to take on new regulations for this new responsibility. These plans are associated with a new SPACE Act, which means Spurring Private Aerospace Competitiveness and Entrepreneurship. One of many areas that are significant in the SPACE Act is the new role of the DOT to collect data, conduct analysis, and report to Congress on the status of the commercial space transportation industry.

Referencing Commercial Travel Safety and the National Transportation Safety Board (NTSB), FAA Safety Regulations for Commercial Travel and Space Flight [18], Congress created the new law in 2015 called the SPACE Act, H.R.2262—US Commercial Space Launch Competitiveness Act [19]. On November 25, 2015, this act became Public Law No: 114-90. The act was introduced by Rep. Kevin McCarthy [R-CA-23] on May 12, 2015.

In November of 2015, the US government updated the US commercial space legislation with the passage of the SPACE Act of 2015. This update to the law allows US citizens to engage in the commercial exploration and exploitation of space resources, such as water and minerals. The SPACE Act includes the extension of indemnification of US launch providers for extraordinary catastrophic third-party losses of as failed launch through 2025. The previous law involving the indemnification of US launch providers was scheduled to expire in 2016. Besides the extension of indemnification of US launch providers, the law also extends the "learning period" restrictions to 2025. These "learning period" restrictions limit the ability of the FAA to enact regulations regarding the safety of spaceflight participants. As of 2015, the indemnification of US launch providers for extraordinary catastrophic third-party losses has been a component of US space law for over 25 years. During this time, this component of the law has never been invoked in any commercial launch mishap [18, 19].

The following text is an excerpt of legislative language from the SPACE Act, Title 1, Sections 101, 102, and 103, as related to launches of commercial space vehicles and the new role of the DOT:

TITLE I--COMMERCIAL SPACE LAUNCH [19]

(Sec. 101) This bill directs the Department of Transportation (DOT) to report periodically to specified congressional committees on the progress of the commercial space transportation industry in developing voluntary consensus standards or any other construction that promotes best practices

DOT shall also report to Congress on the status of the knowledge and operational experience acquired by the commercial space transportation industry while providing flight services for compensation or hire to support the development of a safety framework.

An independent, private systems engineering and technical assistance organization or standards development organization contracted by DOT shall assess the readiness of the industry and the federal government to transition to a safety framework that may include regulations.

The contracted organization, as part of such review, must evaluate: (1) the progress of the commercial space industry in adopting industry voluntary standards or any other construction as reported by DOT in the interim assessments in the knowledge and operational report; and (2) the knowledge and operational experience obtained by the commercial space industry while providing services for compensation or hire.

(Sec. 102) DOT shall provide the committees a plan to update the methodology used to calculate maximum probable loss from claims with respect to commercially licensed space launch liability insurance and financial responsibility requirements through the use of a validated risk profile approach. The Government Accountability Office (GAO) shall assess the plan.

(Sec. 103) Certain time constraints of requirements for commercial space launch and reentry experimental permits are repealed. Rockets, reusable launch vehicles that will be launched into a suborbital trajectory, and designs for such vehicles as well as rocket designs are covered. DOT may issue an experimental launch or reentry permit notwithstanding the issuance of any launch or reentry license. Neither shall the issuance of such a license invalidate an experimental permit.

This bill reaffirms that DOT, in overseeing and coordinating commercial launch and reentry operations, should:

- promote commercial space launches and reentries by the private sector;
- facilitate government, state, and private sector involvement in enhancing U.S. launch sites and facilities;
- protect public health and safety, safety of property, national security interests, and foreign policy interests of the United States; and
- consult with another executive agency, including DOD or NASA, as necessary to provide consistent application of commercial space launch licensing requirements.

DOT must consult with DOD, NASA, and other executive agencies to identify and evaluate all requirements imposed to protect health and safety, safety of property, national security interests, and foreign policy interests of the United States relevant to any commercial launch of a launch vehicle or commercial reentry of a reentry vehicle, and:

- determine whether the satisfaction of a requirement of one agency could result in the satisfaction of a requirement of another agency, and
- resolve any inconsistencies and remove any outmoded or duplicative federal requirements or approvals.

The future for commercial space travel looks so bright; you better wear shades (e.g., sunglasses). It is very exciting to witness the birth of the commercial space future and its rapid growth with many new companies being created and existing companies reinventing themselves to enter this new marketplace. As the marketplace develops, we hope that many safety engineering practitioners are brought into the marketplace to apply many of the paradigms and processes, technologies, lessons learned, and experiences described in this book to ensure the commercial space travel flights will be as safe as, if not safer than, today's commercial aviation flights.

18.6 Summary

In conclusion, these special topics show the reader safety data from publicly available Internet locations to help you compare results from different sources to make decisions about the future of new system and product designs. Where a safety design feature works great in one application, consider the possibility of applying that technology to another application that would reap similar benefits. By seeking opportunities to research data, you open possibilities to be innovative. Safety engineers should apply areas of innovation by transitioning technology from one application, such as commercial aviation, where the technology is mature and proven to work, to other applications, such as motor vehicle industry, that could leverage the benefits of a mature technology. With innovation, you may be able to technically solve problems facing the world beyond the point where others have gone. If the effects of these global problems increase the risk of people's safety, consider taking bold and courageous steps to understand the problem and provide innovative solutions. You will become a hero if your technical solutions are proven to save lives. We encourage you to explore the possibilities, use the techniques described in our book, and become that hero.

References

[1] Negroni, C., Why Airplanes Are Safe, Travel and Leisure Magazine, http://www.travelandleisure.com/articles/why-airplanes-are-safe (Accessed on August 9, 2017).

[2] Is Commercial Aviation as Safe and Secure as We're Told? https://www.scientificamerican.com/article/is-commercial-aviation-as-safe-and-secure-as-we-re-told/ (Accessed on August 9, 2017).

[3] Airliner Accident Fatality Data from Aviation Safety Database, https://aviation-safety.net/ (Accessed on August 9, 2017).

[4] AirSafe Website, Key Information for Air Travelers, http://www.airsafe.com/events/last_15.htm (Accessed on August 9, 2017).

[5] How Safe Is Air Travel? The Statistical Truth, https://thepointsguy.com/2015/02/how-safe-is-air-travel-the-statistical-truth/ (Accessed on August 9, 2017).

[6] Is 2014 the Deadliest Year for Flights? Not Even Close, CNN.com, http://www.cnn.com/interactive/2014/07/travel/aviation-data/ (Accessed on August 9, 2017).

[7] How Safe Is Air Travel? http://www.independenttraveler.com/travel-tips/travelers-ed/how-safe-is-air-travel (Accessed on August 9, 2017).

[8] Boeing (2015) *Statistical Summary of Commercial Jet Airplane Accidents, Worldwide Operations, 1959–2015*, Boeing Commercial Airplanes, Seattle, WA.

[9] Ministry of Defense (2006) JSP553 Military Airworthiness Regulations, Edition 1, Change 5.

[10] Purton, L. and Kourousis, K. (2014) Military Airworthiness Management Frameworks: A Critical Review. *Procedia Engineering*, 80, 545–564.

[11] Insurance Institute for Highway Safety (IIHS) Highway Loss Data Institute (HLDI) Website, http://www.iihs.org/iihs/topics/t/general-statistics/fatalityfacts/state-by-state-overview (Accessed on August 9, 2017).

[12] National Highway Traffic Safety Administration (NHTSA) Website, https://www.nhtsa.gov/research-data (Accessed on August 9, 2017).

[13] Fatalities in the United States, NHTSA's National Center for Statistics and Analysis, September 16, 2016, https://crashstats.nhtsa.dot.gov/#/ (Accessed on August 9, 2017).

[14] National Highway Traffic and Safety Association (NHTSA), Preliminary Statement of Policy Concerning Automated Vehicles, May 13, 2013.

[15] Ni, R. and Leung, J., (2014) *Safety and Liability of Autonomous Vehicle Technologies*, Massachusetts Institute of Technology (MIT), Boston, MA.

[16] Hackers Remotely Kill a Jeep on the Highway—With Me in It, http://www.wired.com/2015/07/hackers-remotely-kill-jeep-highway/(Accessed on August 9, 2017).

[17] Ong, E.C. and Leveson, N.G., Fault Protection in a Component-Based Spacecraft Architecture. Proceedings of the International Conference on Space Mission Challenges for Information Technology, Pasadena, July 2003.

[18] FAA commercial space transportation regulations, Chapter III, Parts 400 to 460, of Title 14 Code of Federal Regulations (CFR), Federal Aviation Administration (FAA), Washington, DC.

[19] The Space Act, H.R.2262, U.S. Commercial Space Launch Competitiveness Act SPACE Act of 2015 (reference Wikipedia).

Appendix A: Hazards Checklist [1]

Electrical

_____ Shock
_____ Burns
_____ Overheating
_____ Ignition of combustibles
_____ Inadvertent activation
_____ Power outage
_____ Distribution backfeed
_____ Unsafe failure to operate
_____ Explosion/electrical (electrostatic)
_____ Explosion/electrical (arc)

Mechanical

_____ Sharp edges/points
_____ Rotating equipment
_____ Reciprocating equipment
_____ Pinch points
_____ Lifting weights
_____ Stability/topping potential
_____ Ejected parts/fragments
_____ Crushing surfaces

Design for Safety, First Edition. Edited by Louis J. Gullo and Jack Dixon.
© 2018 John Wiley & Sons Ltd. Published 2018 by John Wiley & Sons Ltd.

Pneumatic/hydraulic pressure

_____ Overpressurization

_____ Pipe/vessel/duct rupture

_____ Implosion

_____ Mislocated relief device

_____ Dynamic pressure loading

_____ Relief pressure improperly set

_____ Backflow

_____ Crossflow

_____ Hydraulic ram

_____ Inadvertent release

_____ Miscalibrated relief device

_____ Blown objects

_____ Pipe/hose whip

_____ Blast

Acceleration/deceleration/gravity

_____ Inadvertent motion

_____ Loose object translation

_____ Impacts

_____ Falling objects

_____ Fragments/missiles

_____ Sloshing liquids

_____ Slip/trip

_____ Falls

Temperature extremes

_____ Heat source/sink

_____ Hot/cold surface burns

_____ Pressure evaluation

_____ Confined gas/liquid

_____ Elevated flammability

_____ Elevated volatility

_____ Elevated reactivity

_____ Freezing

_____ Humidity/moisture

_____ Reduced reliability

_____ Altered structural properties
(e.g., embrittlement)

Radiation (ionizing)

_____ Alpha

_____ Beta

_____ Neutron
_____ Gamma
_____ X-ray

Radiation (non-ionizing)
_____ Laser
_____ Infrared
_____ Microwave
_____ Ultraviolet

Fire/flammability—presence of:
_____ Fuel
_____ Ignition source
_____ Oxidizer
_____ Propellant

Explosives (initiators)
_____ Heat
_____ Friction
_____ Impact/shock
_____ Vibration
_____ Electrostatic discharge
_____ Chemical contamination
_____ Lightning
_____ Welding (stray current/sparks)

Explosives (effects)
_____ Mass fire
_____ Blast overpressure
_____ Thrown fragments
_____ Seismic ground wave
_____ Meteorological reinforcement

Explosives (sensitizes)
_____ Heat/cold
_____ Vibration
_____ Impact/shock
_____ Low humidity
_____ Chemical contamination

Explosives (conditions)
_____ Explosive propellant present
_____ Explosive gas present
_____ Explosive liquid present

_____ Explosive vapor present
_____ Explosive dust present

Leaks/spills (material conditions)
_____ Liquid/cryogens
_____ Gasses/vapors
_____ Dusts—irritating
_____ Radiation sources
_____ Flammable
_____ Toxic
_____ Reactive
_____ Corrosive
_____ Slippery
_____ Odorous
_____ Pathogenic
_____ Asphyxiating
_____ Flooding
_____ Runoff
_____ Vapor propagation

Chemical/water contamination
_____ System cross connection
_____ Leaks/spills
_____ Vessel/pipe/conduit rupture
_____ Backflow/siphon effect

Physiological (see Ergonomic)
_____ Temperature extremes
_____ Nuisance dusts/odors
_____ Baro pressure extremes
_____ Fatigue
_____ Lifted weights
_____ Noise
_____ Vibration (Raynaud's syndrome)
_____ Mutagens
_____ Asphyxiants
_____ Allergens
_____ Pathogens
_____ Radiation (see Radiation)
_____ Cryogens
_____ Carcinogens
_____ Teratogens
_____ Toxins

_____ Irritants

Human factors (see Ergonomic)

_____ Operator error
_____ Inadvertent operation
_____ Failure to operate
_____ Operation early/late
_____ Operation out of sequence
_____ Right operation/wrong control
_____ Operated too long
_____ Operate too briefly

Ergonomic (see Human factors)

_____ Fatigue
_____ Inaccessibility
_____ Nonexistent/inadequate "kill" switches
_____ Glare
_____ Inadequate control/readout differentiation
_____ Inappropriate control/readout location
_____ Faulty/inadequate control/readout labeling
_____ Faulty work station design
_____ Inadequate/improper illumination

Control systems

_____ Power outage
_____ Interferences (EMI/ESI)
_____ Moisture
_____ Sneak circuit
_____ Sneak software
_____ Lightning strike
_____ Grounding failure
_____ Inadvertent activation

Unannunciated utility outages

_____ Electricity
_____ Steam
_____ Heating/cooling
_____ Ventilation
_____ Air conditioning
_____ Compressed air/gas
_____ Lubrication drains/slumps
_____ Fuel
_____ Exhaust

Common causes
_____ Utility outages
_____ Moisture/humidity
_____ Temperature extremes
_____ Seismic disturbance/impact
_____ Vibration
_____ Flooding
_____ Dust/dirt
_____ Faulty calibration
_____ Fire
_____ Single-operator coupling
_____ Location
_____ Radiation
_____ Wear-out
_____ Maintenance error
_____ Vermin/varmints/mud daubers

Contingencies (emergency responses by system/operators to "unusual" events)
_____ "Hard" shutdowns/failures
_____ Freezing
_____ Fire
_____ Windstorm
_____ Hailstorm
_____ Utility outages
_____ Flooding
_____ Earthquake
_____ Snow/ice load

Mission phasing
_____ Transport
_____ Delivery
_____ Installation
_____ Calibration
_____ Checkout
_____ Shake down
_____ Activation
_____ Standard start
_____ Emergency start
_____ Normal operation
_____ Load change
_____ Coupling/uncoupling
_____ Stressed operation

_____ Standard shutdown
_____ Shutdown emergency
_____ Diagnosis/troubleshooting
_____ Maintenance

Reference

[1] Goldberg, B. E., *System Engineering "Toolbox" for Design-Oriented Engineers*, NASA Reference Publication 1358, National Aeronautics and Space Administration, Marshall Space Flight Center, Huntsville, AL, 1994.

Appendix B: System Safety Design Verification Checklist [1]

Section 1: Electrical safety	Verify	Remarks
♦ Commercial off-the-shelf equipment ♦		
1.1 Are Commercial Off-the-Shelf (COTS) equipment listed or certified by a Nationally Recognized Testing Laboratory (NRTL)?		
1.2 Are listed COTS used in accordance with the manufacturer's manuals, in the intended environment, and within the limitations of the listing?		
1.3 Have any modifications to the equipment been reevaluated by the listing NRTL?		
1.4 Is maintenance not required on listed COTS equipment beyond that specified in the manufacturer's manuals?		
♦ Protection against shock and arcing ♦		
1.5 Are personnel suitably protected from access to hazardous voltages (>30 V between live parts and ground) when setting up, operating, or tearing down the equipment?		
1.6 Are personnel suitably protected from accidental contact with hazardous voltages (>30 V between live parts and ground) during maintenance and when maintenance covers are opened?		

(Continued)

Design for Safety, First Edition. Edited by Louis J. Gullo and Jack Dixon.
© 2018 John Wiley & Sons Ltd. Published 2018 by John Wiley & Sons Ltd.

Section 1: Electrical safety	Verify	Remarks

1.7 If the answer to Question 1.6 is **NO**, is a bypassable safety interlock incorporated to kill all power within the compartment once the maintenance cover is removed?

1.8 Are enclosures or guards that protect terminals or like devices exhibiting 30–500 V marked "WARNING, XXX Volts" in black on an orange background?

1.9 Are portions of assemblies operating at potentials above 500 V completely enclosed from the remainder of the assembly, and is this enclosure provided with non-bypassable safety interlocks?

1.10 Are enclosures for potentials, which exceed 500 V, marked "DANGER, HIGH VOLTAGE, XXX VOLTS" in white on a red background?

1.11 Are all terminals, conductors, etc., capable of supplying greater than 25 A, protected against accidental short circuit by tools, removable conductive panels and assemblies, etc.?

1.12 Are all high voltage circuits (>500 V) and capacitors (>30 V or >20 J energy) reliably and automatically discharged to less than 30 V/20 J within 2 seconds after power is removed?

1.13 Are all test points, required to be measured by maintainers, limited to less than 300 V (between test points and accessible dead metal/ground)?

1.14 If voltage dividers are used to reduce test point potentials, are two resistors used between the test points and neutral (not ground)?

1.15 Where test point voltages are to be measured through holes in protective barriers, is the maximum voltage labeled?

1.16 Is sufficient space provided between live parts and dead metal parts to prevent shorting or arcing?

1.17 Are parts and components suitably affixed to prevent loosening or rotation that could lead to shorting or arcing?

1.18 If a tool is required to make adjustments while equipment is powered, is spacing and insulation adequate to prevent contact with energized parts by the tool?

◆ **Connectors and plugs** ◆

1.19 Have connectors, used for multiple electric circuits/voltages, been selected to preclude mismating?

1.20 Has the use of similar configuration connectors in close proximity avoided?

1.21 Are plugs and receptacles coded and marked to clearly indicate mating connectors, where those of similar configuration are in close proximity?

1.22 Are plugs and receptacles designed to preclude electrical shock and burns while being disconnected?

1.23 Are male plugs de-energized when disconnected?

Section 1: Electrical safety	Verify	Remarks

1.24 Is the operator protected from potential arcing if accidentally disconnecting RF power cables?

1.25 Are all receptacles marked with their voltage, amperage, phase, and frequency characteristics where these ratings differ from the standard ratings?

♦ Wiring ♦

1.26 Is the wiring and insulation suitable for the intended load and operating voltage?

1.27 Is the wiring insulation suitable for the anticipated environment, temperature, and/or possible exposure to fuel, grease, or other chemicals?

1.28 Are wires and cables supported, protected, and terminated in a manner that prevents shock and fire?

1.29 Is wiring protected when passing though openings, near sharp edges, and near hot surfaces?

1.30 Is suitable strain relief provided for conductors and cords at their terminations to prevent stress from transmitting to terminals, splices, or internal wiring?

1.31 Where the user has access to wiring that carries hazardous voltage/current, does the wiring have a second barrier of protection (i.e., jacketed cord, conduit, etc.)?

1.32 Are single-phase line conductors color-coded black or otherwise clearly identified?

1.33 Are three-phase line conductors color-coded as follows: A is black, B is red, and C is blue, or otherwise clearly identified?

1.34 Are DC power conductors color-coded red for positive polarity and black for negative polarity?

♦ Grounding ♦

1.35 Are all equipment noncurrent-carrying metal parts and surfaces at ground potential when the equipment is powered (excluding self-powered equipment)?

1.36 Does self-powered equipment have all external surfaces at the same potential?

1.37 Is the path from various equipment points to ground continuous and permanent (hinges and slides not relied upon as the ground path)?

1.38 Are the noncurrent-carrying parts of internal components grounded where they can be accessed by maintainers?

1.39 Are panels and doors containing meters, circuit breakers, etc. grounded in a reliable manner, whether in a closed or open/ removed position ($<0.1\,\Omega$)?

(Continued)

Section 1: Electrical safety	Verify	Remarks

1.40 Does the grounding path have capacity to safely conduct any currents that might be imposed thereon?

1.41 Is the impedance of the grounding path sufficiently low to limit the potential drop and to allow overcurrent devices to clear quickly?

1.42 Does the path from the equipment tie point to ground have sufficient mechanical strength to minimize accidental grounding disconnection?

1.43 Do cables that carry a grounded conductor (neutral) also carry an Equipment Grounding Conductor (EGC) that terminates in the same manner as the other conductors?

1.44 Are insulated grounding wires color-coded green with or without yellow stripes?

1.45 Are neutral/grounded conductors color-coded white or natural gray?

1.46 Is green and white color-coding applied ONLY to grounding and grounded conductors, respectively?

1.47 Do power attachment plugs automatically ground equipment?

1.48 When the grounded power plug is mated with the receptacle, does the ground pin contact make first/break last?

1.49 Are noncurrent-carrying metal parts, grounding wires, etc. (except RF cable shields) not used to complete electrical circuits?

1.50 Is the grounding wire separate from electrical circuits, that is, not tied to neutral other than at the power source?

1.51 If a neutral/ground bond point is provided at the equipment's secondary supply circuit, is it isolated from the primary power source neutral/ground bond point in order to prevent ground loops?

1.52 On transmitting equipment, is a grounding stud provided that permits attachment of a portable shorting rod?

1.53 Is a ground stud provided on equipment intended to be interconnected to remote systems via long lengths of signal cables?

1.54 Has a test been conducted to verify that the equipment (as well as equipment systems) allows <5 Measurement Indication Units (MIU) of residual leakage current to flow to ground under the most adverse conditions of input voltage/frequency (3.5 MIU if the system can be powered from GFCI protected circuits)?

1.55 Where equipment has excessive leakage current, are redundant EGCs provided?

◆ **Power disconnects and switches** ◆

1.56 Is a means provided so that power can be cut off while installing, replacing, or servicing a complete system or any Line Replaceable Unit (LRU)?

1.57 If a main power switch is provided, does it cut off all power to the complete system?

Section 1: Electrical safety	Verify	Remarks

1.58 Is the switch located on the front panel and clearly identified?

1.59 Are power and control switches selected and located to prevent accidental actuation or stopping of the equipment?

1.60 Are switches provided to deactivate mechanical drive units without disconnecting other parts of the equipment?

1.61 Are power/maintenance switches provided at equipment that can be powered or controlled remotely?

1.62 Can lockout/tagout devices be applied to switches that are relied upon to deactivate power during maintenance?

1.63 Is protection provided against accidental contact with the supply side of the main power switch?

1.64 Are emergency controls readily accessible and clearly identified?

◆ Interlocks ◆

1.65 Where safety interlocks are used, is the interlock actuator recessed or otherwise protected against contact?

1.66 Are safety interlock circuits designed to be fail-safe?

1.67 Are live parts of safety interlocks protected from contact?

1.68 Where bypassable safety interlocks are used, do they automatically reset once the cover or guard is replaced?

1.69 Are battle short interlocks provided with an indicator to show when active?

◆ Overload and overcurrent protection ◆

1.70 Is equipment that is designed to have multiple-input power capabilities, or powered by a generator with multiple-voltage output capabilities, protected from damage when connected to incorrect input power/voltage levels?

1.71 Are overcurrent and/or overload protective devices provided for primary circuits?

1.72 If overcurrent protective devices are provided in series with any conductor grounded at the power source, does this device simultaneously open all other load conductors in the circuit?

1.73 Are multi-pole circuit breakers provided for multi-phase circuitry that will open all phases during a fault in any one?

1.74 If circuit breakers are used to power up/down equipment, have they specifically been designed for this purpose?

1.75 Do circuit breakers provide a visual indication when tripped?

1.76 Can fuses be removed safely (no exposed live parts) and without the use of tools?

1.77 Are fuse replacement types and ratings labeled?

1.78 Is surge protection incorporated to protect the user and the equipment?

(*Continued*)

Section 2: Mechanical safety	Verify	Remarks

◆ Enclosures and guards ◆

2.1 Are equipment enclosures suitably designed to protect the equipment and personnel when considering the anticipated environment and rough handling?

2.2 Are equipment openings and vents sized and located to prevent access to hazardous parts, as well as to prevent objects from falling inside and contacting hazardous parts?

2.3 Are fasteners and methods of securing doors and peripheral components sufficiently strong to prevent breakaway during normal use?

2.4 Are snag hazards due to exposed gears, cams, fans, belts, guy wires, and other moving parts avoided?

2.5 Does the equipment enclosure material and any enclosure openings limit fire propagation?

2.6 Are switches and other electrical components adequately protected against water entry due to rain or equipment washdown?

2.7 Is the equipment designed to provide personnel adequate and safe access (free of obstructions) during installation, operation, and maintenance?

2.8 Are "no step" markings provided at necessary locations to prevent injury and equipment damage?

◆ Stops, limits, and interlocks ◆

2.9 Are self-locking or other fail-safe devices incorporated into expandable and collapsible structures, such as shelters, jacks, masts, and tripods, to prevent accidental or inadvertent collapsing or falling?

2.10 Are reliable stops/limits integrated to protect moving parts from damage due to overextension or by being driven into fixed parts?

2.11 Where pins or latches are applied during equipment stowage, transportation, or maintenance to secure moveable components (i.e., motorized antenna dish, etc.), is damage prevented if the pins are left in and the drive mechanism activated?

2.12 Are doors and drawers and associated hinges, supports, slides, and stops positively locked or otherwise secured to prevent unintended movement when in the open or closed position?

◆ Pinch points and sharp edges ◆

2.13 Are telescoping ladders and assemblies provided with adequate clearance between rungs/parts to prevent pinch points?

2.14 Are hinged brackets and such devices designed and located so that fingers are not exposed to pinch points during adjustment?

2.15 Are sharp corners, edges, and projections avoided?

Section 2: Mechanical safety	Verify	Remarks

2.16 Is the installed equipment free of overhanging edges and corners that may cause injuries?

2.17 Are door and cover edges not at eye level when in an open position?

◆ Handling ◆

2.18 Is the equipment weight limited to permit safe handling by the anticipated user/maintainer crew size per the criteria below?

No. of soldiers	Weight (lbs)
1	37
2	74
3	102
4	130

2.19 Is a caution label specifying weight and lifting requirements affixed to equipment exceeding the single soldier handling criteria?

2.20 Are suitable carrying handles or hand grasp areas provided?

2.21 Does the equipment's size and weight distribution allow for easy handling, moving, and positioning?

2.22 Is the temperature of all exposed parts subject to momentary contact <60°C for metal, 68°C for glass, or 85°C for plastic/wood at an ambient temperature of 25°C, regardless of the condition of operation?

2.23 If the answer to Question 2.22 is NO, are the hot surfaces adequately labeled and protected against accidental contact?

2.24 Where prolonged contact is required (handles, controls, etc.), are surface temperatures <49°C for metal, 59°C for glass, or 69°C for plastic/wood at an ambient temperature of 25°C, regardless of the condition of operation?

◆ Miscelaneous ◆

2.25 Is the equipment likely to remain upright under normal use and in strong wind, considering its means of support, center of gravity, and slope?

2.26 Is the weight bearing capacity of hoists, jacks, and other such equipment suitable for the expected loading conditions and is the load capacity labeled?

2.27 Are pressurized systems or components provided with relief valves that will vent in a safe direction and manner?

2.28 Are positive means provided to prevent mismating of fittings; couplings; fuel, oil, hydraulic, and pneumatic lines; and mechanical linkages?

2.29 Are there provisions to prevent injury from the implosion of cathode ray tubes?

2.30 Is all glass of the non-shatterable type?

(*Continued*)

Section 3: Other safety	Verify	Remarks

3.1 Is the system designed to preclude injury or equipment damage due to operator induced error?

3.2 Is equipment designed to prevent accidental ignition when used in hazardous atmospheres (applicable to equipment that is intended for use in atmospheres of explosive gas or vapors, combustible dusts, or ignitable fibers and flyings)?

3.3 Are emergency controls readily accessible and clearly identified?

3.4 Are switches, indicators, panel instruments, and control devices adequately labeled to prevent confusion that could lead to a hazard?

3.5 Is an audible/visual warning device provided to warn personnel of impending danger or to indicate malfunction that could cause injury or equipment damage?

3.6 Is proper color-coding provided for safety-critical indicators (green, power on, ready; amber, caution; red, danger; white, info)?

3.7 Is adequate separation provided between critical warning lights and other lights?

3.8 Are audible warning signals distinguishable from other sounds under normal operating conditions?

3.9 Is the display lighting of aircraft electronics (avionics) compatible to the use of night vision goggles?

3.10 Have all equipment-related mechanical, electrical, chemical, and health hazards been suitably addressed through warning labels?

3.11 Are guards, covers, and barriers marked to indicate the hazard that may be present upon removal of such devices?

3.12 Are labels sized and placed so that the associated hazard is identified before the user is exposed to the hazard?

3.13 When possible, are labels located such that they are not removed when the barrier or access door is removed?

3.14 Do warning labels comply to the marking, design, and color requirements detailed in the system specification?

3.15 Are warning labels capable of lasting for the normal life expectancy and operational environments of the equipment to which they are affixed?

3.16 Is PMCS established for safety-critical circuits such as safety interlocks, voltage dividers, capacitor discharge circuits, etc.?

3.17 Are all maintenance procedures within the qualifications of the designated Military Occupational Specialty (MOS)?

Section 4: Health hazards	Verify	Remarks

4.1 Are noise levels less than the below listed limits at both operator and maintainer locations?

Steady state, 8 hours TWA: 85 dBA
Steady state, 16 hours TWA: 82 dBA
Impulse: 140 dBP

4.2 Where safe noise levels can be exceeded during operation or maintenance, are appropriate warning labels provided on the equipment?

4.3 If headsets or earphones are to be used with the equipment, are labels provided on the equipment to warn users to keep the volume at the lowest, useable level?

4.4 Is the equipment (considering operation, maintenance, storage, and/or disposal) free from hazardous or potentially hazardous materials?

4.5 Have nonhazardous substitute materials been utilized as much as possible?

4.6 Are potential exposures to hazardous materials controlled to levels below the Occupational Safety and Health Administration (OSHA) Permissible Exposure Limit (PEL) and/or American Conference of Governmental Industrial Hygienists (ACGIH) Threshold Limit Values (TLV), which ever is the more stringent requirement?

4.7 Is the release of toxic, corrosive, or explosive fumes or vapors prevented?

4.8 Is the equipment free of advanced composite materials (e.g., textile glass fiber, carbon/graphite fiber, aramid fiber, ceramic fiber, composite matrix)?

4.9 Are the outer coverings of cables, wires, and other components free of glass fiber materials?

4.10 Are personnel not required to occupy the shelter during normal operations for extended periods? If the answer is NO, answer Questions 4.11–4.13.

4.11 Is an environmental control unit provided that maintains temperatures within the shelter between 65 and 85°F to prevent heat or cold stress?

4.12 Do shelter air temperatures at the floor level and head level differ by <10°F?

4.13 Is adequate ventilation provided within the shelter (20 cfm/person of fresh air)?

(Continued)

Section 4: Health hazards	Verify	Remarks

4.14 Where generators and vehicles are to be operated within the vicinity (<25 ft) of the shelter, has air sampling been conducted to ensure compliance with the diesel exhaust PELs listed below? Permissible Limits (PPM)

Substance	8 hours TWA	STEL
Carbon monoxide	25	N/A
Formaldehyde	—	0.3
Sulfur dioxide	2	5
Acrolein	0.1	0.3
Nitric oxide	25	N/A
Nitrogen dioxide	3	5

4.15 Is the system free of insulating materials (e.g., asbestos, fibrous glass, mineral wool, polystyrene foam, polyurethane foam)? If the answer is **NO**, are appropriate warnings and/or safeguards provided on the equipment and in the technical manuals?

4.16 Is a fixed type fire suppression system provided? If **YES**, specify type, concentration by volume, and answer Questions 4.17 and 4.18.

4.17 Is an audible or visual alarm activated prior to release of the fire suppression agent?

4.18 Is there a time delay prior to release of the fire suppression agent?

4.19 Is the system free of all other health related hazards (vibration, shock, trauma, biological hazards, etc.)?

Section 5: Environmental impact	Verify	Remarks

5.1 Is the item or component free of hazardous or potentially hazardous materials as defined by the Federal Standard 313, EPA (40 CFR), DOT (49 CFR), OSHA ACGIH, or other federal law, regulation, or standard?

5.2 Is the item or component free of reactive or flammable chemicals such as solvents, thinners, or diluents?

5.3 Is the item or component free of toxins and carcinogens (e.g., polychlorinated biphenyls, elemental mercury, beryllium oxide, asbestos, etc.)?

5.4 Is it free of Ozone-Depleting Chemicals (ODC, i.e., Ozone-Depleting Substances (ODS)), refrigerant gasses, chlorofluorocarbons, etc.?

5.5 If the answer to any of Questions 5.1 through 5.4 is "no," has every effort been made to substitute nonhazardous materials?

Section 5: Environmental impact	Verify	Remarks

5.6 Have Material Safety Data Sheets for any hazardous materials been completed and submitted to the government?

5.7 Does the equipment avoid the use of batteries? If NO, complete Section 8, Battery Safety

5.8 Is the system free of ionizing radiation sources (radioactive isotopes, etc.)? If NO, complete Section 6, Radiation Safety. *All radioactive isotopes, regardless of quantity, must be reported to Department of the Army*

5.9 Is the system free of non-ionizing radiation sources (radio frequency, laser, etc.)? If NO, complete Section 6, Radiation Safety

5.10 Have all components that are routinely replaced in the course of maintenance been selected so as not to require special handling or disposal?

5.11 Have those components having a significant economic recovery or reclamation potential been identified (components with precious metals, special alloys, etc.)?

5.12 Have all materials that have the potential for the evolution or release of hazardous gasses, vapors, or fumes in violation of federal, state, or local regulations been eliminated?

5.13 Has the system been designed so as not to release combustion products, emit objectionable odors, or create airborne particulates?

5.14 Has the potential for the release of toxic or hazardous substances onto the soil or to surface or subsurface water been eliminated?

Section 6: Radiation safety	Verify	Remarks

6.1 Are warning labels provided that indicate the hazardous range of microwave emissions for components that produce a power density in excess of the following limits?

Frequency (f) MHz	Power density mW/cm^2
<100	See DODI 6055.11
100–300	1.0
300–3,000	$f/300$
3,000–300,000	10

6.2 For transmitting equipment, where antennas can develop RF currents on nearby dead metal objects, is the maximum current through an impedance equivalent to that of the human body for conditions of grasping the dead metal object limited to the following values: $I = 1000f$ mA for $(0.003 < f \leq 0.1\,\text{MHz})$; $I = 100\,\text{mA}$ for $(0.1 < f \leq 100\,\text{MHz})$?

(Continued)

Section 6: Radiation safety	Verify	Remarks

6.3 Have all devices that exceed 10,000 V been evaluated
for X-radiation?

6.4 Are X-ray producing devices shielded to reduce personnel
exposure to <2.0 mR/hour and no more than 50 mR/year?

6.5 Are X-ray producing devices and the components in which
they are located labeled with an X-radiation hazard warning
symbol?

6.6 Has the use of any amount of radioactive material in the design
and manufacture of any part or component been avoided?

6.7 If the answer to Question 6.6 is **NO**, has the manufacturer
identified the physical form, isotope, and quantity of ANY
radioactive material utilized in each component/system?

6.8 If the answer to Question 6.6 is **NO**, does the manufacturer
have the appropriate authorization (NRC license or CECOM
authorization) for radioactive material?

6.9 Are optical products (lenses, mirrors, windows, fiber optics,
etc.) free of ANY amounts of radioactive material?

6.10 Are radiation markings and labels affixed to all parts or
components containing radioactive material?

6.11 Are filters, goggles, or other protective devices identified and/
or provided, and are warning signs posted, for all sources of radio
frequency, ultraviolet, infrared, high-energy visible, laser, and any
other type of hazardous radiant energy?

6.12 If lasers are used, has output power been limited to the
lowest power density that could meet the performance
requirements?

6.13 Are warning labels affixed near the beam exit port and the
laser fire button (as applicable) for all class III and IV lasers?

6.14 Do lasers conform to the Code of Federal Regulations
requirements as detailed in the system specification? If the answer
is **NO**, answer Questions 6.15–6.17.

6.15 Has a military exemption been approved through the
contracting office?

6.16 Do exempt laser systems comply with MIL-STD-1425A?

6.17 Are exempt laser systems provided with a permanent caution
label notifying of such?

Section 7: Antenna and mast safety	Verify	Remarks

7.1 Are antenna terminals insulated to prevent RF burns?

7.2 Are antenna tips designed to prevent puncture wounds?

7.3 Are labels provided to warn against contact with overhead
electrical lines?

Section 7: Antenna and mast safety	Verify	Remarks

7.4 Where design permits, are antennas provided with blocking capacitors and/or coated with dielectric material to insulate against overhead electrical lines?

7.5 Are lockout devices provided for remotely operated antennas posing a mechanical RF hazard to maintainers?

7.6 Are winches, collapsible parts, tensioners, and other similar devices provided with safety latches or the like to prevent unintended collapse, freewheeling, or uncontrolled release of guy cable?

7.7 If the answer to Question 7.6 is YES, are the safety latches designed to prevent accidental or intentional bypass?

7.8 For masts >45 ft in height, is a means provided (pulley and rope) to raise any warning beacons that may be required at a particular locality?

7.9 Is a means provided to ensure that the mast is level?

7.10 Are tripping and "clothes-hanger" hazards due to guy wires minimized?

7.11 Can the designated crew size safety set up and tear down the antenna mast?

7.12 Are alternative methods of recovering the mast during emergencies, component failure, ice buildup, or jamming safe?

7.13 Are maximum wind speed limits identified for safe mast assembly, removal, and maintenance?

7.14 Are stakes suitably sized to prevent pullout in all soil conditions for worst-case wind load conditions?

7.15 Are tripods designed so that adjustments can be safely made at any time during erection of the mast should any of the legs sink?

Questions 7.16 through 7.27 pertain to Lightning Protection Adequacy. Note: If the mast is electrically continuous, treat it as the down-conductor

7.16 If antenna acts as an aerial terminal, does conductivity equal or better that of #3 AWG solid copper?

7.17 If the answer to Question 7.16 is NO (e.g., dish antenna), is the antenna contained within a 45° cone from the tip of a provided air terminal?

7.18 Is down-conductor equivalent to #3 AWG solid copper with a minimum strand size of #17 AWG?

7.19 Are joints mechanically strong and corrosion resistant?

7.20 Are resistance of joints less than that of 2 ft (0.6 m) of down-conductor ($R = 0.002\,\Omega$ or less—negligible resistance)?

7.21 Will the down-conductor remain free of bends or kinks after repeated use?

7.22 Is down-conductor straight as possible with any turns not <90° with 8 inch radius of turn?

(Continued)

Section 7: Antenna and mast safety	Verify	Remarks

7.23 Is ground rod at least 1/2-inch-diameter, 8-ft-long copper clad steel or equivalent?

7.24 Is ground rod free of paint?

7.25 Does antenna mast configuration during erection, storage, take-down, or operation prevent any component of the lightning protection system from mechanical damage or wear?

7.26 If mast is electrically continuous and is acting as the down-conductor, is the ground stud adequate?

7.27 Is a safety tip cap provided for the air terminal?

Section 8: Battery safety	Verify	Remarks

8.1 For each battery type, identify the manufacturer, model number, chemistry, its purpose, and the quantity used

8.2 Are the batteries in the government inventory? If **YES**, indicate the battery's nomenclature (BA-xxx, BB-xxx, etc.) and NSN

8.3 Does the equipment prevent the charging of non-rechargeable batteries when installed?

8.4 Does the equipment incorporate a voltage cutoff to prevent overdischarge of the battery during usage or long-term storage?

8.5 Are design features incorporated to prevent charging of rechargeable batteries at high temperatures?

8.6 Are design features incorporated to prevent overcharging of rechargeable batteries?

8.7 Is the battery enclosed and protected from mechanical shock and the environment?

8.8 Are battery compartments/enclosures adequately vented to prevent the buildup of explosive gasses?

8.9 Is the battery compartment designed to prevent any leaking liquid/gas from entering the main equipment?

8.10 Is the use of conductive battery covers avoided?

8.11 Where battery boxes house non-rechargeable lithium batteries, are they designed to prevent injury or damage in the event of a violent battery venting or rupture IAW CECOM TB 7, battery box design?

8.12 Has a test been conducted on a CECOM-certified test apparatus to verify Question 8.11?

8.13 Are battery compartments oriented so that in the event of a battery venting, gasses or liquids are directed away from the user's face and body?

8.14 Is adequate spacing/guarding provided around battery terminals so that tools cannot create an electrical short while working near the batteries or disconnecting the battery cables?

Section 8: Battery safety	Verify	Remarks

8.15 If the equipment utilizes two batteries in parallel, is electrical circuitry incorporated to prevent reverse or parallel battery charging and to limit any imbalance in current draw?

8.16 Are mechanical or electrical features incorporated to prevent equipment damage due to the insertion of different batteries having similar dimensions?

Section 9: Generators	Verify	Remarks

9.1 Is a main circuit breaker provided and located in an easily accessible location?

9.2 Are the following protective devices present with suitable indicators to safeguard against operator injury and/or equipment damage: overspeed, over-temperature, overvoltage, overload and short circuit, low oil pressure, and low fuel?

9.3 Is the battle short switch provided and located on the main control panel?

9.4 Are all supply connection points clearly marked with terminal information and polarity?

9.5 Are all convenience receptacles provided with overcurrent protection as well as ground fault circuit interrupter protection?

9.6 Is a suitable grounding terminal lug provided and identified for connection to an earth grounding electrode?

9.7 Is an Army-approved grounding system fielded with the generator set, and is a storage location provided for it?

9.8 Are components, conductors, and shielding appropriately located such that overheating, arching, shorting, and contact with moving parts are avoid?

9.9 Are battery terminals and cables marked for polarity and provided with nonconductive guards to prevent accidental shorting?

9.10 Are tools to be used near high voltages, such as load terminal wrench, adequately insulated?

9.11 Are fuel lines adequately supported and separated from live wires and cables?

9.12 Are fuel lines projecting through metal apertures protected by grommets and secured to framing members?

9.13 Is thermal and sound insulating material treated with fire retardant, free from noxious fumes, unaffected by battery electrolyte or petroleum derivatives, capable of maintaining it shape position and consistency inherently or by retaining methods, and replaceable?

(Continued)

Section 9: Generators	Verify	Remarks

9.14 Where safe noise levels can be exceeded during operation or maintenance, are appropriate warning labels provided on the equipment?

9.15 Is a type B/C dry chemical extinguisher provided with the generator? Specify size

9.16 Is CARC paint applied only to surfaces that will not exceed 400°F?

9.17 Is the generator exhaust located and directed away from operator designated areas?

9.18 Is the air intake at a sufficient distance from the exhaust?

9.19 Is the fuel tank designed and located in a manner that will not allow spills or overflow to run into the engine, exhaust, or electrical equipment?

9.20 Is the fuel tank equipped with a float valve to prevent fuel from overflowing when the set is being fueled from the auxiliary fuel connection?

9.21 Where an auxiliary refueling system is integrated, is a fuel line and jerry can adapter provided for connection to the external fuel container?

9.22 Is the center of gravity and weight of the set distinctly marked?

9.23 Are tie-downs and lifting positions clearly marked?

9.24 Are lifting rings, slings, and forklift eyes provided?

Section 10: Equipment integration of shelters and trailers	Verify	Remarks

10.1 Is the vehicle weight properly distributed and is the vehicle laterally stable?

10.2 Does the shelter/equipment Center of Gravity (COG) fall within the prime mover COG envelope?

10.3 Is the center of gravity and equipment weight distinctly marked?

10.4 Does the system weight (including crew gear and trailer pintle weight) not exceed the load capacity of the prime mover?

10.5 Has the vehicle satisfactorily passed road worthiness testing (e.g., Munson road test)?

10.6 Have no vehicle speed restrictions been placed on the prime mover as a result of system integration?

10.7 Are adequate instructions provided for placement of detached trailers?

10.8 Are safety chains provided to prevent the trailer from detaching from the towing vehicle?

10.9 Will the lifting rings support the total weight of the shelter and the installed equipment?

10.10 Are entries and exits free of obstructions?

Section 10: Equipment integration of shelters and trailers	Verify	Remarks
10.11 Do the entryway ladders or steps allow safe entrance and exit?		
10.12 Is an emergency exit provided and marked?		
10.13 Is the emergency exit readily accessible and simple to operate in a high stress, zero visibility situation?		
10.14 Where extended operations are required on top of the shelter, are ladders, nonslip surfaces, and guardrails or chains provided for the shelter roof?		
10.15 Does the floor surface prevent slipping?		
10.16 Is adequate illumination provided in all areas?		
10.17 Are wall, floor, and ceiling fastenings sufficient to prevent equipment from breaking away, falling, or accidentally dislodging?		
10.18 Are accessories secured or stowed to prevent damage when the vehicle is moving?		
10.19 Is equipment that is designed to have multiple-input power capabilities or powered by a generator with multiple-voltage output capabilities protected from damage when connected to incorrect input power/voltage levels?		
10.20 Is an Army-approved earth grounding system (ground rod, SWGK, etc.) provided?		
10.21 Do the grounded and grounding circuits remain isolated throughout the shelter, including at the supply side of the power panel?		
10.22 Where a switch is provided to switch between different power sources, is the grounded/neutral conductor also switched to avoid ground loops?		
10.23 Is a ground stud provided at the power entry box and is it suitably identified?		
10.24 Are no parts of the vehicle/shelter enclosure or frame used as the AC ground path?		
10.25 Are the ground pins of the convenience outlets hard wired to the shelter/system ground point?		
10.26 Is lightning surge protection provided at the power and signal entry panels for all cables?		
10.27 Are all outdoor receptacles Ground Fault Circuit Interrupter (GFCI) protected?		
10.28 If the answer to Question 10.27 is **NO**, is the socket configuration of each outdoor receptacle that is not connected to a GFCI unique to its special application and unusable for other applications or as a convenience outlet?		
10.29 Has the amount of residual leakage current to ground for the entire system been verified through test to be <5 mA (3.5 mA if system can be powered from a GFCI protected circuit)? If **YES**, indicate the amount of current that was measured		
10.30 Is a main power switch provided at the shelter entrance?		

(*Continued*)

Section 10: Equipment integration of shelters and trailers	Verify	Remarks

10.31 Are safety switches provided at remotely located assemblies to protect maintainers?

10.32 Are terminals, plugs, and other exposed parts located within power distribution panels that may exhibit over 30 V or 20 A guarded against accidental contact if exposed during maintenance?

10.33 Where transmitting equipment exists and room permits, are shorting rods provided?

10.34 Are fuel lines that are inside the shelter made as short as possible?

10.35 Is there a heater fuel shutoff valve inside the shelter?

10.36 Is a fuel line and jerry can adapter provided for connection to the external fuel tank or container?

10.37 Are fuel lines and fuel sources suitably protected from potential damage and sources of heat?

10.38 Are battery compartments designed to prevent gas buildup within the shelter (i.e., forced air ventilated to the outside)?

10.39 Is a warning device provided to indicate when either the battery vent lid or door is closed or when the ventilation fan is inoperable?

10.40 Is the vehicle exhaust sufficiently separated from shelter openings to avoid an accumulation of carbon monoxide in the shelter?

10.41 Is a type B/C carbon dioxide or dry chemical extinguisher provided for electrical equipment and located near shelter exit? Specify size

10.42 Are ceilings, walls, and other surfaces adjacent to aisles free of electrical components and switches that are vulnerable to breakage by accidental collision?

10.43 Are controls, connectors, or other parts that project into walkways or located a foot level protected from mechanical damage?

10.44 Are climbing rings, handholds, rails, etc. provided where needed?

10.45 Are handles recessed rather than extended where they might be hazardous?

10.46 Can EMI emitted by the equipment cause any degraded or erratic operation of other equipment?

10.47 Where equipment is installed on platforms with weapons or turrets, has a test been conducted to ensure that EMI cannot cause uncontrolled turret movement or weapons misfire?

♦ Operation on-the-move ♦

10.48 Is the shelter to be occupied and operated only when the system is stationary (no on-the-move operations)? If NO, answer Questions 10.49–10.59

10.49 Have suitable seats and restraints been provided for the required number of users?

Section 10: Equipment integration of shelters and trailers	Verify	Remarks
10.50 Is equipment positioned so that personnel will not bump or rub against it when riding over rough terrain?		
10.51 Where equipment requires frequent viewing and access, is the operator not required to twist in his seat, which could diminish seat belt effectiveness in an accident?		
10.52 Is the equipment suitably mounted to prevent loosening or dislodging when the vehicle is driven over rough terrain or in the event of an accident or rollover?		
10.53 Can users in the shelter maintain reliable communication with the driver at all times?		
10.54 Can operators access fire suppression systems, communication systems, and other critical controls while in a seated position?		
10.55 Is adequate airflow and temperature control maintained and can the user access the ECU controls while in a seated position?		
10.56 Is the use of batteries that can vent poisonous gasses, such as lithium sulfur dioxide batteries, avoided?		
10.57 Is emergency power and lighting available in the event of primary power loss?		
10.58 Is adequate noise protection provided for the users?		
10.59 Do critical commands that can be accidentally keyed require a second confirming action?		

Section 11: Equipment integration of vehicle cabs	Verify	Remarks
11.1 Does placement of equipment or interconnecting cables avoid interference with preexisting controls, indicators, or other equipment and panels requiring access?		
11.2 Are trip, snag, and obstruction hazards by the equipment or interconnecting cables avoided in both primary and secondary egress paths?		
11.3 Does the placement of equipment not interfere with step locations?		
11.4 Is the equipment located so that it will not be stepped on or otherwise damaged by personnel during ingress or egress?		
11.5 Are equipment controls and switches protected from accidental activation?		
11.6 Do critical commands that can be accidentally keyed require a second confirming action?		
11.7 Are conductors that supply equipment power properly fused?		
11.8 Are sharp bends in cables avoided?		
11.9 Are sharp or protruding edges, surfaces, or corners avoided that can cause injury with equipment in operating or stowed position?		
11.10 Is equipment positioned so that personnel will not bump or rub against it when riding over rough terrain?		

(Continued)

Section 11: Equipment integration of vehicle cabs	Verify	Remarks

11.11 Does the equipment location prevent increased personnel injury in the event of a vehicle accident or rollover?

11.12 Is the equipment suitably mounted to prevent loosening or dislodging when the vehicle is driven over rough terrain or in the event of an accident or rollover?

11.13 Is the mounting hardware designed and installed such that when the equipment is removed, the remaining mounting hardware does not pose a mechanical hazard to personnel?

11.14 Is Electromagnetic Interference (EMI) that can cause any degraded or erratic operation of other equipment avoided?

11.15 Where equipment is installed on platforms with weapons or turrets, has a test been conducted to ensure that EMI cannot cause uncontrolled turret movement or weapons misfire?

11.16 Is all equipment suitably grounded to chassis?

11.17 Are interconnecting cables run neatly and tied down to avoid any tripping and snag hazards?

11.18 Are cable ties and adhesives suitable to withstand the environment and rough handling?

11.19 Do hinged or adjustable mounting hardware avoid pinch hazards?

11.20 Are mounting screws and bolts properly sized to prevent projections?

11.21 Does the driver's field of view through the driver and passenger windshield and windows remain unobstructed with the equipment installed and adjusted in any position? Can the driver clearly see all mirrors?

11.22 Where the driver's view can be obstructed by adjustable displays or equipment, can the equipment be aligned with existing vehicle obstructions to minimize impact? Are labels provided to warn the driver to do so? Are the adjustable mounts designed so they don't progressively move/shift out of position due to vehicle motions and vibrations?

11.23 Where the equipment requires frequent viewing or access, is twisting or other motion by the operator avoided, which could cause repetitive motion stress?

11.24 Where equipment requires frequent viewing and access, is the operator not required to twist in his seat, which could diminish seat belt effectiveness in an accident?

11.25 Where night vision goggles are used, is direct or reflected light from indicators or displays avoided where it can interfere with driver night vision driving?

11.26 Is equipment heat load dissipated adequately so that cab temperatures will not increase significantly?

Section 11: Equipment integration of vehicle cabs	Verify	Remarks

11.27 Is the use of batteries that can vent poisonous gasses, such as lithium sulfur dioxide batteries, avoided where the equipment cannot be readily jettisoned from the vehicle?

11.28 Is the equipment so located that it will not divert the driver's attention?

Section 12: Transit case mounted equipment	Verify	Remarks

12.1 Where COTS is mounted in the transit cases, is additional protection against the elements provided for when the case is open?

12.2 Is GFCI protection integrated into the power cord?

12.3 Are ground studs and bonding straps provided to permit bonding between cases?

12.4 Is an earth grounding system supplied if it can be powered more than 25 ft from the power source?

12.5 Is overcurrent protection incorporated, located on the front panel, and clearly identified?

12.6 Can the system be stacked or secured in a manner that will prevent it from tipping?

12.7 Does the system prevent tipping if weight is applied to a drawer or cover that can be extended?

12.8 Are leveling systems provided for setup in uneven terrain?

12.9 Are legs or other similar mechanisms provided with pads to prevent sinking in soft soil conditions?

12.10 Are adequate handles and warning labels provided for repositioning the equipment with the transit box covers removed?

12.11 Where transportation or storage orientation is important, is the transit case labeled "This Side Up?"

12.12 Are pressure relief valves provided for the transit cases?

Section 13: Software safety	Verify	Remarks

13.1 Is the system or equipment free of software that (a) could create a hazard, (b) controls hazardous processes or outputs, or (c) controls information upon which the operator must rely in order to make safe decisions? If **YES**, skip Section 13

13.2 Does the software adequately control all hazardous routines and outputs?

13.3 Does the software allow the operator to take control over the hardware at any time? If the answer is **YES**, then skip Question 13.4

13.4 Does the software allow the operator to take control over the hardware when hazardous routines or outputs are involved?

(Continued)

Section 13: Software safety	Verify	Remarks

13.5 Will operator have information needed in order to make safe decisions without reliance upon information generated by the software? If **YES**, skip Question 13.6.

13.6 Is the probability that the software will fail to provide information needed by the operator in order to make safe decisions at an acceptably low level?

13.7 Is the probability that the software will induce a critical hazard at an acceptably low level?

13.8 Can the failure of any input or output device cause a critical hazard?

13.9 Does the system assume or revert to a safe state upon a power failure or upon the failure of any hardware component, such as the primary computer? Is the corruption of safety-critical data prevented in such an event?

13.10 Has all legacy software (including operating system and commercial software) been tested as part of the overall system to ensure that its integration does not create critical hazards?

13.11 Have all safety-critical software functions been identified for specific software testing in and apart from testing of other software functions?

13.12 Have the results of all software safety analyses, testing, and hazard abatement been formally documented for present consideration and for future changes/upgrades?

13.13 Where systems display situational awareness data, is common warfighting symbology used as depicted in MIL-STD-2525? Is the modification of any symbols avoided?

13.14 Does the display interface adequately address the limitations of color-blind users?

13.15 Do nonreversible/destructive actions (zeroize) require two discrete entries?

13.16 Are operators adequately notified of safety-critical messages and alerts?

Reference

[1] US Army Communications-Electronics Command (CECOM), System Safety Design Verification Checklist, SEL Form 1183, February 2001.

Index

Design for Safety, First Edition. Edited by Louis J. Gullo and Jack Dixon.
© 2018 John Wiley & Sons Ltd. Published 2018 by John Wiley & Sons Ltd.